METHODS IN MOLECULAR BIOLOGY

Series Editor
John M. Walker
School of Life and Medical Sciences
University of Hertfordshire
Hatfield, Hertfordshire, AL10 9AB, UK

For further volumes:
http://www.springer.com/series/7651

Myogenesis

Methods and Protocols

Edited by

Sissel Beate Rønning

Nofima AS, Ås, Norway

Humana Press

Editor
Sissel Beate Rønning
Nofima AS
Ås, Norway

ISSN 1064-3745 ISSN 1940-6029 (electronic)
Methods in Molecular Biology
ISBN 978-1-4939-8896-9 ISBN 978-1-4939-8897-6 (eBook)
https://doi.org/10.1007/978-1-4939-8897-6

Library of Congress Control Number: 2018958511

This Humana Press imprint is published by the registered company Springer Science+Business Media, LLC part of Springer Nature.
The registered company address is: 233 Spring Street, New York, NY 10013, U.S.A.

Preface

The skeletal muscles account for up to half of the human body mass and are mainly composed of postmitotic, multinucleated muscle fibers. The skeletal muscle comprises more than 600 individual muscles, and is not only important for movement but is also a major site of fatty acid oxidation and carbohydrate metabolism. Skeletal muscle function is highly dependent on the ability to regenerate. Fifty years ago, Mauro first suggested that satellite cells were involved in skeletal muscle regeneration. Since the first discovery of these cells, numerous reports have identified these stem cells as primary contributors to the postnatal growth, maintenance, and repair of skeletal muscles. The satellite cells are located between the basal lamina and sarcolemma (plasma membrane) of skeletal muscle fibers and are normally quiescent in the adult muscle before they become activated upon exercise, injury, or disease. Then the cells have a remarkable ability to self-renew, expand, grow, or undergo myogenic differentiation to fuse and restore damaged muscle. The conversion of mononuclear muscle precursors (myoblasts) into multinucleated myotubes is a complex process and is still not fully characterized.

This volume of the *Methods in Molecular Biology* series therefore provides a collection of many of the common experimental approaches used to study myogenesis. The chapters are all written in the same format, describing the basic theory of the method, listing all the chemicals, reagents, equipment, and materials necessary. Then the method section is written as a step-by-step description of every protocol. Most importantly, the chapters give an idea of problems that can be encountered, how these can be identified, and, finally, how to overcome the difficulties.

The book covers subjects ranging from isolation and purification protocols, manipulation of muscle cells, transcriptomics and proteomics, metabolism and exercise, and tissue engineering. Presented methods involve different species, including human, bovine, Atlantic salmon, rats, mice, larval zebra fish, and *Drosophila melanogaster*. The book includes advanced techniques for manipulation of the muscle cells including lentivirus-mediated RNAi, siRNA gene silencing, and CRISPR/Cas9-mediated targeting, as well as transcriptomics and proteomics. The book also includes protocols important for tissue engineering such as co-culture with other cells, trans-differentiation, use of serum-free media, and bioengineering skeletal muscle in bioreactors. Together, I hope that this volume will be an essential part of many laboratory libraries and assist researchers throughout the world to help reveal the unknowns of myogenesis.

I would like to thank Prof. John M. Walker, the series editor, for his help, advice, and guidance in making this book.

Enjoy!

Ås, Norway *Sissel Beate Rønning*

Contents

Contributors

VIGDIS AAS • *Department of Life Sciences and Health, Faculty of Health Sciences, Oslo Metropolitan University, Oslo, Norway*

MARIKO AKAHOSHI • *Department of Bioresource Sciences, Faculty of Agriculture, Kyushu University, Fukuoka, Japan*

EDWARD B. ARIAS • *Muscle Biology Laboratory, School of Kinesiology, University of Michigan, Ann Arbor, MI, USA*

VÉRONIQUE BLANQUET • *INRA, PEIRENE EA7500, USC1061 GAMAA, Université de Limoges, Limoges, France*

KHALED BOUHOUCHE • *INRA, PEIRENE EA7500, USC1061 GAMAA, Université de Limoges, Limoges, France*

ALEXANDER D. BROWN • *Exercise Metabolism and Adaptation Research Group (EMARG), Research Institute for Sport and Exercise Sciences (RISES), Liverpool John Moores University, Liverpool, UK*

ANTON L. BRYANTSEV • *Department of Molecular and Cellular Biology, Kennesaw State University, Kennesaw, GA, USA*

GREGORY D. CARTEE • *Muscle Biology Laboratory, School of Kinesiology, University of Michigan, Ann Arbor, MI, USA; Department of Molecular and Integrative Physiology, University of Michigan, Ann Arbor, MI, USA; Institute of Gerontology, University of Michigan, Ann Arbor, MI, USA*

LIZZET CASTILLO • *Department of Biology, University of New Mexico, Albuquerque, NM, USA*

VISHAL CHATURVEDI • *School of Pharmacy and Biomedical Science, Curtin Health Innovation Research Institute, Faculty of Health Science, Curtin University, Bentley, WA, Australia*

MARIA B. CHECHENOVA • *Department of Molecular and Cellular Biology, Kennesaw State University, Kennesaw, GA, USA*

JIE CHEN • *Department of Cell and Developmental Biology, University of Illinois at Urbana-Champaign, Urbana, IL, USA*

JINGJUAN CHEN • *Department of Animal Science, Purdue University, West Lafayette, IN, USA*

GRAEME L. CLOSE • *Exercise Metabolism and Adaptation Research Group (EMARG), Research Institute for Sport and Exercise Sciences (RISES), Liverpool John Moores University, Liverpool, UK*

DEIRDRE R. COOMBE • *School of Pharmacy and Biomedical Science, Curtin Health Innovation Research Institute, Faculty of Health Science, Curtin University, Bentley, WA, Australia*

PETER D. CURRIE • *Australian Regenerative Medicine Institute, Monash University, Clayton, VIC, Australia; European Molecular Biology Laboratory Australia Melbourne Node, Monash University, Clayton, VIC, Australia*

LINDA DESENDER • *Tissue Engineering Laboratory, Department of Development and Regeneration, KU Leuven campus Kulak, Kortrijk, Belgium*

TRACY E. DOHN • *Department of Molecular and Cellular Biology, Kennesaw State University, Kennesaw, GA, USA; Department of Biology, University of New Mexico, Albuquerque, NM, USA*
TOMASZ DOMORADZKI • *Department of Physiological Sciences, Faculty of Veterinary Medicine, Warsaw University of Life Sciences (SGGW), Warsaw, Poland*
DANIELLE E. DYE • *School of Pharmacy and Biomedical Science, Curtin Health Innovation Research Institute, Faculty of Health Science, Curtin University, Bentley, WA, Australia*
CLAIRE E. EYERS • *Department of Biochemistry, Institute of Integrative Biology, University of Liverpool, Liverpool, UK; Centre for Proteome Research, Institute of Integrative Biology, University of Liverpool, Liverpool, UK*
NILS J. FAERGEMAN • *Department of Biochemistry and Molecular Biology, Villum Center for Bioanalytical Sciences, University of Southern Denmark, Odense M, Denmark*
SAMANTHA FERRIES • *Department of Biochemistry, Institute of Integrative Biology, University of Liverpool, Liverpool, UK; Centre for Proteome Research, Institute of Integrative Biology, University of Liverpool, Liverpool, UK*
MELANIE GERARD • *Tissue Engineering Laboratory, Department of Development and Regeneration, KU Leuven campus Kulak, Kortrijk, Belgium*
DACHA GHOLOBOVA • *Tissue Engineering Laboratory, Department of Development and Regeneration, KU Leuven campus Kulak, Kortrijk, Belgium*
KATARZYNA GRZELKOWSKA-KOWALCZYK • *Department of Physiological Sciences, Faculty of Veterinary Medicine, Warsaw University of Life Sciences (SGGW), Warsaw, Poland*
EMANUELA GUSSONI • *Division of Genetics and Genomics, Boston Children's Hospital, Boston, MA, USA; Department of Pediatrics, Harvard Medical School, Boston, MA, USA; The Stem Cell Program at Boston Children's Hospital, Boston, MA, USA*
GEMMA E. HARDMAN • *Department of Biochemistry, Institute of Integrative Biology, University of Liverpool, Liverpool, UK; Centre for Proteome Research, Institute of Integrative Biology, University of Liverpool, Liverpool, UK*
JESPER F. HAVELUND • *Department of Biochemistry and Molecular Biology, Villum Center for Bioanalytical Sciences, University of Southern Denmark, Odense M, Denmark*
VIBEKE HØST • *Nofima AS, Aas, Norway*
FIONA K. JONES • *Department of Biochemistry, Institute of Integrative Biology, University of Liverpool, Liverpool, UK*
ANDREAS M. KASPER • *Exercise Metabolism and Adaptation Research Group (EMARG), Research Institute for Sport and Exercise Sciences (RISES), Liverpool John Moores University, Liverpool, UK*
AYNUR KAYA-ÇOPUR • *Aix Marseille Univ, CNRS, IBDM, Marseille, France*
DONGWOOK KIM • *Department of Cell and Developmental Biology, University of Illinois at Urbana-Champaign, Urbana, IL, USA*
BEVERLEY F. KINNEAR • *School of Pharmacy and Biomedical Science, Curtin Health Innovation Research Institute, Faculty of Health Science, Curtin University, Bentley, WA, Australia*
SHIHUAN KUANG • *Department of Animal Science, Purdue University, West Lafayette, IN, USA; Center for Cancer Research, Purdue University, West Lafayette, IN, USA*
TYANNA L. LOVATO • *Department of Biology, University of New Mexico, Albuquerque, NM, USA*
LAETITIA MAGNOL • *INRA, PEIRENE EA7500, USC1061 GAMAA, Université de Limoges, Limoges, France*

ALICJA MAJEWSKA • *Department of Physiological Sciences, Faculty of Veterinary Medicine, Warsaw University of Life Sciences (SGGW), Warsaw, Poland*
NEIL R. W. MARTIN • *Musculoskeletal Biology Research Group, School of Sport, Exercise, and Health Sciences, Loughborough University, Loughborough, UK*
YUJI MATSUYOSHI • *Department of Bioresource Sciences, Faculty of Agriculture, Kyushu University, Fukuoka, Japan*
FREA MEHTA • *Department of Physiology, Maastricht University, Maastricht, The Netherlands*
WATARU MIZUNOYA • *Department of Bioresource Sciences, Faculty of Agriculture, Kyushu University, Fukuoka, Japan*
MARTA MORENO-TORRES • *Department of Biochemistry and Molecular Biology, Villum Center for Bioanalytical Sciences, University of Southern Denmark, Odense M, Denmark*
MARK MURPHY • *School of Pharmacy and Biomolecular Sciences, Liverpool John Moores University, Liverpool, UK*
MAKO NAKAMURA • *Department of Bioresource Sciences, Faculty of Agriculture, Kyushu University, Fukuoka, Japan*
NATAŠA NIKOLIĆ • *Department of Pharmaceutical Biosciences, School of Pharmacy, University of Oslo, Oslo, Norway*
SANDY T. OAS • *Department of Biology, University of New Mexico, Albuquerque, NM, USA*
TONE-KARI K. OESTBYE • *Nofima AS, Ås, Norway*
ANNA PAKULA • *Division of Genetics and Genomics, Boston Children's Hospital, Boston, MA, USA; Department of Pediatrics, Harvard Medical School, Boston, MA, USA*
ALEXIS PARENTÉ • *INRA, PEIRENE EA7500, USC1061 GAMAA, Université de Limoges, Limoges, France*
MARK W. PATAKY • *Muscle Biology Laboratory, School of Kinesiology, University of Michigan, Ann Arbor, MI, USA*
MONA E. PEDERSEN • *Nofima AS, Aas, Norway*
LUCE PÈRIÉ • *INRA, PEIRENE EA7500, USC1061 GAMAA, Université de Limoges, Limoges, France*
ADDOLORATA PISCONTI • *Department of Biochemistry, Institute of Integrative Biology, University of Liverpool, Liverpool, UK*
MARK J. POST • *Department of Physiology, Maastricht University, Maastricht, The Netherlands*
VINAY KUMAR RAO • *Department of Physiology, Yong Loo Lin School of Medicine, National University of Singapore, Singapore, Singapore; Temasek Life Sciences Laboratory, National University of Singapore, Singapore, Singapore*
DHANUSHIKA RATNAYAKE • *Australian Regenerative Medicine Institute, Monash University, Clayton, VIC, Australia; European Molecular Biology Laboratory Australia Melbourne Node, Monash University, Clayton, VIC, Australia*
ADRIANA REYES-ORDOÑEZ • *Department of Cell and Developmental Biology, University of Illinois at Urbana-Champaign, Urbana, IL, USA*
FRANK SCHNORRER • *Aix Marseille Univ, CNRS, IBDM, Marseille, France*
ROBERT A. SEABORNE • *Institute for Science and Technology in Medicine (ISTM), Keele University School of Medicine, Keele University, Staffordshire, UK; Exercise Metabolism and Adaptation Research Group (EMARG), Research Institute for Sport and Exercise Sciences (RISES), Liverpool John Moores University, Liverpool, UK*
SHILPA RANI SHANKAR • *Department of Physiology, Yong Loo Lin School of Medicine, National University of Singapore, Singapore, Singapore*

JANET SHANSKY • *Department of Pathology and Laboratory Medicine, Brown University, Providence, RI, USA*
ADAM P. SHARPLES • *Institute for Science and Technology in Medicine (ISTM), Keele University School of Medicine, Keele University, Staffordshire, UK; Exercise Metabolism and Adaptation Research Group (EMARG), Research Institute for Sport and Exercise Sciences (RISES), Liverpool John Moores University, Liverpool, UK*
JANELLE M. SPINAZZOLA • *Division of Genetics and Genomics, Boston Children's Hospital, Boston, MA, USA; Department of Pediatrics, Harvard Medical School, Boston, MA, USA*
CLAIRE E. STEWART • *Exercise Metabolism and Adaptation Research Group (EMARG), Research Institute for Sport and Exercise Sciences (RISES), Liverpool John Moores University, Liverpool, UK*
RESHMA TANEJA • *Department of Physiology, Yong Loo Lin School of Medicine, National University of Singapore, Singapore, Singapore*
RYUICHI TATSUMI • *Department of Bioresource Sciences, Faculty of Agriculture, Kyushu University, Fukuoka, Japan*
LISANNE TERRIE • *Tissue Engineering Laboratory, Department of Development and Regeneration, KU Leuven campus Kulak, Kortrijk, Belgium*
RUUD THEUNISSEN • *Department of Physiology, Maastricht University, Maastricht, The Netherlands*
LIEVEN THORREZ • *Tissue Engineering Laboratory, Department of Development and Regeneration, KU Leuven campus Kulak, Kortrijk, Belgium*
DANIEL C. TURNER • *Institute for Science and Technology in Medicine (ISTM), Keele University School of Medicine, Keele University, Staffordshire, UK*
HERMAN VANDENBURGH • *Department of Pathology and Laboratory Medicine, Brown University, Providence, RI, USA*
EVA VEISETH-KENT • *Nofima AS, Ås, Norway*
CHAO WANG • *Department of Animal Science, Purdue University, West Lafayette, IN, USA*
ELISABETH YTTEBORG • *Nofima AS, Ås, Norway*

Chapter 1

Purification of Myogenic Progenitors from Human Muscle Using Fluorescence-Activated Cell Sorting (FACS)

Anna Pakula, Janelle M. Spinazzola, and Emanuela Gussoni

Abstract

Primary myoblasts derived from human tissue are a valuable tool in research of muscle disease and pathophysiology. However, skeletal muscle biopsies, especially from diseased muscle, contain a plethora of non-myogenic cells, necessitating purification of the myogenic cell population. This protocol describes techniques for dissociation of cells from human skeletal muscle biopsies and enrichment for a highly myogenic population by fluorescence-activated cell sorting (FACS). We also describe methods for assessing myogenicity and population expansion for subsequent in vitro study.

Key words Skeletal muscle, Myoblast isolation, Tissue dissociation, Fluorescence-activated cell sorting (FACS), CD82, CD56, Immunostaining, Pax7

1 Introduction

Myogenic cells derived from muscle biopsies are a valuable resource for modeling human muscle disease in vitro. They can be utilized to assess many cell-based and pharmacological therapies for translational research by assaying myoblast proliferation, differentiation, and fusion, which are often compromised in diseased states. However, human skeletal muscle biopsies, especially those affected by disease, often contain extensive populations of non-myogenic cells including adipocytes and fibroblasts that may confound experimental results. Thus, it is important to isolate a myogenic population for accurate in vitro study of skeletal muscle development and disease.

Early studies of muscle disease through the 1970s largely involved the use of tissue explants or unpurified dissociated cells [1–4], which were used to determine in vitro conditions for culture and differentiation of human muscle cells [5, 6]. In the 1980s, Blau

Anna Pakula and Janelle M. Spinazzola contributed equally to this chapter.

Sissel Beate Rønning (ed.), *Myogenesis: Methods and Protocols*, Methods in Molecular Biology, vol. 1889,
https://doi.org/10.1007/978-1-4939-8897-6_1, © Springer Science+Business Media, LLC, part of Springer Nature 2019

and Webster introduced a pre-plating technique to remove fibroblasts [7], and this group later described more specific and efficient isolation of myoblasts from a dissociated human sample utilizing human neural cell adhesion molecule (NCAM), also known as CD56 [8], a cell surface antigen shown to be expressed on myogenic cells [9] using fluorescence-activated cell sorting (FACS).

Here, we describe an effective technique for dissociation of mononuclear cells from human muscle biopsies, and purification of a highly myogenic population utilizing FACS to detect the cell surface markers CD56 in combination with tetraspanin CD82 (*see* **Note 1**). We recently demonstrated that CD82 is an excellent myogenic marker in both human fetal and adult skeletal muscle that is also retained on activated and differentiating myogenic progenitors [10]. This protocol also describes methods to confirm enrichment of muscle stem cells via Pax7 immunostaining; to culture the purified progenitors and confirm the presence of a myogenic population with an in vitro fusion assay. Isolation and expansion of these cells from normal individuals and from individuals with muscle disorders will help accelerate the development of therapies for human disorders such as muscular dystrophies.

2 Materials

2.1 Dissociation of Primary Cells

1. Protected disposable scalpels with stainless steel blade size #10.
2. Sterile 10 cm tissue culture-treated plastic dishes.
3. Assorted sterile 5, 10, and 25 mL pipettes.
4. Sterile 0.22 μm Polyethersulfone (PES) low protein binding membrane filters, 250 and 500 mL volumes.
5. 0.22 μm Cellulose nitrate (CN) filter unit, 500 mL volume.
6. Sterile 15 and 50 mL conical centrifuge tubes.
7. BD Falcon sterile nylon cell strainers, 100 μm and 40 μm pore sizes.
8. 10× Hank's Balanced Saline Solution (HBSS), free of calcium chloride, magnesium chloride, and magnesium sulfate, diluted to 1× with double distilled water and filter sterilized with a 0.22 μm CN filter. This solution can be stored at 4 °C or room temperature.
9. Complete growth medium: High glucose Dulbecco's Modified Eagle's Medium (DMEM), 20% fetal bovine serum (FBS), 1% Penicillin-Streptomycin-Glutamine (PSG 100× stock, Thermo Fisher 10378016). Mix 395 mL of high glucose DMEM with 100 mL of FBS (*see* **Note 2**) and 5 mL of 100× PSG. Sterilize by filtering the solution through a 500 mL 0.22 μm PES filter unit. Store at 4 °C and use within 1 month.

10. Sterile HEPES buffered saline solution, without phenol red.
11. 1M calcium chloride solution ($CaCl_2 \cdot 2H_2O$, FW 147). Dissolve 1.47 g powder in 10 mL of double distilled water. Store at 4 °C.
12. Dispase II 2.4 U/mL stock solution. Dissolve 1 g powder dispase II (Roche Applied Science) in 100 mL HEPES buffered saline to generate a stock solution of 2.4 U/mL. Check the total units for the specific dispase lot and add the appropriate volume of DMEM to reach the 2.4 U/mL concentration. Filter-sterilize the solution through a PES 500 mL filter; aliquot into 15 mL conical tubes (10 mL/tube) and store aliquots at −20 °C.
13. Collagenase D stock solution: dissolve 2.5 g powder collagenase D (Roche Applied Science) in 250 mL solution of 1× HBSS supplemented with 1.25 mL of 1 M $CaCl_2$. Sterilize by filtering through a PES filter unit. The filtered solution can be dispensed in 15 mL conical tubes (10 mL/tube) and stored at −20 °C.
14. Sterile freezing medium: 90% FBS and 10% dimethyl sulfoxide (DMSO). Prepare freezing medium and immediately store on ice. Unused sterile freezing medium can be stored at 4 °C for up to 4 weeks.
15. Sterile red blood cell lysis solution, stored at room temperature (such as Qiagen RBC Lysis solution).
16. Sterile 1.8 mL CryoTube™ vials.
17. Bench top centrifuge.
18. Hemocytometer.
19. Sterile laminar flow biosafety cabinet.
20. −150 °C freezer, liquid nitrogen storage tank.
21. Humidified 5% CO_2 incubator set to 37 °C.

2.2 Purification of Myoblasts

2.2.1 Thawing of Cryopreserved Sample

1. Sterile 0.22 μm PES filter, 500 mL volume.
2. Sterile 50 mL conical centrifuge tubes.
3. Sterile tissue culture-treated plastic dishes, 10 or 15 cm size.
4. CryoTube™ vial containing dissociated unpurified primary cells.
5. Sterile complete growth medium, as above.
6. Sterile laminar flow biosafety cabinet.
7. Water bath set to 37 °C.
8. CO_2 incubator, as above.

2.2.2 Preparation of Sample for FACS

1. Sterile 0.22 μm PES filter, 50 mL volume.
2. Sterile 15 and 50 mL conical centrifuge tubes.
3. Sterile 5 mL round bottom test tubes with cell strainer caps.
4. Dissociated unpurified primary cells.
5. Sterile 1× HBSS (diluted from 10× stock).
6. Sterile 1× Dulbecco's Phosphate Buffered Saline (DPBS, diluted from 10× stock).
7. TrypLE Express™ dissociation enzyme with phenol red.
8. Antibodies (all stored at 4 °C, protected from light): APC anti-CD56 antibody, Clone HCD56 (BioLegend, catalog number: 318310); PE anti-CD82 antibody, Clone ASL-24 (BioLegend, catalog number: 342103).
9. Calcein blue (1 mg vial): Resuspend in 200 μL dimethyl sulfoxide (DMSO). Aliquot in 25 μL aliquots and store at −20 °C (stock). Use 0.5 μL stock calcein/10^6 cells.
10. Sterile 0.5% Bovine Serum Albumin (BSA) in HBSS. Add 2.5 g BSA to 1× HBSS. Sterilize by filtering the solution through a 500 mL 0.22 μm PES filter unit. Store at 4 °C.
11. Sterile laminar flow biosafety cabinet.
12. Bench top centrifuge.
13. Inverted microscope.
14. Hemocytometer.

2.2.3 Fluorescence-Activated Cell Sorting (FACS)

1. FACS or 5 mL round-bottom tubes.
2. Cell sorting machine.
3. Cell sorting software.

2.3 In Vitro Culture of Myoblasts

2.3.1 In Vitro Cell Culture

1. Sterile 50 mL conical centrifuge tubes.
2. Sterile 10 cm tissue culture-treated plastic dishes.
3. Sterile 0.22 μm PES filter, 50 mL volume.
4. Sterile 1× DPBS.
5. TrypLE™ Express Dissociation Enzyme with Phenol Red (Invitrogen).
6. Sterile complete growth medium, as above.
7. Differentiation medium (50 mL): Mix 48.5 mL of low glucose Dulbecco's Modified Eagle's Medium (DMEM) with 1 mL of horse serum (HS) and 0.5 mL of 100× Penicillin-Streptomyocin-Glutamine (PSG). Sterilize by filtering the solution through a 150 mL 0.22 μm PES filter unit. Store at 4 °C and use within 1 month.
8. 0.1% gelatin: Add 0.5 g gelatin to 500 mL of double distilled water, do not shake. Sterilize the solution by autoclaving for 20 min and store at 4 °C.

9. Sterile laminar flow biosafety cabinet.
10. Water bath set to 37 °C.
11. Humidified 5% CO_2 incubator set to 37 °C.
12. Bench top centrifuge.
13. Inverted microscope.
14. Hemocytometer.

2.3.2 Immunofluorescent Staining of Pax7 and In Vitro Fusion Assay

1. Aluminum foil.
2. Microscope slides, 25 × 75 mm, 1 mm thick, such as Tissue Tack microscope slides from Polysciences.
3. Cytospin funnels, such as EZ Single Cytofunnel (Shandon).
4. Cytospin centrifuge, such as Cytospin 4 (Thermo Shandon).
5. 4-well chamber slides, Nunc Lab-Tek II Permanox.
6. 10× Phosphate Buffered Saline (PBS), diluted to 1x with double distilled water. Store at room temperature.
7. 4% paraformaldehyde (4% PFA) in PBS. Dilute 16% paraformaldehyde with 1× PBS. USE CAUTION as paraformaldehyde is extremely toxic; it is recommended that paraformaldehyde be used in a fume hood for safety. Aliquot and store at −20 °C. Aliquots should not be repeatedly frozen and thawed; discard unused PFA after initial use.
8. Permeabilization solution: PBS 0.5% Triton-X100. Mix 50 μL of Triton X-100 with 10 mL of 1× PBS.
9. Blocking solution: 10%FBS, 0.1% Triton-X100 in PBS. Mix 1 mL of fetal bovine serum (FBS), 10 μL of Triton X-100, and 9 mL of 1× PBS.
10. Antibodies (all stored at 4 °C): Anti-Pax7 (Developmental Studies Hybridoma Bank, PAX7 concentrate supernatant); AffiniPure $F(Ab')_2$ Alexa Fluor 594 Donkey anti-Mouse IgG (H + L) (Jackson ImmunoResearch, catalog number: 715-586-150), protect from light.
11. Inverted microscope with epi-fluorescence capabilities including ultraviolet/DAPI, FITC/GFP and Rhodamine/TRITC filter sets.

3 Methods

3.1 Dissociation of Primary Cells (See Note 3)

All the steps in this protocol should be performed in a sterile laminar flow biosafety cabinet using the sterile tissue culture technique. Human skeletal muscle can only be obtained following approval from the Institutional IRB. We obtained de-identified, discarded skeletal muscle tissue under a protocol approved by Boston Children's Hospital IRB.

1. Pre-weigh one 10 cm tissue culture plate, and place the tissue sample to be dissociated in a second (non pre-weighed) 10 cm tissue culture plate.
2. Using sterile scalpels, remove any connective tissue from the muscle tissue. Tissue should be kept moist. Add a few drops of sterile 1× HBSS to tissue as necessary to prevent it from drying out. Place muscle tissue in the pre-weighed 10 cm tissue culture plate, replace the lid, and weigh the plate again. Subtract from this number the tare of the empty plate to calculate the amount of muscle tissue to be dissociated.
3. Thaw frozen aliquots of dispase II and collagenase D in a 37 °C water bath. The solutions will be added at a volume of 3.5 mL each per gram of muscle tissue to be dissociated. Thaw only the amounts of collagenase D and dispase II necessary for dissociation. If an excess of enzymes is thawed, it can be refrozen once and reused.
4. Using sterile scalpels, mince muscle tissue until it resembles a fine paste. During mincing, add a few drops of sterile 1× HBSS to prevent exposed tissue from drying out. Tissue should always appear moist, but with no excess of liquid.
5. After tissue is finely minced, add equal amounts of the thawed dispase II and collagenase D solutions. The final concentration will be 5 mg/mL for collagenase D and 1.2 U/mL for dispase II in this solution. Pipette minced tissue and enzyme solution up and down through a sterile wide-bore 25 mL pipette a few times.
6. Incubate the plate in a humidified 5% CO_2 incubator set to 37 °C for 15 min.
7. Pipette the digestion solution up and down through a sterile 25 mL pipette a few times and incubate again for 15 min. Repeat this step an additional 1–2 times, until the slurry easily passes through a sterile 5 mL pipette and all tissue chunks are dissolved. The total digestion time will range between 45 min and 1 h 15 min.
8. Add 2 volumes of complete growth medium (based on the total volume of dispase II and collagenase D) to the digested slurry and filter the digestion solution through a 100 μm cell strainer over a 50 mL conical tube. Change cell strainer if it appears clogged.
9. Pellet cells for 10 min at 1100 × *g*, room temperature.
10. Resuspend the pellet in 1 volume of complete growth medium (i.e., 3 mL) and add 7 volumes (i.e., 21 mL) of red blood cell lysis solution. Invert the tube a few times and then filter the solution through a 40 μm cell strainer over a 50 mL conical tube.

11. Count cells using a hemocytometer, then pellet the cells for 10 min at 1100 × *g*, room temperature. Expect approximately 10^7 cells/gram tissue from postnatal skeletal muscle and 10^8 cells/gram tissue from fetal skeletal muscle. Cell numbers vary among individuals.
12. Freeze cells at a concentration of 1 × 10^7 cells/mL in ice-cold freezing medium. Store cryovials at −80 °C overnight, then transfer them to −150 °C where they can be permanently stored until necessary. Cell freezing is not required if all reagents and FACS equipment are immediately available. In this case, proceed to Subheading 3.2.2, **step 11**.

3.2 Purification of Myoblasts

All the steps in this protocol except for cell sorting (Subheading 3.2.3) should be performed in a sterile laminar flow hood using the sterile tissue culture technique. Cell sorting should be performed in as clean an environment as possible.

3.2.1 Thawing of Cryopreserved Sample Prior to FACS

Cryopreserved cells should be carefully thawed and plated 1 day prior to cell sorting. This allows the cells to recover from the freezing process before undergoing FACS.

1. Pre-warm complete growth medium in a water bath set to 37 °C. Then, pipette 10 mL pre-warmed medium into a sterile 50 mL conical tube.
2. Coat sterile tissue culture-treated plates (10 cm plate) with 10 mL 0.1% gelatin for 1 h at 37 °C, then remove the gelatin solution by aspiration.
3. Let the plates dry briefly in the biosafety cabinets and replace the lid.
4. Carefully and quickly thaw a vial of cryopreserved, dissociated cells in a 37 °C water bath and transfer the cells into the 50 mL conical tube with 15 mL pre-warmed proliferation medium using a 1 mL pipette. Rinse the inside of the cryovial with fresh complete growth medium to remove as many cells as possible. This step should be performed very quickly as the DMSO used during the cryopreservation process is toxic to the cells at room temperature.
5. Plate the cells in the pre-warmed medium onto sterile, tissue-culture treated plates at approximately 0.5–1 × 10^7 cells/10 cm plate or 1.5–3 × 10^7 cells/15 cm plate. If using a 15 cm plate, add 15 mL pre-warmed medium to bring the total medium volume to 25 mL.
6. Incubate the cells in a humidified 5% CO_2 incubator set to 37 °C overnight. If plating cells several days in advance, change the growth medium every other day and do not allow confluence to exceed 80%.

3.2.2 Preparation of Sample for FACS

1. Pre-warm complete growth medium and 1× DPBS in a 37 °C water bath. Place the 0.5% BSA/HBSS on ice.
2. Check cells under a phase contrast microscope with 10× magnification (*see* **Note 4**). Ensure that there is no contamination and that the cells look healthy.
3. Wash the cells with 5 mL (10 cm plate) or 10 mL (15 cm plate) of 1× DPBS 2–3 times.
4. Pipette 2 mL (10 cm plate) or 5 mL (15 cm plate) TrypLE Express™ dissociation enzyme onto the plate of washed cells and incubate in a humidified 5% CO_2 incubator set to 37 °C for 2 min.
5. Check under the microscope if the cells have lifted and are now floating freely in the medium. If the cells still adhere to the plate, gently tap the bottom of the plate to loosen the cells and return the plate to the incubator for another minute.
6. After incubation, remove the cells by gently swirling the medium, pipetting the cells a few times and pooling the cells to one side of the plate by tilting it at an angle (~45°), then carefully pipette the medium into the 50 mL conical tube (*see* **Note 5**).
7. Repeat **step 6** with growth medium (5 mL for 10 cm plate and 10 mL for 15 cm plate) to collect any remaining cells and quench the TrypLE™ (*see* **Note 6**).
8. Check the plate under a phase contrast microscope at 10× magnification for the presence of cells. There should be very few cells on the surface of the plate after this process.
9. Centrifuge the 50 mL conical tubes containing the cells and wash at 1100 × *g* at 4 °C for 10 min to pellet the cells.
10. Remove the supernatant, and resuspend the cells in 10 mL 5% FBS/HBSS.
11. Determine the cell concentration using a hemocytometer or other cell counting device.
12. For FACS controls, use 5 mL round-bottom test tubes and set aside 2.5×10^5 cells in 500 μL 5% FBS/HBSS for each of the following controls: "Unstained" control; Calcein Blue single color control (live cells); CD56 single color control; CD82 single color control.
13. Pipette the "unstained" control sample through the strainer cap of a 5 mL round-bottom test tube (*see* **Note** 7). Keep on ice.
14. Centrifuge the remaining cells (to be labeled with both CD56 and CD82 antibodies or single color controls) for 10 min at 1100 × *g* at 4 °C.

15. Resuspend cells at a concentration of 1×10^7/mL in 5% FBS/HBSS.
16. Primary antibody incubation: add CD56 and CD82 antibodies to the appropriate cell solutions at a concentration of 5 μL per 1×10^6 cells (as recommended by the manufacturer).
17. To gate for live cells, add calcein blue at a concentration of 0.5 μL per 1×10^6 cells to the appropriate cell solutions. Gently mix and place on ice protected from the light for 30 min (*see* **Note 8**).
18. After this incubation, wash the cells 1× in 2 mL of 5% FBS/HBSS.
19. Centrifuge the cells for 10 min at 1100 × *g* at 4 °C.
20. Resuspend the CD56 and CD82 single color controls in 500 μL of 5% FBS/HBSS, and pipette through the strainer cap of a 5 mL round bottom test tube. Store on ice in the dark.
21. Resuspend the CD56/CD82/calcein blue stained cells in 1 mL of 5% FBS/HBSS, and pipette through the strainer cap of a 5 mL round-bottom test tube. Store on ice in the dark.
22. Prepare collection tube for $CD56^+CD82^+$ sorted cells by pipetting 500 μL of growth medium into a new tube. Store on ice.

3.2.3 Fluorescence-Activated Cell Sorting

It is beyond the scope of this chapter to review FACS or flow cytometry in detail. Gating specifications are briefly indicated.

1. Determine optimal excitation voltages and compensation values using the "no stain" and single color controls (Fig. 1a).
2. Determine the live cell population gating for calcein blue positive cells (Fig. 1b).
3. Determine the double positive (DP) $CD56^+/CD82^+$ and double negative (DN) populations. Gate and sort for the DP cell population (Fig. 1c).

3.3 In Vitro Culture of Myoblasts

3.3.1 In Vitro Cell Culture

All the steps in this protocol except immunofluorescent staining (Subheading 3.3.5) should be performed in a sterile laminar flow hood using the sterile tissue culture technique.

1. Coat sterile 10 cm tissue culture-treated plates with 10 mL 0.1% gelatin for 1 h in a humidified 5% CO_2 incubator set to 37 °C, then remove the gelatin solution by aspiration. Let the plates dry briefly in the biosafety cabinets and replace the lid.
2. Pre-warm complete growth medium in a water bath set to 37 °C.
3. Resuspend sorted CD56/CD82 double positive cells at 0.5–1×10^6 cells/10 mL complete growth medium and plate on coated plates. Gently rock plate(s) to evenly distribute

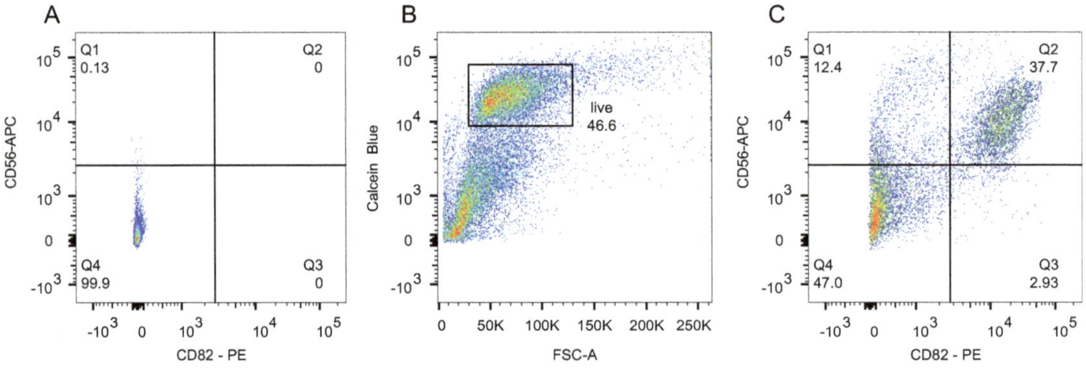

Fig. 1 Gating of myogenic cells double positive for CD56 and CD82 from dissociated human skeletal muscle following FACS analysis. (**a**) Unstained control; (**b**) Gating of live cells based on Calcein blue uptake and (**c**) gating of double positive cells (Q2) that will be sorted

cells, and then place in a 5% CO_2 incubator set to 37 °C. Sorted cells will be small and have a bright, rounded appearance and should attach within 1 day post-sorting.

4. Propagate the cells to 60–75% confluency (*see* **Note 9**). This should take approximately 2–3 days; however, if necessary, replace the medium with fresh growth medium every 2 days until the plate is at 60–75% confluency.

3.3.2 Cell Passaging

1. Coat sterile tissue culture-treated plates with 0.1% gelatin as in Subheading 3.3.1, **step 1**.
2. Pre-warm the following in a water bath set to 37 °C: 1× DPBS, TrypLE™ Express dissociation enzyme, and complete growth medium.
3. Remove the medium from the plate by aspiration and wash the cells twice with 10 mL (10 cm plate) 1× DPBS. Remove DPBS by aspiration.
4. Pipette 2 mL TrypLE™ Express onto the plate and incubate in a humidified 37 °C CO_2 incubator for 2–3 min. Gently remove the cells from the plate by pipetting up and down a few times before transferring cells into a sterile conical tube. Wash any remaining cells from the surface of the plate with additional complete growth medium.
5. Centrifuge the cells at 1100 × *g* at room temperature for 10 min.
6. Resuspend the cells in 10 mL fresh complete growth medium.
7. Determine the cell concentration using a hemocytometer and plate the cells at 0.5–1 × 10^6 cells in 10 mL complete growth medium/10 cm plate.
8. Cells should be passaged every 2–3 days and should not be grown past 75% confluency.

3.3.3 Cell Freezing

1. Remove the medium from the plate by aspiration and wash the cells twice with 10 mL (10 cm plate) 1× DPBS. Remove DPBS by aspiration.
2. Pipette 2 mL TrypLE™ Express onto the plate and incubate in a humidified 5% CO_2 incubator set to 37 °C for 2–3 min.
3. Gently remove the cells from the plate by pipetting up and down a few times before transferring cells into a sterile conical tube. Wash any remaining cells from the surface of the plate with additional complete growth medium.
4. Centrifuge the cells at 1100 × *g* at room temperature for 10 min.
5. Resuspend cells in ice-cold freezing medium (10% DMSO in 90% fetal Bovine Serum) at desired cell concentration (10^6–10^7/mL).
6. Store cryovials at −80 °C overnight and then transfer to −150 °C where they can be permanently stored until necessary.

3.3.4 In Vitro Fusion Assay

1. Coat 4-well chamber slides with 0.1% gelatin.
2. Trypsinize the cells with TrypLE Express™ Dissociation Enzyme with Phenol Red and determine the cell concentration as described above, then plates 20,000 cells in 500 μL complete growth medium/well.
3. Incubate the cells in a humidified 5% CO_2 incubator set to 37 °C until the cells are ~80% confluent.
4. When the cells are ~80% confluent, remove the growth medium from each well and replace with 500 μL pre-warmed differentiation medium (*see* **Note 10**).
5. Incubate the cells in a humidified 5% CO_2 incubator set to 37 °C overnight.
6. Replace the differentiation medium in each well daily during the course of the fusion assay.
7. Monitor the differentiation of the cells using a phase contrast microscope at 10× or 20× magnification. Fusion and robust formation of myotubes should occur within 1 week of exposure to differentiation medium (*see* examples in Fig. 2).

3.3.5 Immunofluorescent Staining for PAX7 (See Note 11)

1. Thaw 4% PFA at room temperature.
2. Following cell sorting, set aside an aliquot of the sorted cells for cytospins. Estimate to use between 2000 and 3000 cells for each cytospin.
3. Resuspend the cells for cytospins at a concentration of 2000 cells/150 μL in 4% PFA solution. Fix cells for 20 min at room temperature; then cytospins 150 μL cells in each cytofunnel. Spin at 600 rpm for 5 min, then remove slides, draw a

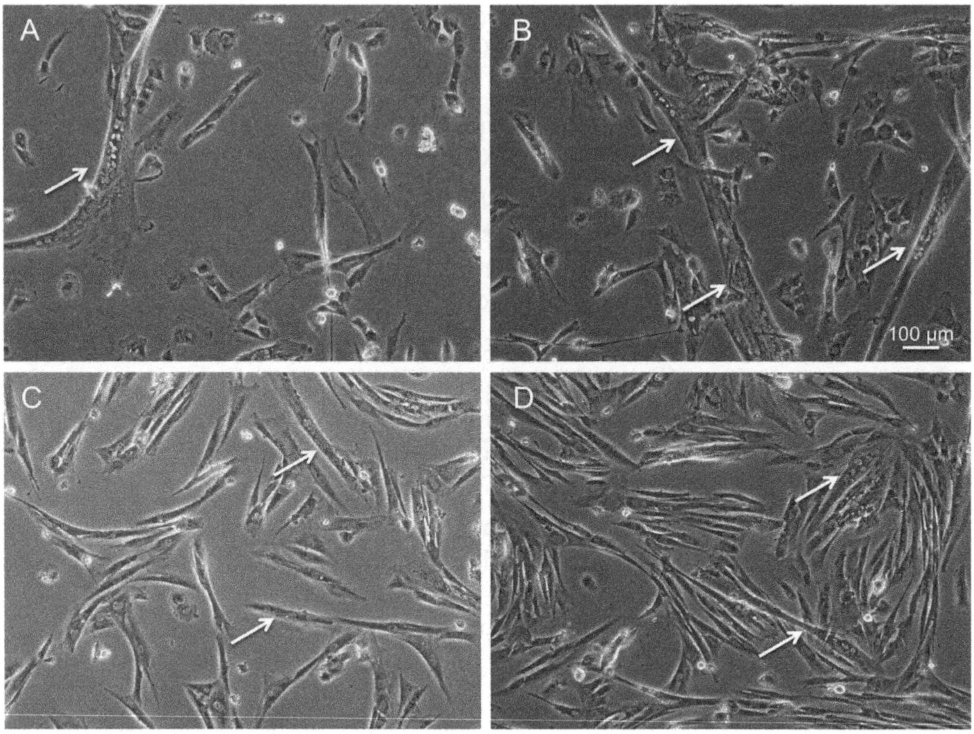

Fig. 2 Examples of human fetal (**a**, **b**) and human adult (**c**, **d**) $CD56^{+}CD82^{+}$ sorted cells showing myotube formation. Scale bar: 100 μm. Arrows point to myotubes in the fields

hydrophobic barrier around the cytospun cells, and place the slides in 1× PBS (*see* **Note 12**).

4. Permeabilize the cells with 200 μL of permeabilization solution for 3 min at room temperature.
5. Remove permeabilization solution using a pipette, and then block the cells for 30 min at room temperature with 200 μL of blocking solution.
6. Prepare the primary antibody solution by diluting the anti-PAX7 concentrate supernatant antibody 1:100 in fresh blocking solution. Incubate the cells with primary antibody solution overnight at 4 °C.
7. Wash the cells three times with 1× PBS for 5 min at room temperature. The slides may be gently agitated on a rotating shaker.
8. Prepare the secondary antibody solution by diluting the anti-mouse antibody 1:1000 in blocking solution. Incubate the cells in the dark with secondary antibody solution for 1 h at room temperature.
9. Wash the cells three times with 1× PBS for 5 min at room temperature in the dark. The plate may be gently agitated on a rotating shaker.

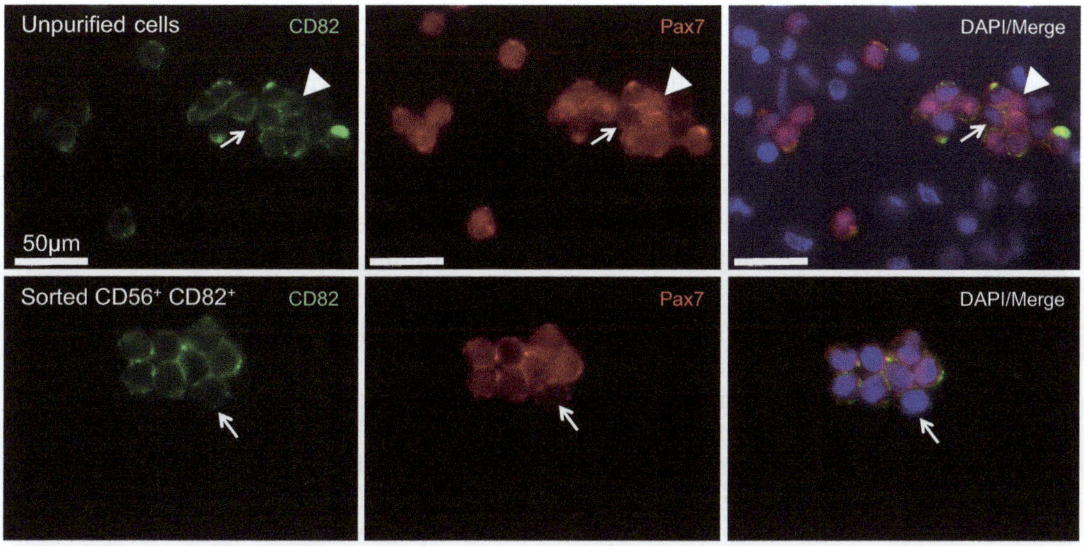

Fig. 3 Immunostaining of CD82 and Pax7 in unpurified cells (upper panels) and in sorted $CD56^+$ $CD82^+$ progenitors (lower panels). CD82 is expressed in >90% of Pax7+ cells (arrowhead), although $CD82^+$ $Pax7^-$ cells are also present (arrows). Scale bars: 50 μm

10. Mount the cells with Vectashield with DAPI and store at 4 °C protected from light with aluminum foil.
11. Visualize the cells by fluorescent microscopy using ultraviolet/DAPI and TRITC filter sets for DAPI and PAX7, respectively. Following cert sorting >85% of the cells should be expressing PAX7 when using fetal tissue (*see* Fig. 3).

3.3.6 In Vitro Fusion Assay Fusion Index Calculation

1. Count the following in each of five random fields per well: Number of total nuclei; number of nuclei within myotubes..
2. Calculate fusion index (%) as: number of nuclei within myotubes/number total nuclei (× 100).
3. Average the fusion index of the five fields.
4. Compare the fusion index of sorted myoblasts versus unsorted.

4 Notes

1. We would like to note that FACS with either CD56 (NCAM) or CD146 (MCAM) in conjunction with CD82 as enriching markers is a highly effective method for isolating human fetal myogenic progenitors. We refer the readers to the following protocol describing use of MCAM as a positive selection marker in cells sorted from human fetal tissue [11]. For adult skeletal muscle, CD82 should be used in conjunction with CD56, since endothelial cells express MCAM.

2. FBS varies considerably between companies and even lot to lot from the same company. Therefore, several different FBS samples should be tested using the in vitro methods described in Subheading 3.3 to determine which lot/company works best.
3. Institutional review and protocol approval are required prior to collection and processing of human tissue. All personnel handling human tissue must receive appropriate safety and human subject education training.
4. There will be many floating, live cells in the culture, which is normal for dissociated human skeletal muscle. It is also likely that there will be small clumps of cells in the culture, and the number of clumps will vary. These clumps will be filtered out prior to cell sorting. Additionally, the dissociation process results in a large amount of debris in addition to cells. This will make the culture appear "dirty" (i.e., little black specks, etc.), but again, this is normal and should not be considered contamination. This debris will be removed during the FACS sample preparation process.
5. When the plate is tilted at an angle, the cells can be seen on the surface of the plate as a light opaque coating. Repeatedly rinse gently the cells off the plate using the trypsin solution (TrypLE Express™ Dissociation Enzyme with Phenol Red) until this coating is no longer visible.
6. Prolonged exposure of the cells to undiluted TrypLE Express™ may negatively affect the health of the cells. Addition of growth medium to the trypsinized cells will quench this effect.
7. Some FACS machines may require tubes that are different in diameter/size from the tube specified in this protocol. Check in advance that the tubes fit in the FACS machine.
8. Calcein blue is a cell viability dye and is used in this protocol to discriminate live from dead cells/debris during the FACS.
9. Cells should never reach 100% confluency when proliferating, as they will begin to differentiate and fuse on contact. The high serum growth medium will lower in serum concentration over time and will not be able to prevent fusion (*see* **Note 10**).
10. Low serum medium induces differentiation and fusion of myoblasts in culture [12].
11. This immunofluorescent staining protocol can also be utilized for the detection of other myogenic markers of proliferating or differentiating cells.
12. It is useful to draw a hydrophobic barrier around the spot where the cells were deposited on the slide following cytospins using a PAP pen. This will minimize the amount of antibody solution to be used during the immunostaining process.

Acknowledgments

This work is supported by a grant from the Muscular Dystrophy Association #479606 (EG) and by the National Institute of Arthritis and Musculoskeletal and Skin Diseases of the National Institutes of Health under Award Number 1R01AR069582-01 (EG). The content is solely the responsibility of the authors and does not necessarily represent the official views of the National Institutes of Health. This protocol was modified from previous work, specifically from the listed references [10, 11].

References

1. Geiger RS, Garvin JS (1957) Pattern of regeneration of muscle from progressive muscular dystrophy patients cultivated in vitro as compared to normal human skeletal muscle. J Neuropathol Exp Neurol 16(4):523–543
2. Herrmann H, Konigsberg UR, Robinson G (1960) Observations on culture in vitro of normal and dystrophic muscle tissue. Proc Soc Exp Biol Med 105:217–221
3. Goyle S, Kalra SL, Singh B (1967) The growth of normal & dystrophic human skeletal muscle in tissue culture. Neurol India 15(4):149–151
4. Bishop A, Gallup B, Skeate Y, Dubowitz V (1971) Morphological studies on normal and diseased human muscle in culture. J Neurol Sci 13(3):333–350
5. Witkowski JA (1977) Diseased muscle cells in culture. Biol Rev Camb Philos Soc 52 (4):431–476
6. Hauschka SD (1974) Clonal analysis of vertebrate myogenesis. II. Environmental influences upon human muscle differentiation. Dev Biol 37(2):329–344
7. Blau HM, Webster C (1981) Isolation and characterization of human muscle cells. Proc Natl Acad Sci U S A 78(9):5623–5627
8. Webster C, Pavlath GK, Parks DR, Walsh FS, Blau HM (1988) Isolation of human myoblasts with the fluorescence-activated cell sorter. Exp Cell Res 174(1):252–265
9. Walsh FS, Ritter MA (1981) Surface antigen differentiation during human myogenesis in culture. Nature 289(5793):60–64
10. Alexander MS, Rozkalne A, Colletta A, Spinazzola JM, Johnson S, Rahimov F, Meng H, Lawlor MW, Estrella E, Kunkel LM, Gussoni E (2016) CD82 is a marker for prospective isolation of human muscle satellite cells and is linked to muscular dystrophies. Cell Stem Cell 19(6):800–807. https://doi.org/10.1016/j.stem.2016.08.006
11. Lapan AD, Gussoni E (2012) Isolation and characterization of human fetal myoblasts. Methods Mol Biol 798:3–19. https://doi.org/10.1007/978-1-61779-343-1_1
12. Yaffe D, Saxel O (1977) A myogenic cell line with altered serum requirements for differentiation. Differentiation 7(3):159–166

Chapter 2

Electrical Pulse Stimulation of Primary Human Skeletal Muscle Cells

Nataša Nikolić and Vigdis Aas

Abstract

Electrical pulse stimulation (EPS) is an in vitro method of inducing contractions in cultured skeletal muscle cells of human and animal origin. Motor neuron activation of muscle fibers can be replaced by applying EPS on differentiated skeletal muscle cells (myotubes) in culture (Thelen et al. Biochemical J 321:845–848, 1997, Fujita et al. Exp Cell Res 313:1853–1865, 2007).

Here we describe two protocols for EPS of human myotubes in 6-well plates: acute, high-frequency (single bipolar pulses of 2 ms, 100 Hz for 200 ms every fifth second for 5–60 min, 10–30 V) and chronic, low-frequency (single bipolar pulses of 2 ms, 1 Hz 10–30 V for 48 h) at the end of a 7 days long differentiation.

Key words Skeletal muscle, Myotubes, Contractions, Electrical pulse stimulations

1 Introduction

Skeletal muscle is remarkable in its ability to adapt and remodel in response to contractile activity. Many of the contraction-induced changes in skeletal muscle fibers are well characterized in vivo, but molecular mechanisms underlying these adaptations are still not completely understood. Cell culture systems derived from human muscle biopsies have been used in research for decades and represent a valuable alternative to adult skeletal muscle with respect to genetic background, morphological, metabolic, and biochemical properties [1]. However, differentiated human skeletal muscle cells (myotubes) are quiescent in cultures, and do not typically contract spontaneously, unless being innervated or electrically stimulated [2, 3]. Motor neuron activation can be replaced by electrical pulse stimulation (EPS) in culture [4, 5], and this model is widely applied as an in vitro model of contractions [6–12].

Both commercially available C-Pace culture pacing systems and home-designed pulse generators have been successfully used on different skeletal muscle cell preparations. Several papers have

Sissel Beate Rønning (ed.), *Myogenesis: Methods and Protocols*, Methods in Molecular Biology, vol. 1889,
https://doi.org/10.1007/978-1-4939-8897-6_2, © Springer Science+Business Media, LLC, part of Springer Nature 2019

been published employing C2C12 cell lines, where chronic electrostimulation ("pacing") was used for both de-novo sarcomere formation and a hypertrophic response, rather similar to the signaling programs seen in intact muscle upon exercise in skeletal muscle [5, 8, 13].

Development and application of EPS in human myotubes has mainly been directed toward metabolic and inflammatory responses related to obesity and type 2 diabetes (T2D). By exposing human myotubes to EPS, others and we have been able to induce metabolic adaptations in human myotubes similar to some of the in vivo responses to physical exercise. We have shown that acute, high-frequency EPS has similar effects on glucose uptake in human myotubes in culture as in vivo muscle contractions, with a 30% increase in glucose uptake, resembling an acute high intensity bout of exercise [6, 14]. However, in a clinical setting, beneficial effects of exercise are more profound after regular physical activity, which may be simulated by exposing human myotubes to chronic, low-frequency EPS. Several research groups using human myotubes from different donors have used this type of EPS, with minor variations. Among the observed effects of chronic, low-frequency EPS of human myotubes are reorganization of the cytoskeleton, de novo formation of sarcomeric structures, visible contractions and profound metabolic adaptations reflected in improved oxidative capacity, and increased insulin sensitivity [6, 7, 9]. Importantly, this model of EPS has been an important in vitro tool to identify novel contraction-induced myokines [10].

In this chapter, we describe two EPS approaches to stimulate human myotubes: chronic, low-frequency EPS, and acute, high-frequency EPS.

2 Materials

1. Cells: EPS is performed on adherent, differentiated myotubes. We use human myotubes obtained from human satellite cells.
2. Plating medium: Dulbecco's modified Eagle's medium (DMEM)-Glutamax™ (5.5 mM glucose) supplemented with 2% heat-inactivated FCS, 50 U/ml penicillin, 50 μg/ml streptomycin, and 1.25 μg/ml amphotericin B.
3. Proliferation medium: DMEM- Glutamax™ (5.5 mM glucose) supplemented with 2% heat-inactivated FCS, 50 U/ml penicillin, 50 μg/ml streptomycin, 1.25 μg/ml amphotericin B, 50 mg/ml gentamicin and 25 mM 4-(2-hydroxyethyl)-1-piperazineethanesulfonic acid (HEPES).
4. Differentiation medium: DMEM-Glutamax™ (5.5 mM glucose) supplemented with 2% heat-inactivated FCS, 50 U/ml penicillin, 50 μg/ml streptomycin, 1.25 μg/ml amphotericin B, 50 mg/ml gentamicin, 25 mM HEPES, and 25 pM insulin.

Fig. 1 C-Dish with carbon electrodes used for chronic, low-frequency EPS

5. Corning CellBIND six-well plates (Corning Life-Sciences, Schiphol-Rijk, The Netherlands).
6. BD Matrigel™ Basement Membrane Matrix (Bedford, MA, USA).
7. Pulse generators and C-Dish electrodes: We use either commercially available IonOptix C-Pace pulse generator (IonOptix, Dublin, Irland), or a self-designed pulse generator. In either case, for chronic EPS, we use commercially available C-Dishes (IonOptix, Dublin, Ireland) arranged to precisely place carbon electrode elements to standard 6-well culturing plates (Fig. 1). Self-designed rod carbon electrodes (4 mm diameter), fixed to lid of a culture plate are used for acute, high-frequency stimulation (*see* **Note 1**) (Fig. 2).

3 Methods

All work involving cells and C-Dish electrodes is carried out in a cell culture safety hood using the aseptic working technique. Cells are cultured in a humidified 5% CO_2 atmosphere at 37 °C.

Fig. 2 Rod carbon electrodes fixed to a culture plate lid, used for acute, high-frequency EPS

3.1 Establishment of Differentiated Human Skeletal Muscle Cells (Myotubes)

1. A bank of myoblasts established from human muscle biopsy samples, obtained with informed consent and according to regulations of national ethic committees, as previously described [15] is stored in liquid nitrogen.
2. Human myoblasts are taken out from liquid nitrogen, thawed and seeded at a density of 100,000 cells/well in a corning CellBIND (*see* **Note 2**) 6-well plate, using 1–2 ml/well (*see* **Note 3**) of plating medium. After 24 h, plating medium is replaced by equal volume of proliferation medium, which is then changed every 2–3 days. At approximately 80% confluence (usually after 1–2 weeks), medium is changed to differentiation medium, which induces differentiation of myoblasts into multinucleated myotubes. Differentiation medium is changed every 2–3 days, and cells are differentiated for 7 days.
3. Myotubes should be attached to the surface of the culturing plate and fully differentiated (at least 4 days of differentiation) prior to EPS. Inspect the cells frequently under a microscope, both during culturing, before the onset of, and during EPS (*see* **Note 4**).

3.2 Electrical Pulse Stimulation (EPS)

1. Chronic, low-frequency EPS should be applied on day 5 of the differentiation period. Onset of acute, high-frequency EPS is on day 7 of the differentiation.
2. For chronic, low-frequency EPS, use C-Dishes from Ionoptix. For acute, high-frequency EPS, use rod carbon electrodes.
3. Sterilize the C-Dishes and rod carbon electrodes immediately before use by spraying with 70% ethanol, and allow them to air dry completely.
4. Fresh, warm (37 °C) differentiation medium should be added to the cells before EPS is started.
5. Fit the C-Dish on top of the culturing plate so that C-Dish electrodes descend into the medium. The cell culture plate lid will fit the groove on the top of the C-Dish. Rod carbon electrodes are already fixed to the lid of a culturing plate.
6. Transfer the assemblies of culturing plates, C-Dishes/rod carbon electrode lids into the incubator.
7. Connect the assemblies to the pulse generator with the appropriate cables, run the cables through the door opening of incubator, and apply the stimulation protocol as anticipated. For chronic, low-frequency stimulation, we apply single, bipolar pulses of 2 ms, 30 V and 1 Hz continuously for the last 24 or 48 h of a 7 days differentiation period (*see* **Note 5**). For acute, high-frequency stimulation, we generally apply single bipolar pulses of 2 ms, 100 Hz for 200 ms every fifth second for 5–60 min, 10–30 V, on day 7 of the differentiation.
8. Change culturing medium once per 24 h during chronic, low-frequency EPS. This is carried out by carefully removing the C-Dish from the culturing plate. Fresh medium is then added to the cells. During medium change, C-Dish electrodes are again sterilized by spraying with 70% ethanol, allowed to dry in the safety hood, before being placed back on the top of cell culturing plates.
9. After ended stimulation, cells are harvested and C-Dishes and rod carbon electrode lids are cleaned and sterilized as described in the following section.

3.3 Cleaning and Sterilization of C-Dishes and Rod Carbon Electrode Lids

Electrodes must be thoroughly cleaned and sterilized after each use, following next procedure (*see* **Note 6**).

3.3.1 Cleaning

1. After ended application, soak the entire C-Dish or rod carbon electrode lid in distilled water for a couple of days with daily changing the water. The water should be changed until the distilled water stops turning pink. Additionally, measurement

of the water's pH may indicate whether ions or other chemicals still leach from the electrodes.

2. Use a soft toothbrush to scrub the surface of the electrodes under rinsing distilled water to avoid accumulation of absorbed salts (*see* **Note 7**).
3. After the cleaning procedure, electrodes must be sterilized.

3.3.2 Sterilization

1. Autoclave should not be used for this purpose (*see* **Note 8**). Instead, sterilization is performed using either of the following procedures:
   (a) Disinfection with 70% ethanol, followed by placing the C-Dish or rod carbone electrode lid under UV germicidal lamp in the cell culture hood for 15 min.
   (b) Heating the C-Dish or rod carbon electrode lid in a drying oven at 100 °C for 3 h.
2. Cleaned and sterilized C-dishes and rod carbon electrode lids are placed on top of sterile 6-well culturing plates, with the culture plate lid fitted on top, rapped in aluminum foil, and stored at room temperature (*see* **Note 9**).

4 Notes

1. C-Dishes from Ionoptix are, according to producer, specifically designed to enable chronic electrostimulation ("pacing") of cultured cells. We have tested C-Dishes in combination with acute, high-frequency EPS as described in current protocol, which resulted in overheating, detachment from culturing surface and cell death. Thus, as implied by the producer, C-Dishes cannot be used for this type of EPS in our cell culture system. Rod carbon electrodes we use for the purpose of acute, high-frequency EPS, on the other hand, have smaller diameter, and can be used for high-frequency stimulation without damaging the myotubes.
2. The cells should be attached to the surface of cell culturing plate. We generally use Corning CellBIND plates for EPS. However, in our hands, human myotubes originating from different donors may respond differently to EPS, and proper coating should be used to avoid cell detachment. For cells that detach under EPS when cultured on Corning CellBIND plates, we have successfully performed 48 h of EPS by coating the plates with BD Matrigel™ Basement Membrane Matrix. For that purpose, BD Matrigel™ Basement Membrane Matrix solution is thawed in the fridge for 2 h, and diluted 1:50 in DMEM. Cell culture plates are coated for 2 h at room temperature, and BD Matrigel™ Basement Membrane Matrix is removed by aspiration.

3. Electrodes must be immersed in liquid during EPS, and volume of medium is adjusted accordingly. Volume of 2 ml is sufficient when Corning CellBIND 6-well plates and the compatible 6-well C-Dishes are used. For acute, high-frequency EPS, performed using carbon rod electrodes, cells are cultured in 1 ml/well medium. Thus, the volume required will vary depending on the length of electrodes and types of plates. Always check that the electrodes are immersed in medium and adjust volume if necessary.
4. Visualizing contractions in primary human myotubes during EPS may be challenging. Not all cells contract at all times. Avoid vigorous movements when taking the cells out of the incubator under stimulation, as both movements and temperature variations may affect contractions. For optimal visualization of cells during EPS, a microscope within a temperature-controlled chamber may be used. Take time when looking for contractions under a microscope.
5. Preferably, voltage should not be higher than 30 V for the described protocol designed around a 6-well assembly. Higher voltage increases the risk of overheating. For chronic, low-frequency EPS, we have been using voltage from 8 to 30 V, and for acute high-frequency stimulation we have used 10–30 V.
6. Carbon electrodes absorb salts, electrolysis by-products, and proteins from the medium. Toxic chemicals may leach from the electrodes into the medium and could induce unspecific effects or even kill the cells. Although it is time-consuming, performing the entire cleaning and sterilization procedure is an essential part of a successful EPS experiment.
7. Toothbrush will often be black once the rubbing of electrodes is finished, but carbon is a solid material, and this should not be a matter of concern.
8. While Ionoptix strongly discourages using autoclave for sterilization of C-Dishes, it is specified in their manual that if the dishes must be completely sterilized, autoclaving cycles should be limited to 20 min at 121 °C. Temperature should not exceed 145 °C, as it may damage the C-Dishes. Following the described cleaning and sterilization procedure in our protocol, however, we have not experienced contamination, and we do not recommend using autoclave for sterilization.
9. Due to substantial cleaning routines and the fact that carbon is a porous material, it is highly recommended to have multiple C-Dishes and rod carbon electrodes available. This allows the researcher to switch between the electrodes in different experiments, while performing proper maintenance (cleaning and sterilization) after each use.

References

1. Aas V, Bakke SS, Feng YZ, Kase ET, Jensen J, Bajpeyi S, Thoresen GH, Rustan AC (2013) Are cultured human myotubes far from home? Cell Tissue Res. https://doi.org/10.1007/s00441-013-1655-1
2. Nikolic N, Gorgens SW, Thoresen GH, Aas V, Eckel J, Eckardt K (2017) Electrical pulse stimulation of cultured skeletal muscle cells as a model for in vitro exercise - possibilities and limitations. Acta Physiol 220(3):310–331. https://doi.org/10.1111/apha.12830
3. Askanas V, Kwan H, Alvarez RB, Engel WK, Kobayashi T, Martinuzzi A, Hawkins EF (1987) De novo neuromuscular junction formation on human muscle fibres cultured in monolayer and innervated by foetal rat spinal cord: ultrastructural and ultrastructural--cytochemical studies. J Neurocytol 16(4):523–537
4. Thelen MH, Simonides WS, van Hardeveld C (1997) Electrical stimulation of C2C12 myotubes induces contractions and represses thyroid-hormone-dependent transcription of the fast-type sarcoplasmic-reticulum Ca2-+-ATPase gene. Biochem J 321 (Pt 3):845–848
5. Fujita H, Nedachi T, Kanzaki M (2007) Accelerated de novo sarcomere assembly by electric pulse stimulation in C2C12 myotubes. Exp Cell Res 313(9):1853–1865. https://doi.org/10.1016/j.yexcr.2007.03.002 S0014-4827(07)00106-1 [pii]
6. Nikolic N, Bakke SS, Kase ET, Rudberg I, Flo Halle I, Rustan AC, Thoresen GH, Aas V (2012) Electrical pulse stimulation of cultured human skeletal muscle cells as an in vitro model of exercise. PLoS One 7(3):e33203. https://doi.org/10.1371/journal.pone.0033203
7. Feng YZ, Nikolic N, Bakke SS, Kase ET, Guderud K, Hjelmesaeth J, Aas V, Rustan AC, Thoresen GH (2015) Myotubes from lean and severely obese subjects with and without type 2 diabetes respond differently to an in vitro model of exercise. Am J Physiol Cell Physiol 308(7):C548–C556. https://doi.org/10.1152/ajpcell.00314.2014
8. Nedachi T, Fujita H, Kanzaki M (2008) Contractile C2C12 myotube model for studying exercise-inducible responses in skeletal muscle. Am J Physiol Endocrinol Metab 295(5): E1191–E1204. https://doi.org/10.1152/ajpendo.90280.2008 90280.2008 [pii]
9. Lambernd S, Taube A, Schober A, Platzbecker B, Gorgens SW, Schlich R, Jeruschke K, Weiss J, Eckardt K, Eckel J (2012) Contractile activity of human skeletal muscle cells prevents insulin resistance by inhibiting pro-inflammatory signalling pathways. Diabetologia 55(4):1128–1139. https://doi.org/10.1007/s00125-012-2454-z
10. Raschke S, Eckardt K, Bjorklund Holven K, Jensen J, Eckel J (2013) Identification and validation of novel contraction-regulated myokines released from primary human skeletal muscle cells. PLoS One 8(4):e62008. https://doi.org/10.1371/journal.pone.0062008
11. Danilov K, Sidorenko S, Milovanova K, Klimanova E, Kapilevich LV, Orlov SN (2017) Electrical pulse stimulation decreases electrochemical Na+ and K+ gradients in C2C12 myotubes. Biochem Biophys Res Commun 493(2):875–878. https://doi.org/10.1016/j.bbrc.2017.09.133
12. Tarum J, Folkesson M, Atherton PJ, Kadi F (2017) Electrical pulse stimulation: an in vitro exercise model for the induction of human skeletal muscle cell hypertrophy. A proof-of-concept study. Exp Physiol. https://doi.org/10.1113/EP086581
13. Nedachi T, Hatakeyama H, Kono T, Sato M, Kanzaki M (2009) Characterization of contraction-inducible CXC chemokines and their roles in C2C12 myocytes. Am J Physiol Endocrinol Metab 297(4):E866–E878. https://doi.org/10.1152/ajpendo.00104.2009
14. Aas V, Torbla S, Andersen MH, Jensen J, Rustan AC (2002) Electrical stimulation improves insulin responses in a human skeletal muscle cell model of hyperglycemia. Ann N Y Acad Sci 967:506–515
15. Henry RR, Abrams L, Nikoulina S, Ciaraldi TP (1995) Insulin action and glucose metabolism in nondiabetic control and NIDDM subjects. Comparison using human skeletal muscle cell cultures. Diabetes 44(8):936–946

Chapter 3

Transdifferentiation of Muscle Satellite Cells to Adipose Cells Using CRISPR/Cas9-Mediated Targeting of MyoD

Jingjuan Chen, Chao Wang, and Shihuan Kuang

Abstract

Brown adipocytes dissipate energy through non-shivering thermogenesis mediated by UCP1 protein, hence representing a powerful target to overcome obesity due to energy surplus. However, brown adipocytes are scarce in adult humans, especially in obese subjects, urging the development of novel strategies to boost the number of these thermogenic adipocytes from a therapeutical perspective. In this regard, transdifferentiation of myoblasts into brown adipocytes represents a promising approach. Here, we describe a method that we have recently developed to transdifferentiate myoblasts into brown adipocytes through CRISPR/Cas9-medidated targeting of MyoD, the master myogenic regulatory factor.

Key words Satellite cell, Myoblast, Adipocyte, Transdifferentiation, Adipogenesis, CRISPR/Cas9

1 Introduction

Brown adipocytes, previously thought to be absent in adult humans, have recently been detected in the neck, supraclavicular, paravertebral, mediastinal, para-aortic, infradiaphragmatic, and suprarenal regions by PET scans and found to be active [1–6]. More recent studies suggest that human brown adipocytes are similar to the inducible thermogenic adipocytes (namely beige or brite adipocytes) found in subcutaneous white adipose tissues (WAT) of rodents [7]. Innervated by the sympathetic nervous system and in response to cold stimuli, brown adipocytes undergo thermogenesis through uncoupled respiration mediated by uncoupling protein-1 (UCP1), a protein that is uniquely found in brown/beige adipocytes and localized to the inner membrane of mitochondria. UCP1 catalyzes the proton leak across the mitochondria membrane thus uncouples the proton from ATPase, rendering energy dissipation as heat instead of generating ATP. Brown

Jingjuan Chen and Chao Wang contributed equally to this work.

Sissel Beate Rønning (ed.), *Myogenesis: Methods and Protocols*, Methods in Molecular Biology, vol. 1889,
https://doi.org/10.1007/978-1-4939-8897-6_3, © Springer Science+Business Media, LLC, part of Springer Nature 2019

adipose tissue (BAT) has received much attention primarily because it is positively correlated with weight loss but negatively correlated to metabolic disorders including insulin resistance and type 2 diabetes, as well as atherosclerosis [7]. However, the abundance of brown adipocytes is low in adult humans, especially in obese subjects, urging the need to increase the number of brown adipocytes in face of obesity prevention and treatment.

Lineage tracing studies in mice indicate that brown adipocytes are derived from a progenitor cell population that expresses Pax7 and Myf5, which are established markers of skeletal muscle progenitors, suggesting a common developmental origin of brown adipocytes and skeletal muscle cells [8, 9]. During development, while the progenitor cells adopting the skeletal muscle cell fate continue to express Pax7, the progenitor cells destined to the brown adipocyte fate ceased to do so [9] and switch on the expression of PR-domain containing protein 16 (PRDM16), a key regulator of brown adipocyte cell fate [8]. The developmental linkage between brown adipocytes and skeletal muscles opens an opportunity to reprogram (transdifferentiate) muscle progenitors into brown adipocytes, especially considering that the skeletal muscle accounts for ~40% of body mass in the adult humans [10]. Efficient transdifferentiation of skeletal muscle progenitors into brown adipocytes would greatly enhance the application of brown adipocytes in the treatment of metabolic syndromes.

Several mechanisms are involved in the cell fate choice between skeletal muscle and brown fat [11], and corresponding strategies to induce the transdifferentiation from skeletal muscle cells into (brown) adipocytes have been reported (Fig. 1). The first strategy is to induce adipogenesis of muscle progenitors through enforcing or activation of canonical adipogenic pathways. These could be achieved through administration of chemical inducers or manipulating expression of adipogenic genes through genetic or epigenetic approaches (Fig. 1A). Forced expression of PRDM16, or its upstream regulator early B cell factor-2 (EBF2) is sufficient to stimulate brown adipogenesis in C2C12 myoblasts [8, 12]. Consistently, it was reported that a retinoid X receptor (RXR) agonist, bexatotene, can potently reprogram myoblasts into brown adipocyte-like cells by activating RXRα and RXRγ, which subsequently upregulate PRDM16 [13]. Peroxisome proliferator-activated receptor gamma (PPARγ) is a key transcriptional regulator of adipogenesis and its agonists (thiazolidinediones) have been reported to convert myoblasts into adipocyte-like cells [14]. In addition, bone morphogenic protein-2 (BMP2) cooperates with CCAAT-enhancer-binding protein alpha (C/EBP α) to induce adipogenic transdifferentiation of C2C12 myoblasts [15]. Several studies reported that either PPARγ or C/EBP-α is sufficient to induce adipogenesis in myoblast [15–17]. However, Hu et al. reported that ectopic expression of PPARγ or C/EBP-α alone does not

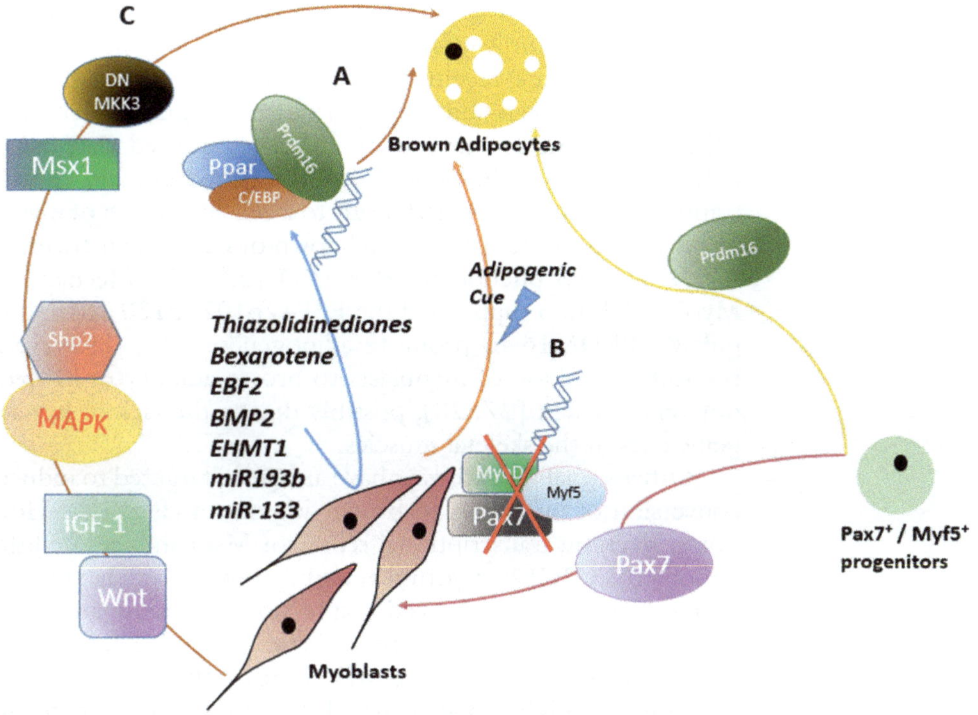

Fig. 1 Overview of the three adipogenic transdifferentiation pathways. The first strategy is to enforce or activate the canonical adipogenic pathway. The ultimate goal is to increase the expression level of adipogenic master genes Prdm16, Pparg as well as C/EBP. It could be directly manipulating these three genes or through indirect activation by chemical compounds (Thiazolidinediones or Bexarotene), molecules (EBF2, BMP2, or EHMT1), or microRNAs (miR-193b or miR-133). The second strategy is to suppress myogenic program in myoblasts. This strategy mostly focuses on the Pax7-MyoD/Myf5 cascade. It has been reported that MyoD acts through E2F4/p107/p130 regulating axis which in turn upregulates PRDM16. The third strategy involves other signaling pathways including DN MKK3, Msx1, Shp2, MAPK, IGF-1, as well as Wnt signaling pathways

significantly change the myogenic potentials in C2C12 cells, but co-expression of PPARγ and C/EBPα induces adipogenic transdifferentiation of myoblasts [18]. Epigenetic regulators such as chromotin modifiers and small noncoding RNA have also been reported to switch the myogenic program of myoblasts to the adipogenic program (Fig. 1). Euchromatic histone-lysine *N*-methyltransferase 1 (EHMT1) directly interacts with PRDM16 to stabilize the PRDM16 transcriptional complex, thus leading to the adipogenic transdifferentiation [19]. Among miRNAs, miR-193b was found to block the myogenic program in C2C12 myoblasts and induce the adipogenic program under adipogenic conditions [20]. MiR-133, a muscle enriched microRNA, was shown to target Prdm16, therefore suppressing the brown adipogenesis in muscle progenitors. Reduction of miR-133 conversely induces the development of brown and beige adipocytes from muscle satellite cells and white adipocyte progenitors [21–23].

The second strategy to induce myoblast to brown adipocyte transdifferentiation is to suppress myogenic program in myoblasts (Fig. 1B). We have recently reported that loss of MyoD converts myoblasts to brown adipocytes through upregulation of Prdm16 [23]. Consistent to our report, An et al. showed that inducible deletion of Pax7, a homeodomain and paired domain containing transcription factor [24–26], leads to a cell fate switch of myoblasts to brown adipocytes [27]. Knockdown of Pax7 downstream genes *MyoD* or *Myf5* phenocopy *Pax7* KO [23, 27]. Mechanistically, MyoD inhibition suppressed the E2F4/p107/p130 axis and upregulated PRDM16 to promote adipogenesis [27]. However, the transdifferentiation of myoblasts to brown adipocytes in vivo was not very efficient [27, 28], possibly due to the lack of key adipogenic cues in the skeletal muscles.

Other signaling pathways have also been targeted to induce fate conversion of myoblasts to brown adipocytes (Fig. 1C). Homeobox containing transcriptional repressor Msx1 induces dedifferentiation of C2C12 myotubes and promotes their adipogenic differentiation [29]. Forced expression of dominant negative mitogen-activated protein kinase kinase 3 (DN MKK3) also transdifferentiates C2C12 cells into adipocytes in the presence of PPAR-γ-activator rosiglitazone [30]. Interestingly, the three major subtypes of mitogen differentiation protein kinases (MAPK) namely ERK (extracellular signal regulated kinases), JNK (c-Jun amino terminal kinases), and p38 have distinct effects on the adipogenic transdifferentiation on C2C12 myoblasts. Qi et al. reported that p38 and JNK were negative regulators while ERK1/2 was a positive regulator of adipogenesis in C2C12 [31]. The negative effect of p38 MAP kinase was substantiated by the observation that Shp2 (nonreceptor tyrosine phosphatase Shp2) promotes adipogenesis through inhibition of p38 [32]. Other signaling pathways including IGF-1 signaling were also reported to induce adipogenesis in C2C12 in the presence of Rho kinase inhibitor [33]. Interrupting Wnt signaling by dominant negative TCF4 also promotes the expression of C/EBP-α and PPARγ, thus switching the myogenic program into adipogenic program [34].

In this article, we describe a protocol to induce the adipogenic transdifferentiation harnessing CRISPR-Cas9 technology to delete MyoD in C2C12 myoblasts. Besides, the adipogenic potential of the primary myoblasts isolated from the MyoD null mice is verified by induced adipogenic differentiation. Both methods manifest a robust switch to the adipogenic program in the myoblasts.

2 Materials

2.1 C2C12 Cell Culture

1. C2C12 cells: C2C12 (ATCC® CRL1772™).
2. Dulbecco's Modified Eagle's Medium (DMEM).
3. Fetal Bovine Serum.
4. Penicillin-streptomycin.
5. Trypsin.
6. 100 mm petri-dish.
7. 24-well culture plate.
8. 48-well culture plate.
9. 50 ml Falcon tube.
10. 15 ml Falcon tube.
11. PBS: 8 g of NaCl, 0.2 g KCl, 1.44 g Na_2HPO_4, 0.24G KH_2PO_4. Adjust the pH to 7.4 with HCl and bring the solution to a total volume of 1 L.
12. C2C12 culture medium: DMEM, 10% FBS, 1% Penicillin/ Streptomycin.

2.2 Generation of Cas9-Expressing C2C12 Myoblasts [35]

1. Plasmid: pSpCas9(BB)-2A-Puro (PX459) V2.0 was a gift from Feng Zhang (Addgene plasmid # 62988).
2. Puromycin.
3. Neon transfection system (Invitrogen).

2.3 Immunofluorescence Staining of C2C12 Cells

1. Fixative (4% paraformaldehyde): ddH_2O, HCl, NaOH (1 N), paraformaldehyde powder, 1× PBS; add 800 ml of 1× PBS to a beaker in a ventilating hood, heat the PBS to approximately 60 °C. Be careful not to boil the PBS. Add 40 g of paraformaldehyde powder to the PBS solution. Slowly add 1 N NaOH until the solution becomes clear. After cooling and filtering the solution, adjust the volume to 1 L and the pH to around 6.9 with HCI.
2. Blocking Buffer: 5% goat serum, 2% BSA, 0.2% Triton X-100, and 0.1% sodium azide in PBS.
3. FLAG antibody (Catalog No. F1804, Sigma).
4. DAPI.
5. Secondary antibody: Goat anti-mouse IgGl, Alexa Fluor 568 (catalog no. A-21124, thermo fisher scientific).

2.4 Production of gRNAs by In Vitro Transcription

1. Q5 hot start polymerase (Catalog No. M0493S, New England BioLabs).
2. MEGAshortscript T7 transcription Kit (Catalog No. AM1354. Thermo Fisher Scientific).

3. MEGAclear Transcription Clean-Up Kit (Catalog No. AM1908, Thermo Fisher Scientific).
4. Cycle Pure Kit (Catalog No. D6492-01, Omega Bio-Tek).

2.5 Primary Myoblasts Isolation from MyoD$^{-/-}$ Mice

1. MyoD$^{-/-}$ mice [36] were generously provided by Dr. Michael Rudnicki (Ottawa Health Research Institute, Ottawa, Ontario, Canada).
2. Type I collagenase and Dispase B (Roche Applied Science).
3. F-10 Ham's medium.
4. 4 ng/ml basic fibroblast growth factor.
5. Primary myoblasts culture medium: F-10 Ham's medium, 20% FBS, 4 ng/ml basic fibroblast growth factor, 1% penicillin-streptomycin.

2.6 Adipogenic Induction of Myoblasts

1. Adipogenic induction medium: DMEM, 10% FBS, 2.85 μM insulin, 0.3 μM dexamethasone, 0.63 mM 3-isobutyl-1-methylxanthine, 1 μM Rosiglitazone.
2. Adipogenic differentiation medium: DMEM, 10% PBS, 200 nM insulin, 10 nM T3, 1 μM Rosiglitazone.
3. Oil red O staining: working solution contains 6 ml Oil Red O (ORO) stock solution (5 mg/ml in isopropyl alcohol, Sigma) and 4 ml ddH_2O.

3 Methods

All the experimental procedures are carried out under room temperature unless otherwise stated. Cells are cultured at 37 °C with 5% CO_2.

3.1 Generation of Cas9-Expressing C2C12 Cells

1. Culture C2C12 myoblasts in 100 mm petri-dish containing 5 ml C2C12 culture medium. When the C2C12 cells reach 70% confluency, aspirate the culture medium and wash cells with 5 ml PBS. Then aspirate the PBS and treat the cells with 1 ml of 0.05% trypsin at 37 °C for 3 min (*see* **Note 1**). Cell detachment can be observed under the microscope. Add 3 ml of culture medium to the culture plate and transfer the cell suspension to a 15 ml Falcon tube. Centrifuge the cell suspension at 300 × *g* for 5 min. Aspirate the supernatant and suspend the cell pellet with 1 ml culture medium (*see* **Note 2**). Cells are counted with hemocytometer. Approximately 10^6 cells are mixed with 1 μg of PX459 plasmid. Transfect the plasmids into C2C12 cells using the Neon Transfection system (Invitrogen) at 1500 V, 10 ms, for 3 times in a 100 μl electroporation tip.

2. After electroporation, seed the C2C12 cells into a 100 mm petri-dish for 2 days in C2C12 culture medium. Then add puromycin (1 μg/ml) into the culture medium for selection of cells transfected with PX459. During the selection, cells are consecutively treated with puromycin. Change the culture medium every 2 days. Non-transfected cells are usually floating in the culture medium after 2 days of puromycin treatment. Transfected cells will form cell clones.
3. Cell clones should appear at 10 days after puromycin selection. To collect cell clones, replace the culture medium with 1 ml of 0.05% trypsin and incubate for 30s (*see* **Note 3**). Immediately aspirate the Trypsin. Suck up the cells in one single cell clone by using a pipette with 10 μl culture medium. Transfer the cells to 12-well plates. The culture medium is still supplemented with puromycin.
4. Passage the cells in each well into a 100 mm petri-dish when they reach 70% confluency in 12-well plates. In this step, seed a small portion of the cells into 24-well plates for immunostaining for the detection of FLAG tag encoded by PX459 plasmid.
5. For immunostaining of FLAG tag, fix the cells grown in 24-well plates with 4% paraformaldehyde (PFA) in PBS (pH 7.4) for 5 min in room temperature. Remove the PFA and wash the cells three times with PBS. Treat the cells with 100 mM glycine for 10 min. Afterward, remove the glycine and wash the cells with PBS once. Incubate the cells in blocking buffer for at least 30 min in room temperature. Remove the blocking buffer and incubate the cells with FLAG antibody (F1804, Sigma) diluted in blocking buffer at 1: 200 overnight at 4 °C. Remove the FLAG antibody and wash the cells with PBS five times for 1 min each. Then incubate the cells with DAPI and secondary antibody diluted in PBS at 1:1000 for 1 h at room temperature. Finally, the cells are washed five times with PBS for 1 min each before checking fluorescence signal under Leica DM 6000B. Only select cell clones expressing FLAG signals.
6. After the positive C2C12 ($C2C12^{Cas9}$) in the 100 mm petri-dish reached 70% confluency, collect the cells and freeze for later use.

3.2 gRNA Selection (See *Note 4*)

1. The gRNA targets of MyoD are designed using an online tool (CRISPR Design, http://crispr.mit.edu/) (Table 1).
2. The gRNAs are produced by following an online protocol (http://www.crisprflydesign.org/grnatranscription/). Briefly, specific primers introducing a T7 promoter and gRNA core sequence are designed and used to amplify PX459. High fidelity polymerase like Q5 hot start polymerase is used in our lab.

Table 1
Sequence of the two sgRNAs designed

	Sequence (5′–3′)	Target region on MyoD Exon1 (5′–3′)
sgRNA-1	GCGCTTTTTGAGGACCTGG	92-139
sgRNA-2	GCGCAACGCCATCCGCTACAT	447-508

After PCR amplification, separate the products on a 1.5% agarose gel and purify with Cycle Pure Kit (Omega Bio-Tek). Transcribe the templates in vitro using Ambion T7 shortscript kit according to the manufacturer's instructions. At last, use the Ambion MEGAclear kit to purify the RNA.

3. Then, mix two gRNAs (50 ng each) with 10^5 C2C12^{Cas9} cells for electroporation (*see* **Note 5**). The condition for electroporation is the same as the condition described in Subheading 3.1, **step 1**. For the control group, C2C12^{Cas9} cells are only given the same electric shocks. After electroporation, seed the cells into 6-well plates.
4. One day after electroporation, collect the cells for DNA extraction using GeneArt Genomic Cleavage Detection Kit according to the manufacturer's instructions. Amplify the region around the target sites by PCR. Clone the PCR products (*see* **Note 6**) and send for sequencing (*see* **Note 7**). By analyzing the sequence, we determine which gRNAs worked with Cas9 to edit MyoD.

3.3 MyoD KO Cell Line Selection (See Note 8)

1. Choose the sgRNA that targets close to the 5 prime of MyoD gene (*see* **Note 9**) and clone into PX459 plasmid. The protocol of cloning of gRNA into PX459 can be found online at Addgene (http://www.addgene.org/62988/). Briefly, primers composing sticky ends for restriction enzyme BbsI for the gRNA are synthesized. Linearize the plasmid with BbsI and purify with Cycle Pure Kit. After annealing the primers, ligate the linearized plasmid and primers using T4 ligase. The plasmid is termed PX459^{MyoD}.
2. Electroporate the wild-type C2C12 cells with either PX459^{MyoD} or PX459 (1 μg/10^6 cells).
3. Two days after electroporation, add 1 μg/ml of puromycin to the culture medium for at least 5 consecutive days. Change the culture medium every 2 days.
4. Collect the remaining cells electroporated with PX459 and survive the selection of puromycin. Freeze the cells as controls for later use.
5. The remaining cells electroporated with PX459^{MyoD} should form cell clones. Collect the cells in cell clones according to

Table 2
The top five potential off targets of sgRNAs

Position	20	19	18	17	16	15	14	13	12	11	10	9	8	7	6	5	4	3	2	1	N	G	G	Locus
MyoD	C	C	A	G	G	T	C	C	T	C	A	A	A	A	A	A	G	C	G	C	A	G	G	Chro:start position
OT1	C	C	T	A	G	T	C	C	T	C	A	A	A	A	A	A	G	C	T	C	A	G	G	Chr1:-106859905
OT2	C	T	A	A	G	T	C	A	T	C	A	A	A	A	A	A	G	C	A	C	T	G	G	Chr13:+78086880
OT3	C	C	A	T	C	T	C	A	T	C	A	A	A	A	A	A	G	C	T	C	A	G	G	Chr19:+50585297
OT4	A	C	A	C	G	T	G	C	T	C	A	A	A	A	A	A	G	C	A	C	G	A	G	Chr1:-39340833
OT5	C	C	A	G	G	T	G	C	T	C	C	A	A	A	A	A	G	C	A	C	A	G	G	Chr12:-113574844

PAM is highlighted in grey.
The different bases between the off target sites and MyoD locus were highlighted in red.

the descriptions in Subheading 3.1, **step 3**. Transfer the cells to 12-well plates.

6. When the cells reach 70% confluency, passage the cells into 6-well plates. Use a small portion of the cells for DNA extraction, PCR amplification and sequencing to confirm correct genomic alteration (*see* **Note 10**). Cells containing predicted genomic alteration are cloned a second time to achieve homogeneity serial dilution.

3.4 Off-Target Detection

1. The gRNA off targets sites are searched using CRISPR Design (http://crispr.mit.edu/).
2. Amplify the top five potential off target sites by PCR using genomic DNA as templates (Table 2).
3. The PCR products are subjected to T7EN1 cleavage assay according to the the manufacturer's protocol. The potential off target sites yielding typical cleavage bands are considered candidates for off target (*see* **Note 11**).

3.5 Adipogenic Induction of MyoD KO Cell Line

1. To examine the adipogenic differentiation potential of MyoD ^{KO} C2C12 myoblasts, treat the cells with induction medium for 5 days and differentiation medium for 5 days (*see* Subheading 1).
2. After adipogenic induction, use Oil Red O staining to detect lipid droplets. Rinse the cells with PBS and fix with 4% paraformaldehyde for 15 min at room temperature. Then stain the cells with working solution of Oil Red O (ORO) for 1 h. Wash the culture plate with distilled water for 10 min and mount with glycerol mounting medium (*see* **Note 12**).
3. Meanwhile, the expression profiles of adipogenic and myogenic specific marker genes are examined using qPCR. The deletion of MyoD downregulated the expression of myogenic genes (such as Myog) and upregulated the expressoin of adipogenic genes (*Fabp4*, *Cebpα*, and *FASN*) (Table 4, also *see* **Note 13**).

3.6 Adipogenic Induction of Primary Myoblasts from MyoD$^{-/-}$ Mice

1. Dissect the Tibialis Anterior (TA) muscle from MyoD$^{-/-}$ mice of the age of 4–6 weeks old. The protocol for isolating myoblasts has been thoroughly characterized by Motohashi et al. [37]. In brief, thoroughly mince the TA muscles and then digest with type I collagenase and Dispase B mixture at 37 °C for 12 min. Triturate the muscle and digest for another 12 min. Add primary myoblast culture medium to stop the digestion. Filter the digestion resultants with a 0.45 μm filter (*see* **Note 14**) and then centrifuge at 300 × *g* for 5 min. Culture the cells in primary myoblast culture medium on collagen-coated cell culture plate (*see* **Note 15**).
2. Adipogenic induction of the primary myoblasts is characterized in Subheading 3.5, **step 1**.
3. The expression levels of adipogenic markers and myogenic markers are examined as described in Subheading 3.5, **step 3** (*see* **Note 16**).

4 Notes

1. Monitor the cell morphology under a light microscope when treating C2C12 with Trypsin as exposing C2C12 myoblasts with Trypsin for too long would induce damages to the cell membranes.
2. Triturate the pellets thoroughly to avoid any cell cluster in this step.
3. The timing of Trypsin treatment is crucial as we want to disassociate the cells from the plate so that the cells could be aspirated yet the clone would remain intact.
4. A schematic overview of validation of sgRNA efficacy is shown in Fig. 2.
5. Two sgRNAs are designed which targeted at exon 1 of MyoD. The sequences of the two sgRNAs are listed in Table 1.
6. After transfection of C2C12^{Cas9} cells with sgRNAs and selection with puromycin, PCR is carried out using the genomic DNA to amplify the region composing the target sites of sgRNAs. Since 2 sgRNAs are transfected, we expect to see several hundred base pairs difference in the PCR product. Indeed, by electrophoresis, two cleaved bands are observed and the ~300 bp band was purified and cloned for sequencing [23].
7. Sequencing results from C2C12^{Cas9} transfected with sgRNA show a 336 bp deletion in MyoD, indicating the efficacy of sgRNA in editing MyoD [23].

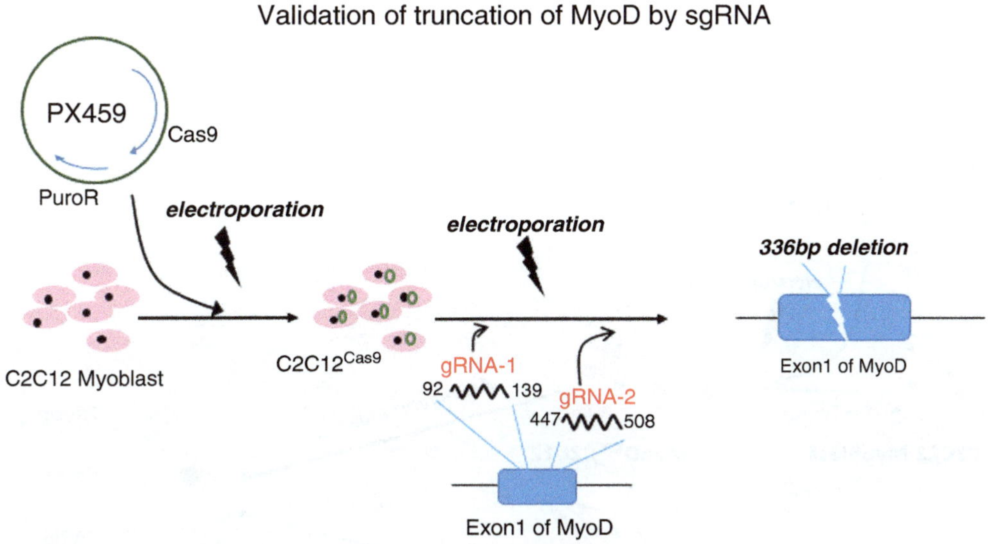

Fig. 2 Schematic view of the experiment validating sgRNA efficacy. In brief, a modified C2C12 cell line is established which expresses Cas9 for later validation of sgRNA. Then sgRNAs are electroporated into the modified C2C12. After sequencing of the target MyoD locus, it is revealed that there is a 336 base pairs deletion, indicating the efficacy of sgRNA in editing MyoD

8. A schematic overview of the adipogenic transdifferentiation of $MyoD^{KO}$ C2C12 myoblasts and primary myoblasts is derived from $MyoD^{-/-}$ mice (*see* Fig. 3).
9. In order to minimize the off target effects, only the sgRNA that targets the proximity of the 5 prime of MyoD exon 1 is cloned to PX459.
10. Of the 21 clones established from MyoD ^{KO} C2C12 cell line, two clones are found out through sequencing that they either have 1 bp or 11 bp frame shifting deletion in both alleles of MyoD (Fig. 4A). MyoD ^{KO} 1 has an 11 bp deletion resulting in the termination of translation at 72nd amino acids and a 1 bp deletion which results in the termination of the translation at 55th amino acids. MyoD ^{KO} 2 has an 11 bp deletion on both alleles resulting in the termination at 72nd amino acids.
11. The potential off targets sites are amplified and examined by electrophoresis. T7EN1 cleavage assay shows there aren't genomic modifications to the potential off target sites (Fig. 4B). The off target sites are listed in Table 2 and the primers for detecting off targeting of sgRNA are listed in Table 3.
12. In control cells, there are only few neutral triglycerides positively stained by Oil Red O whereas in MyoD KO C2C12 cells, there is a substantial region positively stained (Fig. 5B). The

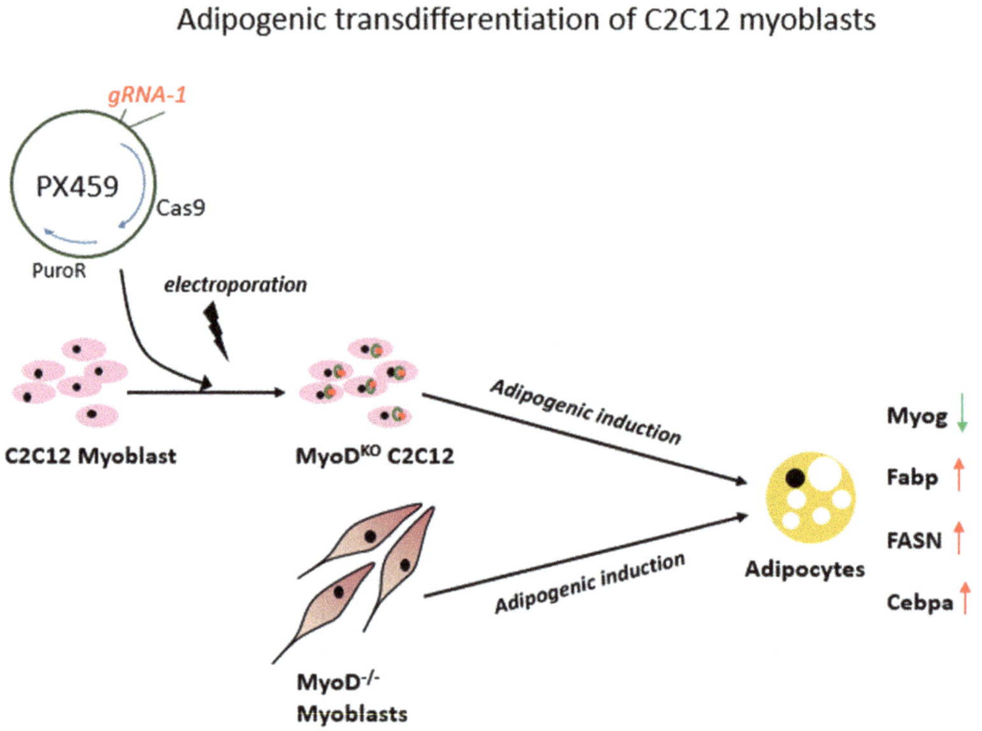

Fig. 3 Schematic overview of adipogenic transdifferentiation of C2C12 myoblasts and primary myoblasts derived from MyoD$^{-/-}$ mice. The sgRNA that targets at the proximity of 5 prime of MyoD is chosen for inducing genomic deletion of MyoD in C2C12 cells. sgRNA-1 is cloned into PX459 as described. Then the plasmid is electroporated into C2C12 myoblasts. The C2C12 cells are then subject to puromycin selection. Both the MyoD^{ko} C2C12 cells and MyoD$^{-/-}$ myoblasts are induced for adipogenesis. Both myoblasts exhibited typical adipogenic features. qPCR of the myogenic and adipogenic marker genes confirmed the shifting of the myogenic program into the adipogenic program, where *Myog* is down regulated yet *Fabp*, *FASN*, and *Cebpa* are upregulated

deletion of MyoD in C2C12 is confirmed by immunostaining using a well-established MyoD antibody (Fig. 5A).

13. The deletion of MyoD leads to the significant downregulation of Myog [23], a myogenic marker for muscle differentiation [38], however, adipogenic-specific genes show significant upregulation indicating a switch from myogenic program to adipogenic program. qPCR primers are listed in Table 4.

14. Passing the mechanically minced cells through a strainer helps to eliminate extra cell debris.

15. During this process, if fibroblasts outgrow myoblasts, an extra step of pre-plating could be carried out to get rid of excessive fibroblasts. The cell culture is carefully disassociated with 1 ml of 0.05% Trypsin for about 3 min. 1 ml of culture medium is added into the cell culture plate to stop trypsinization. The cells are centrifuged at 300 × *g* for 5 min. The supernatant is

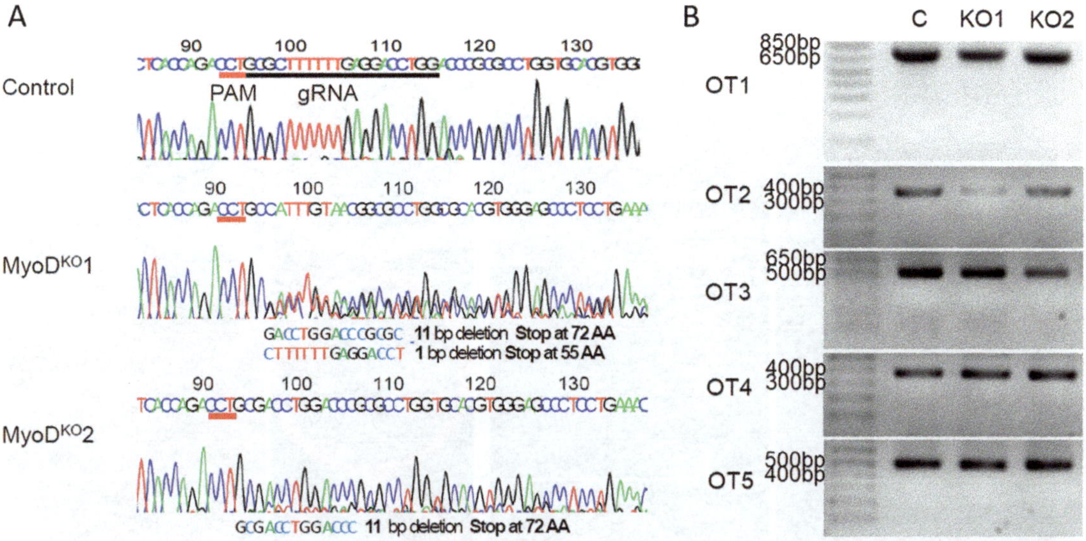

Fig. 4 Correct targeting of the two sgRNAs. (**a**) Two sgRNA are designed targeting the exon 1 of MyoD. The DNA sequence maps show the target locus of the control C2C12 myoblasts and the 2 clones transfected with 2 sgRNAs. $MyoD^{KO}$ 1 has an 11 bp deletion resulting in the termination of translation at 72nd amino acids and a 1 bp deletion which results in the termination of the translation at 55th amino acids. MyoD ^{KO} 2 has an 11 bp deletion on both alleles resulting in the termination at 72nd amino acids. (**b**) Off target detection of the two sgRNAs. The top five potential off target sites subject to T7EN1 assay and amplified by PCR. The electrophoresis result shows that there are not obvious alterations to the off target sites. This figure is adapted from Wang et al., (Ebiomedicine 16:212-223. doi: https:/doi.org/10.1016/j.ebiom.2017.01.015) licensed under CC BY-NC-ND (http:/creativecommons.org/licenses/by-nc-nd/4.0/)

Table 3
Primers for amplifying the off target sites (5′–3′)

Gene	Sequence
OT1	F: GGCTGAAAGGCTTGTACTCG R: GGAGCATTTCAGGCATCAAT
OT2	F: GGTGGAAAGGCTGTTGATGT R: GCCAGAGTGATCGGAATGAT
OT3	F: TCCCTGGCTATGCAAAATTC R: CATTCATGGAGCCTGTGTGT
OT4	F: ACATGTATGCCCAGATGCAA R: GTGCTCCAACGAGTGAGACA
OT5	F: CCCAAGTTCCATGTCCCTAA R: TGAGCAAGGTTTGCAGTTTG
MyoD	F: AAGCGGCCGCATGGAGCTTCTATCGCCGC R: TTTCTAGATCAAAGCACCTGATAAATC

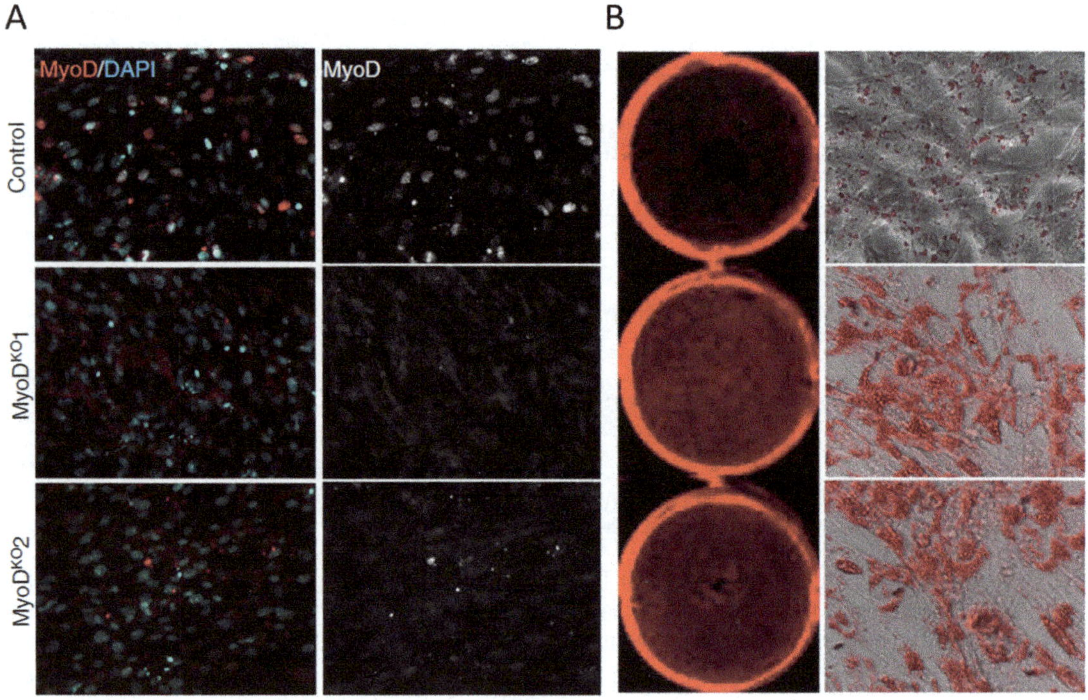

Fig. 5 Transdifferentiation of MyoD^{KO} C2C12 myoblasts into adipocytes. (**a**) Immunostaining of the cultured MyoD^{KO} C2C12 cells using a well-established MyoD antibody. Compared with the control group, the positive staining of MyoD is almost obsolete in MyoD^{KO} cells. (**b**) MyoD^{KO} C2C12 is stained by Oil Red O staining for detection of neutral triglycerides. The MyoD^{KO} C2C12 cells show substantial staining of lipids droplets while the control group shows relatively few. This figure is adapted from Wang et al., (Ebiomedicine 16:212–223. doi: https:/doi.org/10.1016/j.ebiom.2017.01.015) licensed under CC BY-NC-ND (http:/creativecommons.org/licenses/by-nc-nd/4.0/)

Table 4
qPCR primers for myogenic and adipogenic marker genes (5′–3′)

Gene	Sequence
Gadph	F: TCGGTGTGAACGGATTTG R: GGTCTCGCTCCTGGAAGA
Myog	F: GCAATGCACTGGAGTTCG R: ACGATGGACGTAAGGGAGTG
Fabp4	F: ACAAGCTGGTGGTGGAATGTG R: CCTTTGGCTCATGCCCTTT
Fasn	F: GGAGGTGGTGATAGCCGGTAT R: TGGGTAATCCATAGAGCCCAG
Cebpa	F: TGCGCAAGAGCCGAGATAAA R: CCTTCTGTTGCGTCTCCACG

discarded and the cells are suspended with 5 ml primary myoblast culture medium. The suspended cells are seeded in a regular 100 mm culture plate without collagen coating. The cells are incubated in the incubator for 45 min until there is clear morphology of fibroblasts attaching to the culture plate. The culture medium is collected along with the floating cells and transferred to a collagen coated plate. Fibroblasts have better adhesion capacity than do myoblasts, therefore they could easily attach to the culture plate even without collagen coating.

16. The expression profiles of adipogenic and myogenic marker genes are characterized by Wang et al. [23].

Acknowledgments

We thank Jun Wu and Mary Larimore for mouse colony maintenance and members of Kuang laboratory for valuable comments. Jingjuan Chen and Chao Wang have contributed equally to this work.

Funding
This work was supported by a grant from the US National Institute of Health (R01AR071649), National Natural Science Foundation of China (81471070) and CAMS Innovation Fund for Medical Sciences (2016-I2M-1-012).

References

1. Nedergaard J, Bengtsson T, Cannon B (2007) Unexpected evidence for active brown adipose tissue in adult humans. Am J Physiol Endocrinol Metab 293(2):E444–E452. https://doi.org/10.1152/ajpendo.00691.2006
2. Cypess AM, Lehman S, Williams G, Tal I, Rodman D, Goldfine AB, Kuo FC, Palmer EL, Tseng Y, Doria A, Kolodny GM, Kahn CR (2009) Identification and importance of brown adipose tissue in adult humans. New Eng J Med 360(15):1509–1517. https://doi.org/10.1056/NEJMoa0810780
3. Zingaretti MC, Crosta F, Vitali A, Guerrieri M, Frontini A, Cannon B, Nedergaard J, Cinti S (2009) The presence of UCP1 demonstrates that metabolically active adipose tissue in the neck of adult humans truly represents brown adipose tissue. FASEB J 23(9):3113–3120. https://doi.org/10.1096/fj.09-133546
4. Cohade C, Osman M, Pannu HK, Wahl RL (2003) Uptake in supraclavicular area fat ("USA-fat"): description on F-18-FDG PET/CT. J Nucl Med 44(2):170–176
5. Chiba S, Shimada T, Kaneko K, Tomonari K, Satoh Y, Kang M, Fujiwara K, Gotoh K, Masaki T, Katsuragi I, Kakuma T, Hamaguchi K, Yoshimatsu H (2010) Evaluation of human brown adipose tissue using positron emission tomography, computerized tomography and histochemichal studies in association with glucose metabolism. Endocr J 57:S329–S329
6. Bar-Shalom R, Gaitini D, Keidar Z, Israel O (2004) Non-malignant FDG uptake in infradiaphragmatic adipose tissue: a new site of physiological tracer biodistribution characterised by PET/CT. Eur J Nucl Med Mol Imaging 31(8):1105–1113. https://doi.org/10.1007/s00259-004-1506-0
7. Harms M, Seale P (2013) Brown and beige fat: development, function and therapeutic potential. Nat Med 19(10):1252–1263. https://doi.org/10.1038/nm.3361
8. Seale P, Bjork B, Yang WL, Kajimura S, Chin S, Kuang SH, Scime A, Devarakonda S, Conroe HM, Erdjument-Bromage H, Tempst P,

Rudnicki MA, Beier DR, Spiegelman BM (2008) PRDM16 controls a brown fat/skeletal muscle switch. Nature 454(7207):961–U927. https://doi.org/10.1038/nature07182

9. Lepper C, Fan CM (2010) Inducible lineage tracing of Pax7-descendant cells reveals embryonic origin of adult satellite cells. Genesis 48 (7):424–436. https://doi.org/10.1002/dvg.20630
10. Lindle RS, Metter EJ, Lynch NA, Fleg JL, Fozard JL, Tobin J, Roy TA, Hurley BF (1997) Age and gender comparisons of muscle strength in 654 women and men aged 20-93 yr. J Appl Physiol 83(5):1581–1587
11. Farmer SR (2008) Brown fat and skeletal muscle: unlikely cousins? Cell 134(5):726–727. https://doi.org/10.1016/j.cell.2008.08.018
12. Rajakumari S, Wu J, Ishibashi J, Lim HW, Giang AH, Won KJ, Reed RR, Seale P (2013) EBF2 determines and maintains brown adipocyte identity. Cell Metab 17(4):562–574. https://doi.org/10.1016/j.cmet.2013.01.015
13. Nie BM, Nie T, Hui XY, Gu P, Mao LF, Li K, Yuan R, Zheng JS, Wang HX, Tang SB, Zhang Y, Xu T, Xu AM, Wu DH, Ding S (2017) Brown adipogenic reprogramming induced by a small molecule. Cell Rep 18 (3):624–635. https://doi.org/10.1016/j.celrep.2016.12.062
14. Teboul L, Gaillard D, Staccini L, Inadera H, Amri EZ, Grimaldi PA (1995) Thiazolidinediones and fatty-acids convert myogenic cells into adipose-like cells. J Biol Chem 270 (47):28183–28187
15. Fux C, Mitta B, Kramer BP, Fussenegger M (2004) Dual-regulated expression of C/EBP-alpha and BMP-2 enables differential differentiation of C2C12 cells into adipocytes and osteoblasts. Nucleic Acids Res 32(1). https://doi.org/10.1093/nar/gnh001
16. Yu YH, Liu BH, Mersmann HJ, Ding ST (2006) Porcine peroxisome proliferator-activated receptor gamma induces transdifferentiation of myocytes into adipocytes. J Anim Sci 84(10):2655–2665. https://doi.org/10.2527/jas.2005-645
17. Yamanouchi K, Ban A, Shibata S, Hosoyama T, Murakami Y, Nishihara M (2007) Both PPAR gamma and C/EBP alpha are sufficient to induce transdifferentiation of goat fetal myoblasts into adipocytes. J Reprod Dev 53 (3):563–572. https://doi.org/10.1262/jrd.18169
18. Hu ED, Tontonoz P, Spiegelman BM (1995) Transdifferentiation of myoblasts by the adipogenic transcription factors ppar-gamma and C/EBP-alpha. Proc Natl Acad Sci U S A 92 (21):9856–9860. https://doi.org/10.1073/pnas.92.21.9856
19. Ohno H, Shinoda K, Ohyama K, Sharp LZ, Kajimura S (2013) EHMT1 controls brown adipose cell fate and thermogenesis through the PRDM16 complex. Nature 504 (7478):163. https://doi.org/10.1038/nature12652
20. Sun L, Xie HM, Mori MA, Alexander R, Yuan BB, Hattangadi SM, Liu QQ, Kahn CR, Lodish HF (2011) Mir193b-365 is essential for brown fat differentiation. Nat Cell Biol 13 (8):958–U198. https://doi.org/10.1038/ncb2286
21. Yin H, Pasut A, Soleimani VD, Bentzinger CF, Antoun G, Thorn S, Seale P, Fernando P, van Ijcken W, Grosveld F, Dekemp RA, Boushel R, Harper ME, Rudnicki MA (2013) MicroRNA-133 controls brown adipose determination in skeletal muscle satellite cells by targeting Prdm16. Cell Metab 17(2):210–224. https://doi.org/10.1016/j.cmet.2013.01.004
22. Trajkovski M, Ahmed K, Esau CC, Stoffel M (2012) MyomiR-133 regulates brown fat differentiation through Prdm16. Nat Cell Biol 14 (12):1330. https://doi.org/10.1038/ncb2612
23. Wang C, Liu WY, Nie YH, Qaher M, Horton HE, Yue F, Asakura A, Kuang SH (2017) Loss of MyoD promotes fate transdifferentiation of myoblasts into brown adipocytes. EBioMedicine 16:212–223. https://doi.org/10.1016/j.ebiom.2017.01.015
24. Buckingham M, Relaix F (2015) PAX3 and PAX7 as upstream regulators of myogenesis. Semin Cell Dev Biol 44:115–125. https://doi.org/10.1016/j.semcdb.2015.09.017
25. Christ B, Huang RJ, Scaal M (2007) Amniote somite derivatives. Dev Dyn 236 (9):2382–2396. https://doi.org/10.1002/dvdy.21189
26. Kalcheim C (2016) Epithelial-mesenchymal transitions during neural crest and somite development. J Clin Med 5(1). https://doi.org/10.3390/jcm5010001
27. An YT, Wang G, Diao YR, Long YY, Fu XR, Weng MX, Zhou L, Sun K, Cheung TH, Ip NY, Sun H, Wang HT, Wu ZG (2017) A molecular switch regulating cell fate choice between muscle progenitor cells and brown adipocytes. Dev Cell 41(4):382. https://doi.org/10.1016/j.devcel.2017.04.012
28. Borensztein M, Viengchareun S, Montarras D, Journot L, Binart N, Lombes M, Dandolo L (2012) Double Myod and Igf2 inactivation promotes brown adipose tissue development by increasing Prdm16 expression. FASEB J 26

(11):4584–4591. https://doi.org/10.1096/fj.12-208496

29. Odelberg SJ, Kollhoff A, Keating MT (2000) Dedifferentiation of mammalian myotubes induced by msx1. Cell 103(7):1099–1109. https://doi.org/10.1016/s0092-8674(00)00212-9
30. Yeow K, Phillips B, Dani C, Cabane C, Amri EZ, Derijard B (2001) Inhibition of myogenesis enables adipogenic trans-differentiation in the C2Cl2 myogenic cell line. FEBS Lett 506 (2):157–162. https://doi.org/10.1016/s0014-5793(01)02900-3
31. Qi R, Liu H, Wang Q, Wang J, Yang F, Long D, Huang J (2017) Expressions and regulatory effects of P38/ERK/JNK mapks in the adipogenic trans-differentiation of C2C12 myoblasts. Cell Physiol Biochem 44(6):2467–2475. https://doi.org/10.1159/000486169
32. He Z, Zhu HH, Bauler TJ, Wang J, Ciaraldi T, Alderson N, Li SW, Raquil MA, Ji KH, Wang SF, Shao JH, Henry RR, King PD, Feng GS (2013) Nonreceptor tyrosine phosphatase Shp2 promotes adipogenesis through inhibition of p38 MAP kinase. Proc Natl Acad Sci U S A 110(1):E79–E88. https://doi.org/10.1073/pnas.1213000110
33. Sordella R, Jiang W, Chen GC, Curto M, Settleman J (2003) Modulation of rho GTPase signaling regulates a switch between adipogenesis and myogenesis (vol 113, pg 147, 2003). Cell 113(4):547–547. https://doi.org/10.1016/s0092-8674(03)00356-8
34. Ross SE, Hemati N, Longo KA, Bennett CN, Lucas PC, Erickson RL, MacDougald OA (2000) Inhibition of adipogenesis by Wnt signaling. Science 289(5481):950–953. https://doi.org/10.1126/science.289.5481.950
35. Ran FA, Hsu PD, Wright J, Agarwala V, Scott DA, Zhang F (2013) Genome engineering using the CRISPR-Cas9 system. Nat Protoc 8 (11):2281–2308. https://doi.org/10.1038/nprot.2013.143
36. Rudnicki MA, Braun T, Hinuma S, Jaenisch R (1992) Inactivation of myod in mice leads to up-regulation of the myogenic hlh gene MYF-5 and results in apparently normal muscle development. Cell 71(3):383–390. https://doi.org/10.1016/0092-8674(92)90508-a
37. Motohashi N, Asakura Y, Asakura A (2014) Isolation, culture, and transplantation of muscle satellite cells. J Vis Exp 86. https://doi.org/10.3791/50846
38. Olson EN, Klein WH (1994) bHLH factors in muscle development: dead lines and commitments, what to leave in and what to leave out. Genes Dev 8(1):1–8

Chapter 4

Chromatin Immunoprecipitation in Skeletal Myoblasts

Vinay Kumar Rao, Shilpa Rani Shankar, and Reshma Taneja

Abstract

Chromatin immunoprecipitation (ChIP) is a powerful and sensitive technique that is widely used to study DNA-protein interactions. It enables an unbiased genome-wide analysis of transcriptional changes during several biological processes including cellular differentiation. Here, we describe a step-by-step protocol to identify histone modifications, transcription factor, and co-factor binding to chromatin in skeletal myoblasts. We discuss critical steps during cell harvesting, sonication, and immunoprecipitation and provide notes to evade common pitfalls.

Key words Myoblasts, Chromatin modifiers, Histone modifications, Epigenetics, Transcription

1 Introduction

Differentiation of embryonic precursor cells into skeletal muscle is regulated by paired homeobox transcription factors Pax3 and Pax7 and basic helix-loop-helix (bHLH) myogenic regulatory factors (MRFs) [1, 2]. Their transcriptional activity is regulated by association with epigenetic modifiers that are recruited at promoters and enhancers of myogenic genes to mediate dramatic changes in gene expression essential for formation of specialized cell types [3, 4]. In vitro, C2C12 skeletal myoblasts have been widely used to study various steps of muscle differentiation. Using these cells, several studies have employed chromatin immunoprecipitation (ChIP) and its variants to decode genome-wide histone modifications, transcription factor binding and co-factors recruitment during myogenic differentiation [5, 6]. For instance, ChIP assays followed by next-generation sequencing identified genome-wide binding of key transcription factors such as Myogenic determination factor 1 (MyoD), Yin-Yang 1 (YY1) in myoblasts and myotubes [7–9]. Similarly, ChIP analysis of histone modifications such as Histone 3 lysine 9 acetylation (H3K9ac), H3K27me3, has identified global histone modifications during myogenic differentiation [5]. A complete view of the epigenetic landscape, in particular,

Sissel Beate Rønning (ed.), *Myogenesis: Methods and Protocols*, Methods in Molecular Biology, vol. 1889, https://doi.org/10.1007/978-1-4939-8897-6_4, © Springer Science+Business Media, LLC, part of Springer Nature 2019

DNA, and histone modifications is essential to understand transcriptional regulation of myogenesis.

ChIP assay determines if a protein of interest interacts with certain region of the genomic DNA. It also gives an estimate of binding density of the protein at a particular locus/region of the genome [10]. The assay involves crosslinking of DNA with proteins by formaldehyde fixation [11, 12] followed by shearing of DNA to small fragments of 200–500 base pairs (bp) by either sonication or enzymatic digestion [13, 14]. The protein of interest is immunoprecipitated by antibody-bead conjugation followed by reverse crosslinking of DNA and proteins. The binding of protein to specific genomic region is confirmed by polymerase chain reaction (PCR) wherein primers specific to a genomic region of interest are used to assess the enrichment of the DNA in the immunoprecipitated complex compared to control immunoprecipitation. Alternatively, a library is constructed with the immunoprecipitated DNA and high-throughput sequencing is performed to reveal the genome-wide binding of the protein of interest.

Here, we provide a detailed protocol for performing ChIP assay followed by quantitative PCR (ChIP-qPCR) in C2C12 myoblasts. The method consists of six sections that include chromatin crosslinking and cell harvesting; sonication; immunoprecipitation; elution and reverse crosslinking; DNA isolation; real-time quantitative PCR and data analysis (Subheadings 3.1–3.6). We discuss critical steps involved at various stages of the process (Fig. 1).

2 Materials

1. Cell culture: C2C12 myoblasts are cultured in DMEM high glucose containing 20% Fetal Bovine Serum (FBS). Cells should preferably be cultured in 10 cm culture dishes to facilitate collection of cells. Seed 0.5 million cells in a 10 cm dish 1 day before harvesting samples for ChIP. The following day, cells are fixed at 60% confluency (1 million cells). For differentiation, seed 1 million cells, and the following day, switch to differentiation medium (DMEM supplemented with 2% horse serum). Cells should be 80% confluent to initiate differentiation. Cells are cultured in differentiation medium for another 2–5 days depending on the differentiation time point required. In general, 24 h after the addition of differentiation medium, cells should start aligning and cell fusion and myotube formation is apparent from day 2 onward.
2. Formaldehyde: Caution! Very toxic chemical. 37% stock concentration, 1% final concentration.
3. Glycine: 1 M stock concentration, 0.125 M final concentration.

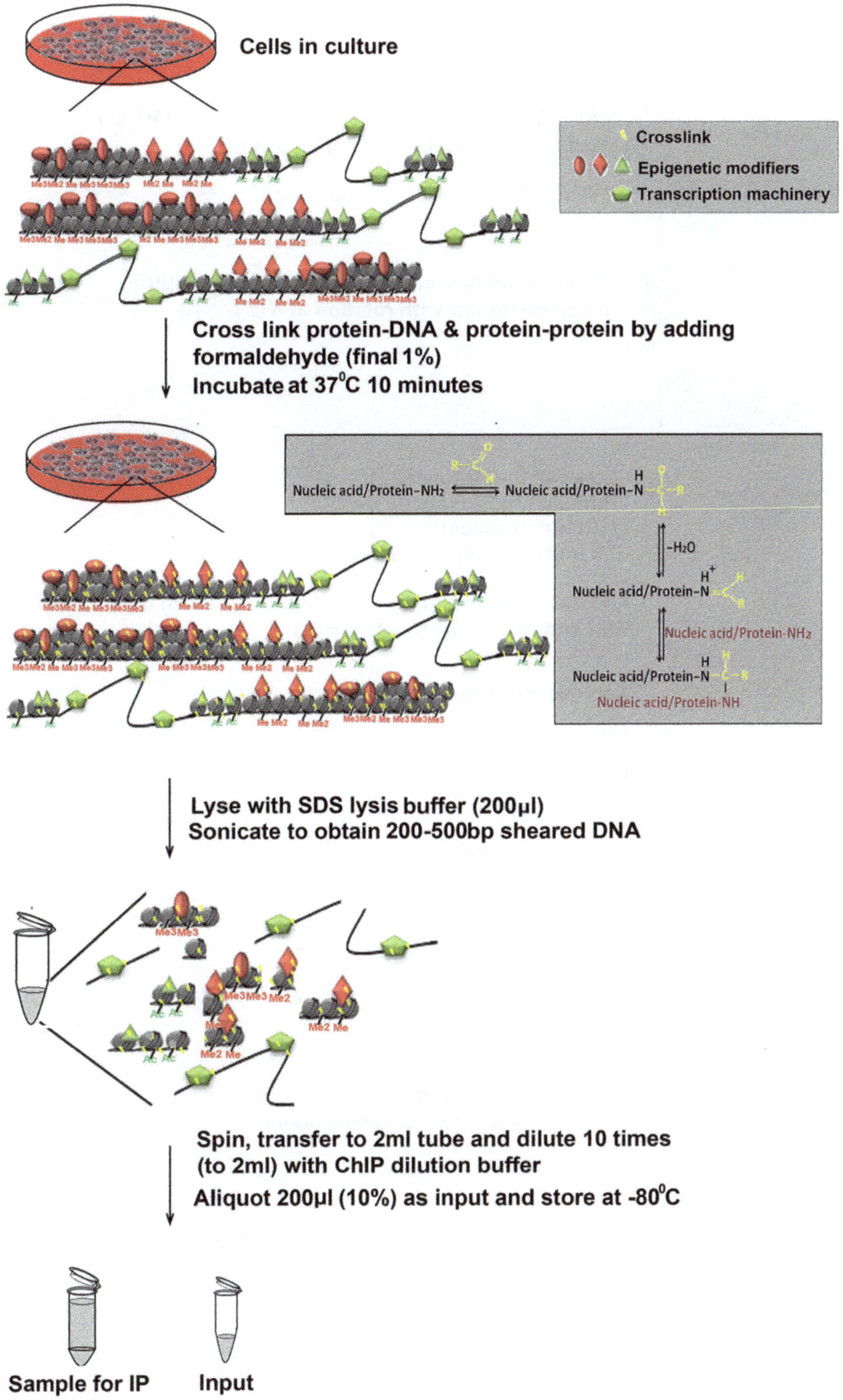

Fig. 1 Schematic representation of the various steps involved in ChIP

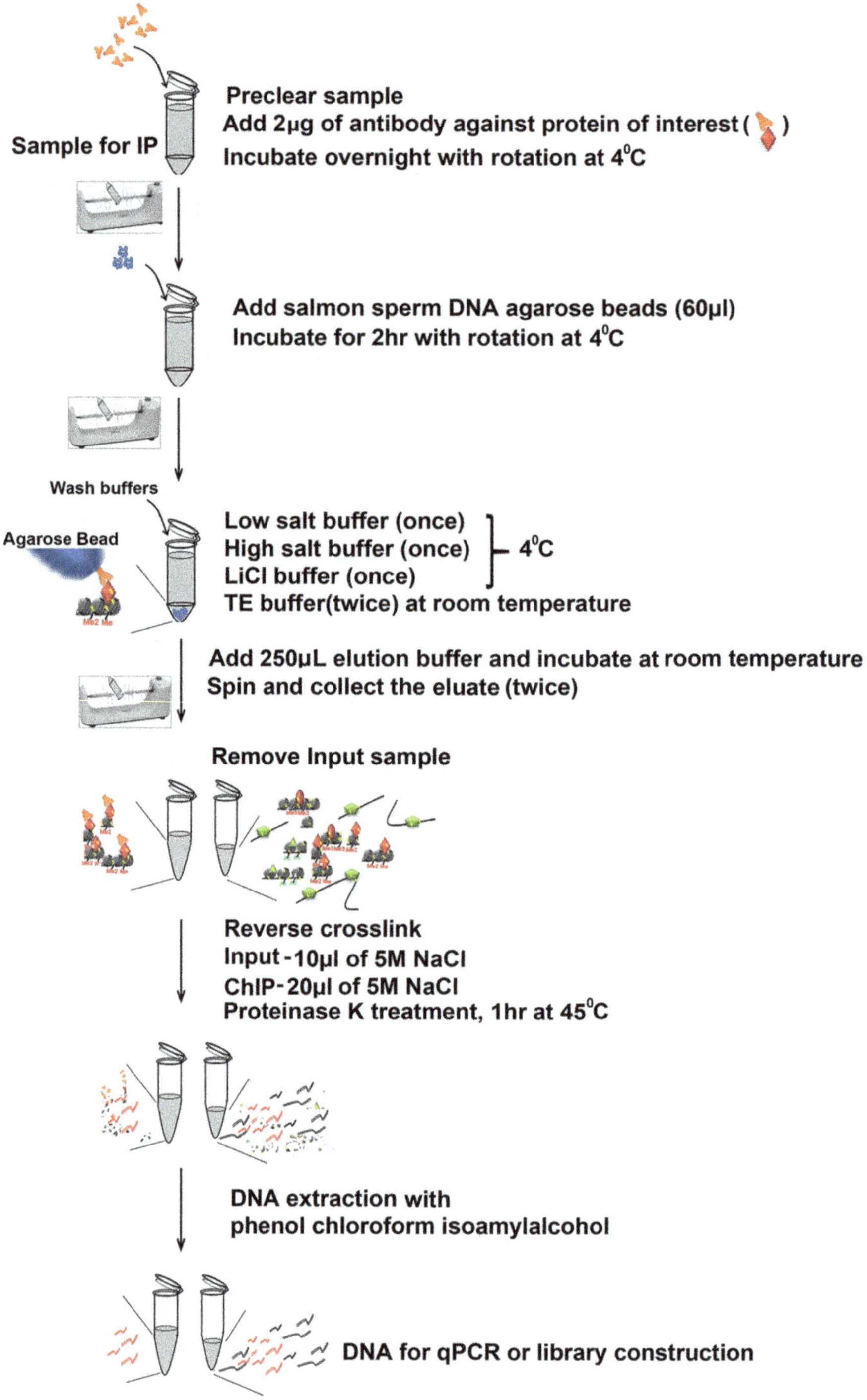

Fig. 1 (continued)

4. 1× Phosphate-Buffered Saline (PBS).
5. Cell lifter.
6. 15 ml Falcon tubes, 1.5 ml and 2 ml Eppendorf tubes.
7. Millipore ChIP assay kit (17-295, Millipore).
8. Buffers:
    - ChIP dilution buffer: 0.01% SDS, 1.1% Triton X100, 1.2 mM EDTA, 16.7 mM Tris–HCl, pH 8.1, 167 mM NaCl.
    - Cell Lysis buffer: 1% SDS, 10 mM EDTA, 50 mM Tris, pH 8.1.
    - Low salt wash buffer: 0.1% SDS, 1% Triton X-100, 2 mM EDTA, 20 mM Tris–HCl, pH 8.1, 150 mM NaCl.
    - High salt wash buffer: 0.1% SDS, 1% Triton X-100, 2 mM EDTA, 20 mM Tris–HCl, pH 8.1, 500 mM NaCl.
    - LiCl wash buffer: 0.25 M LiCl, 1% IGEPAL-CA630, 1% deoxycholic acid (sodium salt), 1 mM EDTA, 10 mM Tris, pH 8.1.
    - TE buffer: 10 mM Tris–HCl, 1 mM EDTA pH 8.0.
    - Elution buffer: 1% SDS and 0.1 M $NaHCO_3$ in water.
9. 1 M Tris–HCl pH 6.8.
10. 0.5 M EDTA.
11. 10 mg/ml RNAse A.
12. 1× Protease inhibitor cocktail.
13. 20 mg/ml Proteinase K.
14. 20 mg/ml Glycogen.
15. Phenol chloroform-isoamylalcohol.
16. SYBR green PCR master mix.
17. Antibodies for protein of interest.
18. Refrigerated centrifuge and micro-centrifuge.
19. Bioruptor (Diagenode).
20. Tube rotator at 4 °C.
21. Heat block.
22. Real time PCR machine.

3 Methods

3.1 Chromatin Crosslinking and Cell Harvesting

1. Culture C2C12 myoblasts at 60% confluency (*see* Subheading 2, **item 1**). Crosslink by adding 1% formaldehyde to the medium. Incubate at 37 °C for 10 min (*see* **Note 1**). It is advised

to perform a cell count from another plate with similar confluency. This protocol utilizes 1 million cells per chromatin IP (*see* **Note 2**).

2. Quench the formaldehyde reaction by adding 0.125 M glycine to the medium. Incubate for 5 min at room temperature.
3. Decant the medium containing formaldehyde and glycine. Rinse cells twice with ice-cold PBS with 1× protease inhibitors.
4. Collect cells using cell scraper into a 15 ml falcon tube (*see* **Note 3**). Ensure that all cells are collected by using PBS. Centrifuge the cells at 500 × *g* for 5 min at 4 °C.
5. Lyse 1 million cells in 200 μl of cell lysis buffer containing freshly added 1× protease inhibitor. Pipette up and down a few times. Transfer the lysate into a 1.5 ml microcentrifuge tube and incubate on ice for 10 min. Henceforth all steps should be performed on ice. Collect one additional lysate ("test sample") containing 1 million cells lysed in 200 μl buffer to test sonication conditions (*see* **Note 4**).

3.2 Sonication

1. Chromatin shearing is performed using a sonciator. Optimal DNA shearing should be between 200 and 500 base pairs (majority of fragments around 500 base pairs) for ChIP-qPCR assays (*see* **Note 5**).
2. Switch on the water bath cooler for at least 15 min for the temperature to reach 4 °C. Sonicate cell lysates under high power setting for 14 cycles of 30 s ON and 30 s OFF. After 7 cycles, take out the tubes from the bioruptor and invert a few times to mix well. Proceed with another 7 cycles. Ensure that the sonicator does not heat up during the procedure.
3. Centrifuge the lysate at 16,000 × *g* for 10 min at 4 °C. Transfer the clear lysate into a fresh microcentrifuge tube.
4. To check sonication efficiency, reverse crosslink the "test sample" by adding 10 μl of 5 M NaCl and incubate at 65 °C for at least 4 h to overnight.
5. Add 2 μl of 10 mg/ml RNAse A and incubate for 30 min at 37 °C.
6. Add 2.5 μl of 20 mg/ml proteinase K for 1 h at 45 °C.
7. Isolate the DNA from the lysate using phenol-chloroform-isoamylalcohol.
8. Load 500 ng to 1 μg of DNA on a 1.5% agarose gel with 100 bp DNA ladder. 200–500 bp DNA shearing should be observed with the conditions mentioned above (*see* **Note 6**) for C2C12 cells (Fig. 2). Samples can be stored at −80 °C at this step.

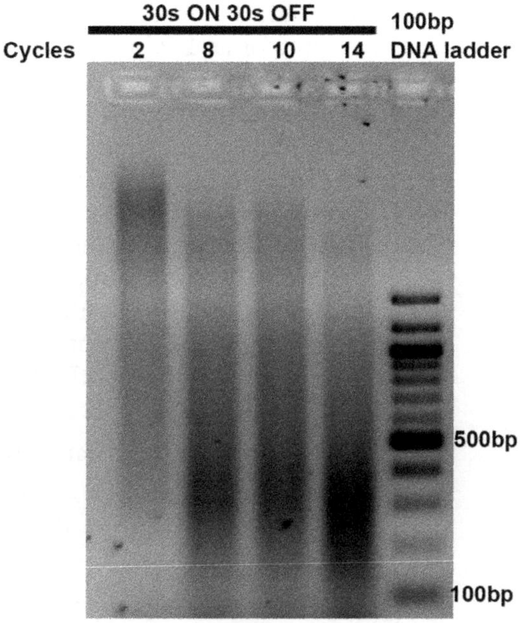

Fig. 2 Image showing chromatin shearing in myoblasts. 1 μg of DNA (sonicated 30 s ON 30 s OFF mode for multiple sonication cycle number as indicated) was analyzed on 1.5% agarose gel. An optimal DNA shearing of 200–500 bp was observed at 14 cycles. 100 bp DNA ladder was used for size approximation

3.3 Immuno-precipitation

1. Use 200 μl of sheared chromatin lysate for each antibody pull-down in 2 ml microcentrifuge tubes. Make up the volume to 2 ml with ChIP dilution buffer with 1× protease inhibitors (*see* **Note 7**). Gently mix the content by inverting tubes 3–4 times. Keep aside 200 μl (10%) of the diluted lysate as input. Store it in a −80 °C freezer until DNA isolation.
2. Preclear the remaining lysate by adding 50 μl of salmon sperm DNA agarose beads provided in the Millipore ChIP kit (make sure to cut the edge of a 200 μl tip to ensure bead integrity). Incubate for 1 h on a rotator at 4 °C.
3. Centrifuge the lysates at 200 × *g* for 2 min. Allow the beads to settle and transfer the upper supernatant to another tube (*see* **Note 8**).
4. Add the desired amount of test antibody to the lysate (1800 μl) (*see* **Note 9**). Additionally, perform the IPs with appropriate control antibody (IgG). Incubate overnight (14–16 h) on a rotator at 4 °C.
5. Add 60 μl of agarose beads from the kit (make sure to cut the edge of tip to ensure bead integrity) and incubate the complex for additional 2 h on rotator at 4 °C.
6. Centrifuge the antibody bead complex at 200 × *g* for 2 min at 4 °C. Retain the beads and discard the clear lysate.

7. Wash the beads with 1 ml of wash buffer for 8 min each on a rotator at 4 °C in the following order: Low salt buffer (once), High salt buffer (once), Lithium chloride buffer (once), and TE buffer (twice). A wash refers to incubating the beads with buffers under rotation and centrifuging the solution at 200 × *g* for 2 min, and discarding the wash buffer (*see* **Note 10**).

3.4 Elution and Reverse Crosslinking

1. Add 250 μl of elution buffer to the beads after the final TE buffer wash. Incubate at room temperature for 15–30 min under rotation. Centrifuge the tube at 200 × *g* for 2 min and collect the clear lysate. Redo the elution step with additional 250 μl of elution buffer for another 15–30 min. Pool the eluted lysates to get a total of 500 μl.
2. Remove the input samples from −80 °C and allow it to thaw.
3. Reverse crosslink by adding 10 μl of 5 M NaCl to the input; and 20 μl of NaCl to the ChIP samples. Incubate at 65 °C overnight.
4. Perform RNAse treatment by adding 2 μl RNAse A for 1 h at 37 °C.
5. Add 20 μl and 10 μl of 1 M Tris–HCl pH 6.8, 8 μl and 4 μl of 0.5 M EDTA to ChIP and input samples respectively. Add 2.5 μl of proteinase K, incubate for 1 h at 45 °C.

3.5 DNA Isolation

1. Add 200 μl and 500 μl of phenol-chloroform-isoamylalcohol mixture to the input and ChIP samples respectively. Vortex for 20 s.
2. Centrifuge the samples at 16,000 × *g* for 10 min at room temperature.
3. Transfer the upper aqueous phase to a new 1.5 ml microcentrifuge tube. Add 500 μl and 800 μl of absolute ethanol to input and ChIP samples respectively. Mix the content of the tube by inverting a few times. Addition of 1 μl of glycogen helps in visualizing the DNA pellet after precipitation.
4. Incubate the tubes at −20 °C for at least 4 h to overnight to allow DNA precipitation.
5. Centrifuge the tubes at 16,000 × *g* for 20 min at room temperature and remove the supernatant. Take care not to discard the pellet.
6. Wash the DNA pellet with 500 μl of 70% ethanol. Centrifuge at 16,000 × *g* for 10 min. Carefully discard the supernatant and air dry the DNA pellet for 5 min at room temperature (*see* **Note 11**).
7. Dissolve ChIP and input DNA in 40 μl and 80 μl of nuclease-free water respectively. Incubate the DNA at 37 °C for 20 min

to ensure that DNA is completely dissolved. DNA can be stored at −20 °C. If the DNA is used for library construction, the quality of the DNA can be checked using a bioanalyzer.

3.6 Real-Time Quantitative PCR and Data Analysis

DNA from input and ChIP samples is used to perform quantitative PCR reaction, with SYBR green master mix.

1. Prepare a 20 μl triplicate master mix (6 μl reaction per well) for each primer pair on a 384-well plate.
2. Each 20 μl master mix for triplicate samples contains 0.8 μl of forward and reverse primer mix from 10 μM primer stock, DNA (2 μl for input, 4 μl for ChIP), 10 μl of 2× SYBR green master mix and rest is nuclease-free water.
3. Run the samples on Light cycler 480 II (Roche) with default settings for 45 cycles (*see* **Note 12**).
4. The ChIP threshold cycles, CT values (from triplicates) are normalized to the corresponding input CT values (from triplicates) to get delta CT (ChIP CT—Input CT). Relative enrichment is calculated using the equation 2^-delta CT.

4 Notes

1. Both the concentration of formaldehyde and the incubation time can affect chromatin shearing efficiency as well as immunoprecipitation with antibodies. Hence care should be taken to quench the reaction with glycine immediately after 10 min fixation. For transcription cofactors that do not bind directly to DNA, cells can be incubated with protein cross-linkers such as (Di-Succinimidly Glutarate, DSG in PBS at a final concentration (2 mM) for 45 min before crosslinking with formaldehyde. This helps stabilize protein interactions.
2. This protocol utilizes 1 million cells in 200 μl of lysis buffer. If performing ChIP with a higher number of cells, increase the lysis buffer volume accordingly to allow proper lysis of the cells. Incomplete lysis may yield low quantity of DNA. Maintain the ratio of 200 μl of lysis buffer for 1 million cells.
3. After fixation, the cells tend to stick to the plate. Collect the cells by adding PBS several times. The cell pellet can be snap frozen in liquid nitrogen and stored at −80 °C. Alternatively, if proceeding with sonication, a mild buffer containing 5 mM PIPES pH 8.0, 85 mM KCl, 0.5% NP-40 with protease inhibitors can be used to harvest the cells. This helps in reducing the stickiness of cells to the plate while keeping the cell nuclei intact.

4. Lysates can be frozen at −80 °C at this step. However we recommend continuing with the sonication before freezing as the efficiency of sonication decreases for frozen pellet (*see* **Note 5**).
5. Chromatin shearing efficiency depends on several factors: the type and amount of cells, lysis buffer content and volume, fresh fixed cells or frozen cell pellet, type of microcentrifuge tubes, and the power setting of the sonicator. Cells that are frozen after formaldehyde fixation require additional sonication cycles [15]. Hence, the sonication conditions have to be empirically determined for the each cell type and conditions used. During standardization, perform the sonication for multiple cycles and choose the shortest sonication time that shows noticeable DNA fragment size around 200–500 bp. Over sonication may break the complexes and denature the antigens while reducing the antibody efficiency. Ensure that the tubes are balanced while sonicating the samples and once standardized keep it constant for all experiments. Ideally sonicate control and test samples at the same time.
6. As each nucleosome is wrapped by 147 bp DNA, chromatin shearing of around 500 bp is ideal for transcription factors or cofactors ChIP. This allows sufficient nucleosome (DNA length) to capture factors bound to chromatin. In case of ChIP for histone modifications, micrococcal nuclease (enzymatic digestion) can be used to digest chromatin to mononucleosomes containing approximately 147 bp DNA. If using the DNA for library construction followed by high-throughput sequencing, the sonication conditions need to be optimized. DNA fragments of around 200–300 bp are ideal for library construction. Make sure the samples are maintained at 4 °C throughout sonication. Sheared chromatin can be stored at −80 °C for a few months. If sonication is performed on differentiated myotubes, the conditions need to be optimized and may require a longer time for chromatin shearing.
7. If cell lysis buffers with a low SDS concentration (0.1%) are used, dilution with ChIP dilution buffer may not be necessary. However, the sonication condition should be standardized with the buffer used. Since this protocol uses 1% SDS lysis, dilution with ChIP dilution buffer is recommended. Lower SDS concentration is ideal for immunoprecipitation assays.
8. Do not centrifuge the beads at high speed (>400 × *g*) as it may rupture the beads.
9. In general, the antibody amount should be in excess to the factor being precipitated. The specificity of the antibody is crucial for the success of ChIP. Certain antibodies may work well for native ChIP but not for crosslinked chromatin. Therefore, the antibody concentration, incubation time, and type of

antibody should be empirically determined. For histone IPs (H3K9me2, H3K9ac, H3K27me3) we typically use 2 μg of antibody and if using whole serum antibodies, 6–8 μl is used per ChIP.

10. After centrifugation, beads tend to form loose pellet. Take extra care not to disturb/lose beads while removing the wash buffers. If required, depending on the antibody, to increase the signal-to-noise ratio, a 3× wash with high salt buffer can be considered.
11. If required, ensure complete ethanol removal by re-centrifugation. Retention of a higher amount of ethanol may interfere with downstream steps as well as dilute the DNA concentration.
12. The PCR reaction should be performed with a specific volume of DNA kept constant across all the samples. This normalizes the effect of a stimulus on the protein being immunoprecipitated. For instance, treatment with a drug can impact expression and hence the amount of protein being pull down may vary compared to controls. Hence, it is important to dissolve DNA in an accurate amount of water across all samples. In addition, care should be taken not to lose beads during washes and handling the ChIP samples.

Acknowledgments

Work in the R.T. laboratory is supported by grants from the National Medical Research Council.

References

1. Kassar-Duchossoy L, Giacone E, Gayraud-Morel B, Jory A, Gomès D, Tajbakhsh S (2005) Pax3/Pax7 mark a novel population of primitive myogenic cells during development. Genes Dev 19:1426–1431. https://doi.org/10.1101/gad.345505
2. Bentzinger CF, Wang YX, Rudnicki MA (2012) Building muscle: molecular regulation of myogenesis. Cold Spring Harb Perspect Biol 4. https://doi.org/10.1101/cshperspect.a008342
3. Bharathy N, Ling BMT, Taneja R (2013) Epigenetic regulation of skeletal muscle development and differentiation. Subcell Biochem 61:139–150. https://doi.org/10.1007/978-94-007-4525-4_7
4. McKinsey TA, Zhang CL, Olson EN (2001) Control of muscle development by dueling HATs and HDACs. Curr Opin Genet Dev 11:497–504
5. Asp P, Blum R, Vethantham V, Parisi F, Micsinai M, Cheng J, Bowman C, Kluger Y, Dynlacht BD (2011) Genome-wide remodeling of the epigenetic landscape during myogenic differentiation. Proc Natl Acad Sci U S A 108:E149–E158. https://doi.org/10.1073/pnas.1102223108
6. Fittipaldi R, Caretti G (2012) Tackling skeletal muscle cells epigenome in the next-generation sequencing era. Comp Funct Genomics 2012:979168. https://doi.org/10.1155/2012/979168
7. Cao Y, Yao Z, Sarkar D, Lawrence M, Sanchez GJ, Parker MH, MacQuarrie KL, Davison J, Morgan MT, Ruzzo WL, Gentleman RC, Tapscott SJ (2010) Genome-wide MyoD binding in skeletal muscle cells: a potential for broad cellular reprogramming. Dev Cell 18:662–674. https://doi.org/10.1016/j.devcel.2010.02.014

8. Blum R, Vethantham V, Bowman C, Rudnicki M, Dynlacht BD (2012) Genome-wide identification of enhancers in skeletal muscle: the role of MyoD1. Genes Dev 26:2763–2779. https://doi.org/10.1101/gad.200113.112
9. Lu L, Sun K, Chen X, Zhao Y, Wang L, Zhou L, Sun H, Wang H (2013) Genome-wide survey by ChIP-seq reveals YY1 regulation of lincRNAs in skeletal myogenesis. EMBO J 32:2575–2588. https://doi.org/10.1038/emboj.2013.182
10. Ohkawa Y, Mallappa C, Vallaster CSD, Imbalzano AN (2012) Isolation of nuclei from skeletal muscle satellite cells and myofibers for use in chromatin immunoprecipitation assays. Methods Mol Biol 798:517–530. https://doi.org/10.1007/978-1-61779-343-1_31
11. Nelson JD, Denisenko O, Bomsztyk K (2006) Protocol for the fast chromatin immunoprecipitation (ChIP) method. Nat Protoc 1:179–185. https://doi.org/10.1038/nprot.2006.27
12. Kuo MH, Allis CD (1999) In vivo cross-linking and immunoprecipitation for studying dynamic protein:DNA associations in a chromatin environment. Methods 19:425–433. https://doi.org/10.1006/meth.1999.0879
13. Nelson JD, Denisenko O, Sova P, Bomsztyk K (2006) Fast chromatin immunoprecipitation assay. Nucleic Acids Res 34:e2. https://doi.org/10.1093/nar/gnj004
14. Thorne AW, Myers FA, Hebbes TR (2004) Native chromatin immunoprecipitation. Methods Mol Biol 287:21–44. https://doi.org/10.1385/1-59259-828-5:021
15. Schoppee Bortz PD, Wamhoff BR (2011) Chromatin immunoprecipitation (ChIP): revisiting the efficacy of sample preparation, sonication, quantification of sheared DNA, and analysis via PCR. PLoS One 6:e26015. https://doi.org/10.1371/journal.pone.0026015

Chapter 5

Exercising Bioengineered Skeletal Muscle In Vitro: Biopsy to Bioreactor

Daniel C. Turner, Andreas M. Kasper, Robert A. Seaborne, Alexander D. Brown, Graeme L. Close, Mark Murphy, Claire E. Stewart, Neil R. W. Martin, and Adam P. Sharples

Abstract

The bioengineering of skeletal muscle tissue in-vitro has enabled researchers to more closely mimic the in-vivo skeletal muscle niche. The three-dimensional (3-D) structure of the tissue engineered systems employed to date enable the generation of highly aligned and differentiated myofibers within a representative biological matrix. The use of electrical stimulation to model concentric contraction, via innervation of the myofibers, and the use of mechanical loading to model passive lengthening or stretch has begun to provide a manipulable environment to investigate the cellular and molecular responses following exercise mimicking stimuli in-vitro. Currently available bioreactor systems allow either electrical stimulation or mechanical loading to be utilized at any given time. In the present manuscript, we describe in detail the methodological procedures to create 3-D bioengineered skeletal muscle using both cell lines and/or primary human muscle derived cells from a tissue biopsy, through to modeling exercising stimuli using a bioreactor that can provide both electrical stimulation and mechanical loading simultaneously within the same in-vitro system.

Key words Skeletal muscle, Tissue engineering, Bioengineering, Myoblasts, Satellite cells, Exercise, Biological scaffolds

1 Introduction

Exercising differentiated multinucleated skeletal muscle cells (myotubes/fibers) in monolayer cultures by means of electrical stimulation [1, 2] and mechanical stretch [3, 4] has partially mimicked responses following resistance [4, 5] and endurance [6, 7] exercise in-vivo. Following pioneering work by Vandenburgh and colleagues three decades ago, culturing in 3-dimensional (3-D) microenvironments may however be considered a more representative

Daniel C. Turner and Andreas M. Kasper contributed equally to this work. Adam P. Sharples is the senior/corresponding author.

Sissel Beate Rønning (ed.), *Myogenesis: Methods and Protocols*, Methods in Molecular Biology, vol. 1889,
https://doi.org/10.1007/978-1-4939-8897-6_5,

model of the in-vivo skeletal muscle niche [8, 9] and with continuing progression, may predominate as an in-vitro tool for investigating the underlying mechanisms of exercise adaptation. Indeed, 3-D tissue engineered skeletal muscle models display highly aligned and differentiated myotubes [10, 11], permitting measures of muscle function [12–16] that are not entirely possible with monolayer cultures where swirling myotube formation results in non-uniaxial force generation. Furthermore, embedding skeletal muscle cells within representative extracellular matrices (e.g. collagen, laminin, fibrin) and allowing their differentiation into myotubes increases the amount of time the myotubes can be kept in culture. Where in monolayer, myotubes may begin to detach, especially following spontaneous contraction due to poorer matrix attachment [10]. The use of self-assembling fibrin "myooids" may also prove advantageous for 3-D skeletal muscle tissue engineering given that the cells can easily proliferate and become confluent prior to differentiation versus the use of a stiffer matrix such as collagen where proliferation maybe inhibited, and therefore fibrin "myooids" usually require lower seeding densities. This is particularly useful if trying to generate bioengineered muscle using limited primary muscle derived cells that eventually senesce. Fibrin myooids also provide superior functionality, such as force production vs. other biological scaffolds and can allow efficient gene transfection [14, 15, 17–20]. Collectively, the advantages of these engineered systems may help delineate the underlying mechanisms of physiological adaptation in response to exercise mimicking stimuli in-vitro. Once fully formed into a cyclindrical-like muscle, constructs undergo electrical [5, 12–15, 17, 21] or mechanical [22, 23] stimulation which can mimic shortening and lengthening contraction of skeletal muscle, respectively. Typically however, only one of these stimulation modes is utilized at once. Therefore, a system that can perform both of these functions simultaneously would be advantageous. Furthermore, delivering an electrical input during the lengthening portion of the movement would also provide a more relevant contraction vs. simply a passive lengthening/stretch, as muscle activity/tension is required to control an eccentric contraction in-vivo. Here, we describe the methodological procedures for creating 3-D bioengineered skeletal muscle constructs, using human skeletal muscle derived cells (SkMDC's) and/or the C2C12 myoblast cell line [24, 25] within a fibrin matrix. We further describe how to undertake stimulatory regimes using a commercially available (with customizations) bioreactor system permitting simultaneous electrical and mechanical stimulation.

2 Materials

2.1 Skeletal Muscle Biopsy Procedure

1. Disposable sterile scalpel.
2. Sterile biopsy instrument: Monopty® disposable core biopsy instrument (e.g., CR Bard, Crawley, UK) or a conchotome biopsy tool (e.g., Gebrüder Zepf Medizintechnik, Dürbheim, DEU) (*see* **Note 1**).
3. 6 × 75 mm steri-strips.
4. 7.5 cm × 5 cm sterile wound dressing plaster with transparent waterproof film.
5. 1.5 ml 0.5% bupivacaine hydrochloride prescribed by a designated physician/medical doctor.
6. Sterile wound cleansing and dressing pack (including swabs, sheet and tray).
7. Fine intramuscular injection needle.
8. Sterile 5 ml syringe.
9. Hydrex surgical scrub (e.g., Ecolab©, UK).
10. Sterile surgical gloves.
11. Disposable hospital razor.
12. 1.5 ml transfer medium: Ham's F-10 medium that includes 1 mM L-glutamine, 0.1% heat inactivated fetal bovine serum (hiFBS) (*see* **Note 2**), 0.1% heat inactivated new born calf serum (hiNBCS) (*see* **Note 2**), 100 U/ml penicillin, 100 μg/ml streptomycin, 2.5 μg/ml amphotericin B. Stored and used at 4 °C.
13. Sterile 2 ml Eppendorf tube containing transfer medium submerged in ice.

2.2 Human Primary Cell Isolation

1. Growth medium: Ham's F-10 medium that includes 1 mM L-glutamine, 10% hiFBS, 10% hiNBCS, supplemented with an additional 2 mM L-glutamine, 100 U/ml penicillin, 100 μg/ml streptomycin, 2.5 μg/ml amphotericin B. Stored at 4 °C.
2. Horse serum (HS) (to neutralize trypsin).
3. 0.5% Trypsin-0.2% EDTA.
4. 0.01 M phosphate buffered saline (1×) (PBS): PBS dissolved in distilled water (dH_20), 100 U/ml penicillin, 100 μg/ml streptomycin, 2.5 μg/ml amphotericin B (*see* **Note 1**). Stored at room temperature.
5. 2× sterile scalpels: either blades (*see* **Note 1**) or disposable sterile scalpel.
6. Magnetic stirring bar (6 × 35 mm) with platform (preheated to 37 °C) (*see* **Note 1**).
7. Disposable sterile petri dish (100 mm diameter).

8. Sterile 100 ml specimen container.
9. 2 mg/ml gelatin from porcine skin (Type A; Sigma-Aldrich, G2500) dissolved in distilled water (dH_20) (*see* **Notes 1** and **3**).

2.3 Cryopreservation of Cells

1. Growth medium for primary human cell culture: Ham's F-10 medium that includes 1 mM L-glutamine, 10% hiFBS, 10% hiNBCS, supplemented with an additional 2 mM L-glutamine and 100 U/ml penicillin, 100 μg/ml streptomycin, 2.5 μg/ml amphotericin B. Stored at 4 °C.
2. Growth medium for C2C12 cell culture: Dulbecco's modified Eagle's medium (DMEM) that includes 4 mM L-glutamine, 10% hiFBS, 10% hiNBCS, supplemented with an additional 2 mM L-glutamine, 100 U/ml penicillin, 100 μg/ml streptomycin. Stored at 4 °C.
3. Cell culture grade dimethyl sulfoxide (DMSO).
4. Cryopreservation freezer container (e.g. "Mr. frosty").
5. PBS (*see* **Note 1**). Stored at room temperature.
6. 0.5% Trypsin-0.2% EDTA.

2.4 Resuscitation of Cryopreserved Cells

1. 2 mg/ml gelatin (*see* **Note 3**). Stored at room temperature.
2. Growth medium for primary human cell culture: Ham's F-10 medium that includes 1 mM L-glutamine, 10% hiFBS, 10% hiNBCS, supplemented with an additional 2 mM L-glutamine and 100 U/ml penicillin, 100 μg/ml streptomycin, 2.5 μg/ml amphotericin B. Stored at 4 °C.
3. Growth medium for C2C12 cell culture: DMEM that includes 4 mM L-glutamine, 10% hiFBS, 10% hiNBCS, supplemented with an additional 2 mM L-glutamine and 100 U/ml penicillin, 100 μg/ml streptomycin. Stored at 4 °C.

2.5 3-D Culture Dishes

1. Easy-Grip 35 mm dishes (BD Falcon®).
2. Sylgard® 184 Elastomer kit (Dow Corning, MI, USA) (*see* **Note 4**).
3. 0.15 mm minutien pins (*see* **Note 5**).
4. Silk suture thread (2.0) (*see* **Note 6**).
5. 70% ethanol.
6. 0.01 M PBS (*see* **Note 1**). Stored at room temperature.

2.6 Fibrin Skeletal Muscle Constructs

1. Fibrinogen from Bovine Plasma (Sigma-Aldrich, F8630): Dissolve fibrinogen (stored at −20 °C) at 20 mg/ml in preheated (37 °C) Ham's F12-K medium and incubate at 37 °C for 2–3 h, swirling every 30 min (*see* **Note 7**). Vacuum filter (0.22 μm; *see* **Note 8**), aliquot and store at −20 °C (*see* **Note 9**).

2. Thrombin from Bovine Plasma (Sigma-Aldrich, T4648): Dissolve thrombin at 200 U/ml in 5 ml of high glucose (4.5 g/l) DMEM (*see* **Note 10**). Sterile filter (0.22 μm), aliquot and store at −20 °C (*see* **Note 9**).
3. Aprotinin from Bovine Lung (Sigma-Aldrich, A3428-10MG): Dissolve total 10 mg aprotinin in 1 ml of distilled water (dH_20) to ensure a stock concentration of 10 mg/ml. Sterile filter (0.22 μm), aliquot and store at −20 °C.
4. 6-Aminocaproic Acid (Sigma-Aldrich, A7824): Dissolve 6-aminocaproic acid at 50 mg/ml in distilled water (dH_20). Sterile filter (0.22 μm), aliquot and store at 4 °C (*see* **Note 11**).
5. L-Ascorbic Acid (Sigma-Aldrich, A4403): Dissolve total 100 mg L-ascorbic acid in 11.36 ml of high glucose (4.5 g/l) DMEM to ensure a stock concentration of 50 mM. Sterile filter (0.22 μm), aliquot and store at 4 °C (*see* **Note 12**).
6. L-Proline (Sigma-Aldrich, P8865): Dissolve 100 mg L-proline in 17.37 ml 0.01 M PBS (*see* **Note 1**) to ensure a stock concentration of 50 mM. Sterile filter (0.22 μm), aliquot and store at 4 °C (*see* **Note 12**).
7. Genipin (Sigma-Aldrich, G4796): Dissolve genipin at 10 mg/ml in DMSO (*see* **Note 13**). Sterile filter (0.22 μm), aliquot and store at −20 °C.
8. Growth medium: High glucose (4.5 g/l) DMEM that includes 4 mM L-glutamine, 10% hiFBS, 10% hiNBCS, supplemented with an additional 2 mM L-glutamine,100 U/ml penicillin, 100 μg/ml streptomycin (*see* **Note 14**), 0.5 mg/ml 6-aminocaproic acid, 50 μM L-ascorbic acid, 50 μM L-proline (*see* **Note 12**). Stored at 4 °C.
9. Differentiation medium: High glucose (4.5 g/l) DMEM that includes 4 mM L-glutamine, 2% HS, supplemented with an additional 2 mM L-glutamine, 100 U/ml penicillin, 100 μg/ml streptomycin (*see* **Note 14**), 1 mg/ml 6-aminocaproic acid, 50 μM L-ascorbic acid, 50 μM L-proline (*see* **Note 12**). Stored at 4 °C.
10. Maintenance medium: High glucose (4.5 g/l) DMEM that includes 4 mM L-glutamine, 3.5% hiFB, 3.5% hiNBCS, supplemented with an additional 2 mM L-glutamine, 100 U/ml penicillin, 100 μg/ml streptomycin (*see* **Note 14**), 1 mg/ml 6-aminocaproic acid, 50 μM L-ascorbic acid, 50 μM L-proline (*see* **Note 12**). Stored at 4 °C.
11. Sylgard coated dishes (*see* **Note 4**).

2.7 TC-3 Bioreactor System (Ebers Medical Technology, ESP) (See Note 15)

1. Control module.
2. 3-D bioreactor cell culture chamber(s) including parts such as clamps and screws (*see* **Note 1**).
3. Laptop/desktop with controlling software.

4. Peristaltic pump/perfusion box.
5. Electrical Stimulation: Electrical stimulation module, output box, electrode anode/cathode splitter, platinum/pure iridium or stainless steel electrodes (*see* **Note 1**).
6. Mechanical Stimulation: Mechanical stimulation box.

2.8 From the Culture Dish to the Bioreactor

1. TC-3 Bioreactor chamber(s) (e.g. Ebers Medical Technology, ESP).
2. 2× sets of angled surgical tweezers (*see* **Notes 1** and **16**).
3. Maintenance medium: High glucose (4.5 g/l) DMEM that includes 4 mM L-glutamine, 3.5% hiFB, 3.5% hiNBCS, supplemented with an additional 2 mM L-glutamine, 100 U/ml penicillin, 100 μg/ml streptomycin (*see* **Note 14**), 1 mg/ml 6-aminocaproic acid, 50 μM L-ascorbic acid, 50 μM L-proline (*see* **Note 12**). Stored at 4 °C.

3 Methods

All procedures must be conducted under fully aseptic conditions. Following relevant ethical approval and informed consent of the participants, the skeletal muscle biopsy procedure must be undertaken by fully trained and insured individuals, usually a physician/clinician. A full medical screen by the physician of the participant is also required and the participants GP/family doctor should be informed. When preparing all biopsy equipment/contents, it is essential that direct contact is only made to the exterior packaging. If transported, the muscle tissue must be immersed in chilled (4 °C) transfer medium and transported to a sterile Class 2 cell culture hood/microbiological safety cabinet (BSC) as soon as possible to undergo subsequent cell isolation procedures. Furthermore, the storage and disposal of any tissue and waste should be in accordance to the relevant legislation, (e.g. in the UK, the Human Tissue Act (2004), and Control of Substances Hazardous to Health (2002) regulations, respectively).

3.1 Skeletal Muscle Biopsy Procedure

1. Ensure participant is laying in a supine position before shaving the desired area (in this case, the skin above the belly of the vastus lateralis) with a disposable razor.
2. Empty the wound dressing pack onto a sterile trolley, avoiding contact with the inner dressing. The biopsy instrument, syringe, needle gauge, scalpel, suture pack and plaster can now be opened and placed on the sheet of the sterile dressing pack (Fig. 1).
3. Pour the surgical scrub into the wound dressing tray and unpackage the Marcain and sterile surgical gloves.

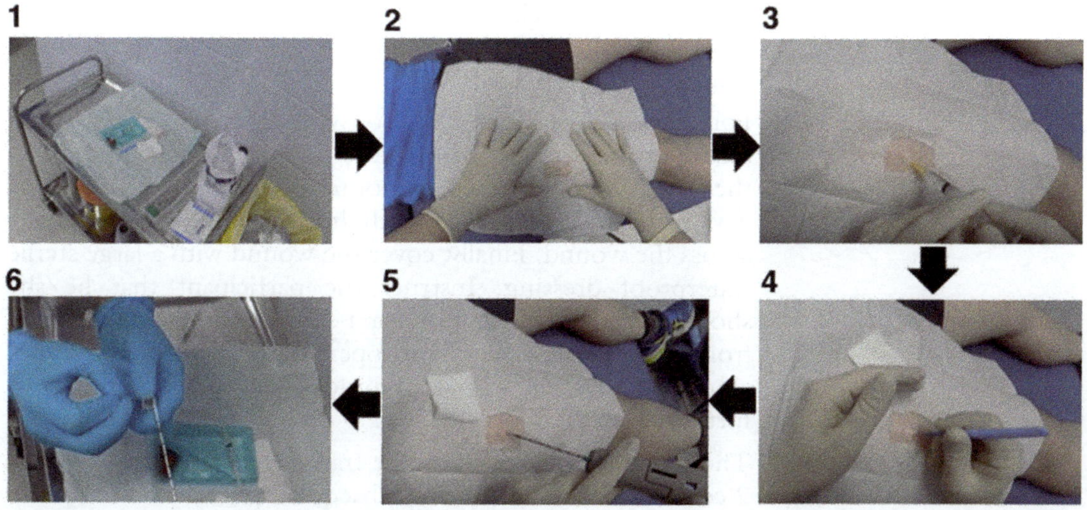

Fig. 1 Depicts (1) removing all packaging of relevant equipment and placing onto a sterile working trolley. (2) Wound dressing sheet placed over the subjects leg with the skin of the biopsy area exposed. (3) Injection of anaesthetic into the deep fascia. (4) Incision of scalpel through the superficial layers and epimysium of the muscle. (5) Insertion of the biopsy needle to obtain muscle tissue which is then (6) transferred to the Eppendorf tube containing transfer medium (4 °C) and submerged in ice

4. Put on the sterile surgical gloves, ensuring the exterior does not make contact with the skin or objects which are non-sterile.
5. Using wound dressing swabs and surgical scrub, wipe the desired biopsy area in order to cleanse the skin. Place the wound dressing sheet/fenestrated drape over the participants leg with the skin of the desired biopsy area exposed (Fig. 1).
6. Assemble the sterile needle and syringe and withdraw the prescribed amount of anesthetic (1.5 ml) ensuring no air bubbles are present. Insert the needle subcutaneously and inject ~50% (~0.75 ml) of until there are signs of peau d'orange (skin resembles orange peel). Insert the needle a little deeper and inject the remaining anesthetic into the deep fascia (Fig. 1) (*see* **Note 17**). To ensure the anesthetic has taken effect (usually ~ 1–2 min post administration), gently apply pressure to the biopsy area with a scalpel (without incision) and ask the participant to verbally confirm a lack of feeling together with numbness.
7. Once anesthetized, using the scalpel, make an incision through the superficial layers and epimysium of the muscle. The tip of the biopsy gun may now be inserted into the wound and muscle tissue may be obtained (Fig. 1). Alternatively, when obtaining muscle tissue using a conchotome instrument, it is important to ensure an incision is made to the fascia to avoid additional pressure and discomfort. You may then insert and rotate (90°) the instrument to cut and extract the muscle tissue [26] (*see* **Note 18**).

8. Using a sterile scalpel, scrape the muscle tissue into a sterile tube containing transfer medium (4 °C) and submerge in ice (Fig. 1).
9. Following 2–3 min (needle biopsy instrument) or 5–10 min (conchotome) of applied pressure to stop the bleeding, clean the area using surgical scrub prior to dressing the wound. Once dry, place the steri-strips both horizontally and diagonally across the wound. Finally, cover the wound with a large sterile waterproof dressing. Instruct the participant that he/she should keep the dressing on for 1–2 weeks and should refrain from swimming, especially in open water or in public pools, alongside covering with waterproof film when showering to further reduce risk of infection.
10. The biopsy sample can now be transferred on ice to the class 2 cell culture hood/BSC for subsequent primary cell isolation.

3.2 Human Primary Cell Isolation

1. Using a modified version of methods previously used [27], empty the contents of the sterile tube into an irradiated sterile petri dish and pipette off the transfer medium (*see* **Note 19**). The biopsy sample should then undergo 3× PBS (preheated to 37 °C) washes and any visible connective and adipose tissue must be removed using a sterile scalpel.
2. Add 5 ml of trypsin-EDTA (preheated to 37 °C) to the sample and mince using two sterile scalpels for at least 1 min.
3. Add the trypsin-EDTA solution and a sterile magnetic stirrer (*see* **Note 1**) to a sterile 100 ml specimen container and place on a heated magnetic stirring platform at 37 °C for 10 min. Alternatively, place the sterile specimen container on a magnetic stirring platform without heating in an oven/incubator set at 37 °C for 10 min.
4. While the sample is digesting, add 1 ml of preheated (37 °C) HS into a separate 15 ml sterile tube. This will neutralize the trypsin in **step 5**.
5. Following 10 mins of heated stirring, remove the trypsin-EDTA solution/supernatant via hand pipetting (*see* **Note 19**) and add to the pre-prepared HS as in **step 4**, leaving the remaining dissected tissue sample in the petri dish. Place the 15 ml tube containing the supernatant and HS into an incubator (37 °C).
6. Add another 5 ml of trypsin-EDTA to the remaining dissected tissue and dissect/mince again for at least 1 min. Place the tissue back on the heated stirrer and repeat as in **step 3.**
7. Once stirred, remove the 5 ml trypsin-EDTA and add to HS/trypsin-EDTA mixture (6 ml) already derived as in **step 5**, yielding approximately 11 ml of solution (1 ml HS and 2 × 5 ml trypsin-EDTA/supernatant) within the 15 ml tube.

8. To isolate the cells from the supernatant, centrifuge the 15 ml tube at 340 × *g* for 5 min at 24 °C.
9. While in the centrifuge, the remaining tissue in the petri dish should be plated on a pregelatinized T25 flask with 7.5 ml of growth medium to derive any remaining cells using an explant technique from the post enzymatically digested tissue derived above (*see* **Note 20**).
10. Following centrifugation after **step 8** above, a cell pellet should be visible at the bottom of the tube. Carefully remove the top 2.5 ml of supernatant (*see* **Note 19**).
11. Resuspend the next 5 ml of supernatant in 5 ml of fresh growth medium in a separate pregelatinized T25 to enable the culture of the isolated cells (*see* **Note 20**).
12. Finally, resuspend the remaining bottom 2.5 ml of supernatant (including cell pellet) in 7.5 ml of fresh growth medium in a separate pregelatinized T25 to enable the culture of cells derived from the above enzymatic digestion (*see* **Note 20**).
13. Allow cells to culture over the ensuing 5+ days without disturbing or removing from the incubator. Additional growth medium should only be added at 7 days to enable full adherence and maintain any cell secreted factors. Cells should then be monitored frequently to assess cell growth and to avoid cells becoming over-confluent and any unwanted cell fusion. Once 80% confluent, cells can be split into larger T75 flasks to generate enough cells for the creation of fibrin bioengineered skeletal muscle (*see* **Note 21**) dependant on the experimental design and the number of fibrin gels required (*see* Sect. 3.6). Cells can also be cryopreserved for later use (*see* Sect. 3.3 below).

3.3 Cryopreservation of Cells

1. Preheat growth medium (primary or C2C12 culture medium), PBS, and trypsin-EDTA to 37 °C.
2. Once PBS is heated, wash the cells 2–3 times, aspirating/removing PBS after each wash.
3. Add 1 ml of trypsin-EDTA and incubate at 37 °C for 5 min. Gently agitate the cells and visualize down an inverted phase contrast microscope to ensure they have all detached from the plastic culture flask surface.
4. Add 4 ml of growth medium to neutralize the trypsin-EDTA solution. Pool cell solution from all flasks of the same participant (primary) or passage number (C2C12) and add to a sterile tube (e.g., 15 ml or 50 ml). Homogenize cell solution slowly (*see* **Note 22**).

5. Count cells using preferred method (i.e. manually using a hemocytometer or electronically using an automated cell counter) and dilute solution to 1 × 10^6 cells/ml using fresh growth medium.
6. Pool solution into a sterile 100 ml specimen container or 50 ml tube and add 10% DMSO (*see* **Note 23**) to prevent formation of ice crystals.
7. Once resuspended, pipette 2 ml of cell solution per prelabeled cryovial (name, cell type, passage number, cell density, and date) and place into a cryopreservation freezer container (i.e. "Mr frosty") filled with isopropanol to ensure a gradual (~1 °C/min) decrease in temperature, further reducing the likelihood of ice crystal formation.
8. Place the cells into a −80 °C freezer for 24 h prior to storing cells in liquid nitrogen.

3.4 Resuscitation of Cryopreserved Cells

1. Preheat growth medium to 37 °C and gelatinize a T25 (primary cells) or T75 (C2C12 cells) flask per vial.
2. Remove the cryovial(s) from the liquid nitrogen Dewar and thaw at room temperature (C2C12 cells) or within an incubator at 37 °C (primary cells).
3. For C2C12 cells, seed cell suspension onto the pregelatanized T75 flask(s) and add 10 ml of growth medium per flask. Gently agitate to ensure even distribution across the culture flask surface and place in an incubator (37 °C) to allow cells to grow to 80% confluency over the ensuing days (typically 72 h).
4. For primary cells, transfer cell suspension into a sterile 15 ml tube and centrifuge at 200 × *g* for 10 min.
5. Remove supernatant and resuspend cell suspension and 6 ml of growth medium onto the pregelatinized T25 flask.
6. Once 80% confluent, resuspend primary cells onto a T75 flask and passage until the desired cell population is attained (*see* **Note 21**).

3.5 Preparation of 3-D Culture Dishes

1. Sylgard coat dishes according to the manufacturer's instructions. Briefly, on a covered surface, add curing agent to elastomer and stir to ensure a consistent silicone solution (*see* **Note 4**). Using a syringe (without needle), add 1.5 ml of silicone to each dish. Allow coated dishes to cure for approximately 24 h. Once cured, replace lids and store for at least 1 week prior to using for muscle cultures (*see* **Note 24**).
2. Prepare 4 × 0.15 mm minutien pins and 2 × 6 mm silk sutures per dish (*see* **Notes 5** and **6**).
3. Using a "fine point" sharpie and ruler, mark the 6, 12, 18, 24, 30 mm points across the center of each dish (*see* **Note 25**).

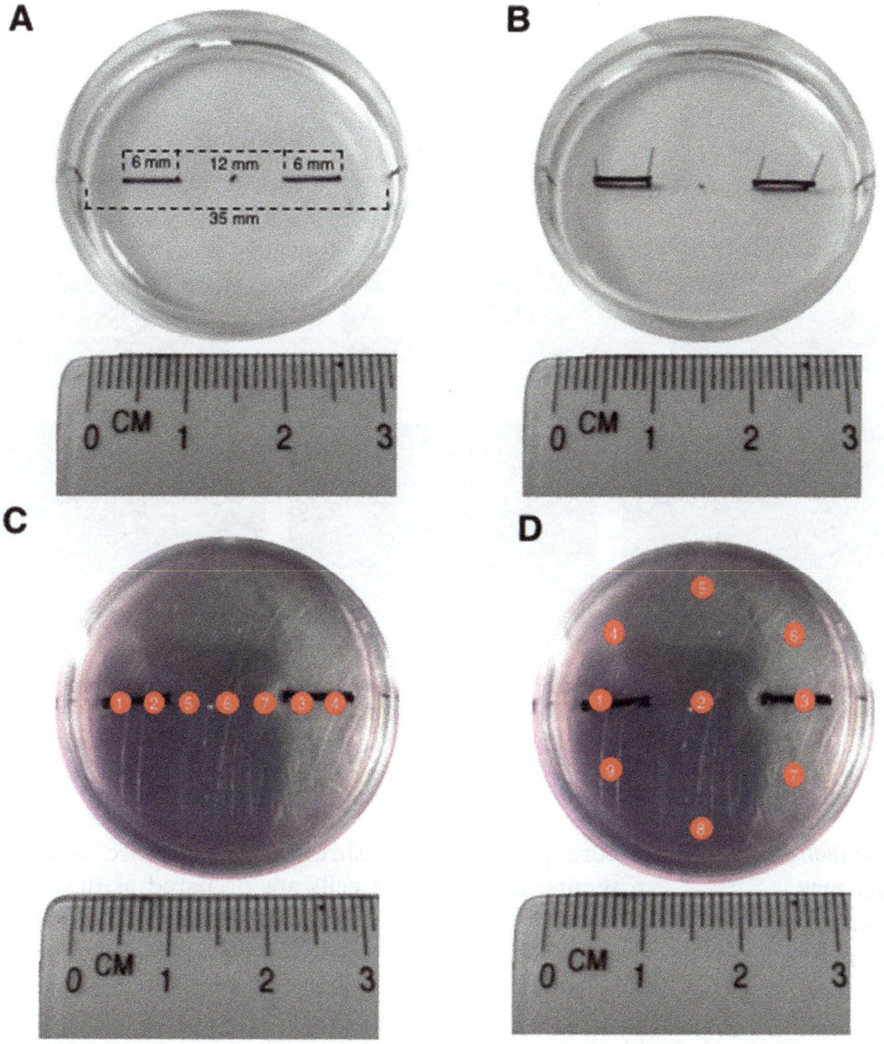

Fig. 2 Depicts (**A**) marked sylgard coated 3-D culture dish with (**B**) silk sutures pinned in place. (**C**) Depicts thrombin solution placement to ensure sutures are saturated and (**D**) indicates pipetting fibrinogen dropwise prior to gentle agitation to ensure even distribution across the cell culture dish surface

Mark a line between the 6–12 and 24–30 mm points to ensure accurate silk suture alignment when pinning in place (Fig. 2).

4. Align silk sutures between 6–12 and 24–30 mm points and pin in place ensuring a 12 mm gap in-between (Fig. 2).

5. Following preparation of 3-D culture dishes, thoroughly spray the inside of the lid and culture dish with 70% immersed methylated spirit (IMS)/ethanol and leave in a Class 2 cell culture hood/BSC under UV to sterilise and air dry overnight (*see* **Note 26**).

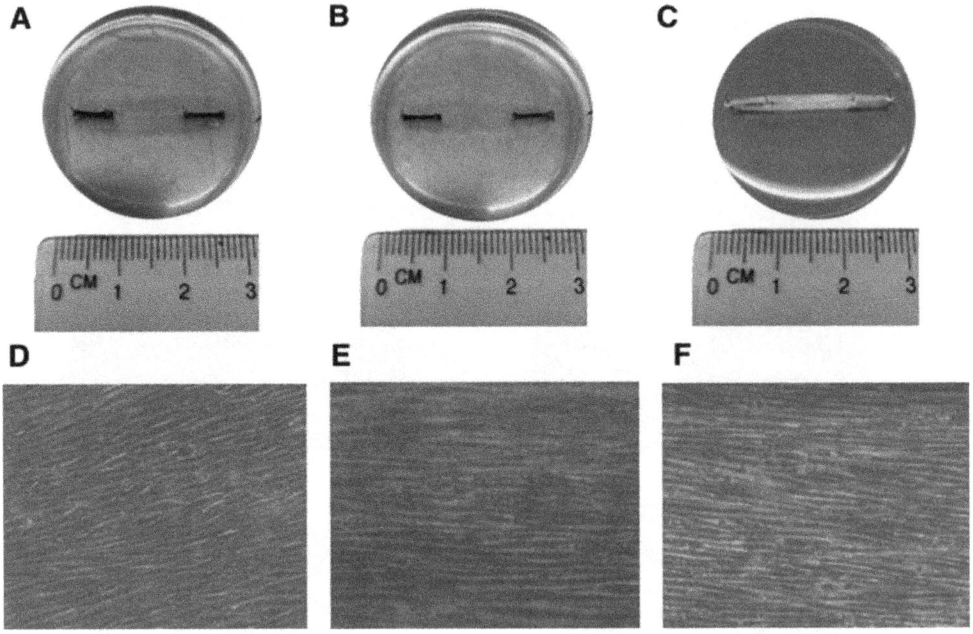

Fig. 3 Macroscopic images of C2C12 3-D fibrin muscle constructs at (**a**) 3 days (**b**) 7 days and (**c**) 14 days in culture. (**d–f**) depict microscopic bright field images (10 × magnification) at 3, 7 and 14 days, respectively

3.6 Preparation and Maintenance of Fibrin Skeletal Muscle Constructs

1. Before proceeding, ensure all 3-D culture dishes, relevant reagents, medium and cells are prepared as described above. For each fibrin muscle construct, thaw 50 μl/ml thrombin, 8 μl/ml aprotinin (*see* **Note 27**), 20 μl/ml genipin (*see* **Note 13**) and 200 μl fibrinogen at room temperature while cell suspension is incubating at 37 °C. Growth medium for thrombin solution should also reach room temperature prior to using.
2. Add thrombin, aprotinin and genipin (*see* **Note 13**) to growth medium to ensure 500 μl of solution per gel (*see* **Notes 27** and **28**).
3. Coat dishes with 500 μl of thrombin/aprotinin/genipin gel solution (*see* **Note 29**) (Fig. 2).
4. Add 200 μl of fibrinogen dropwise around the culture dishes and agitate gently to ensure even distribution with the thrombin solution (Fig. 2).
5. Allow the fibrin gel to polymerize at room temperature for 10 mins prior to incubating at 37 °C for 1 h.
6. Following polymerization, seed 2 ml cell suspension per dish consisting of 100,000–200,000 cells/ml (primary [10]) or 50,000–90,000 cells/ml (C2C12) in growth medium (*see* **Note 30**).

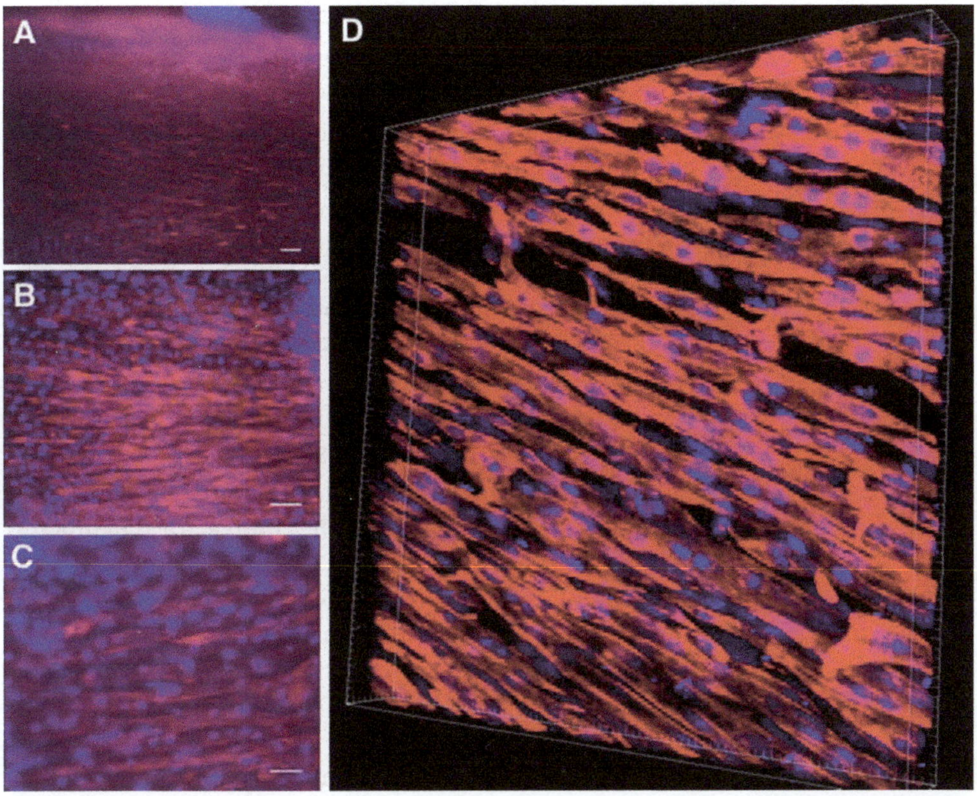

Fig. 4 Immuno-stained microscopic (Nikon, Eclipse Ti-S) images of C2C12 3-D fibrin constructs at (**a**) 10 × (scale bar = 50 μm) (**b**) 20 × (scale bar = 20 μm) and (**c**) 40 × (scale bar = 20 μm) magnification. (**d**) *Z*-stack animated confocal (Olympus, IX83) microscopic image (40 × magnification). Constructs were stained for desmin (red) and nuclei (blue)

7. Gently agitate culture dishes and place in the incubator at 37 °C. Growth medium should be changed every 48 h until cells reach ~90% confluency (Fig. 3).
8. Once approximately 90% confluent, aspirate remaining growth medium and 3× PBS wash (*see* **Note 31**).
9. Transfer to 2 ml of preheated (37 °C) differentiation medium and incubate at 37 °C for 48 h (*see* **Notes 31** and **32**).
10. Remove differentiation media and wash cells as in **step 8** (*see* **Note 31**), transfer to 2 ml of maintenance medium, and incubate at 37 °C (*see* **Note 33**). Be sure to wash cells as in **step 8** (*see* **Note 31**) and top up or change maintenance medium (*see* **Note 34**) for 14 days to allow constructs to mature (Figs. 3 and 4). The gel will contract from the edge of the dish and roll into a cylindrical-like muscle whereby multinucleated myotubes will form in aligned and parallel fashion in line with the direction of the sutures (Figs. 3 and 4).

3.7 From the Culture Dish to the Bioreactor

1. Preheat maintenance medium to 37 °C.
2. Remove bioreactor chamber(s) and parts (screws, clamps, etc.) from the autoclave bag and assemble the bottom the clamps.
3. Manually adjust the distance between the medial edge of each clamp to ensure a 12 mm gap (*see* **Note 35)** and tighten to prevent any unwanted change in resting length.
4. Using sterile angled tweezers (*see* **Note 1**), remove pins from the culture dish and transfer the muscle construct (leaving sutures in if possible) from the cell culture well/dish to the bioreactor chamber(s) (*see* **Notes 16**, **35–37**).
5. Avoiding any stretching or twisting of the muscle construct, clamp the medial end of each suture embedded within either side of the construct (*see* **Notes 36** and **37**).
6. Submerge the clamped construct(s) with 20 ml of maintenance medium by pipetting into the bioreactor chamber(s) (*see* **Notes 31** and **38)**.
7. The muscle construct is now ready to undergo electrical and/or mechanical stimulation.

3.8 Bioreactor Setup

1. Ensure enough free bench space to position all bioreactor equipment (Fig. 5).
2. Connect electrical stimulation and control modules to the mains power supply (ensure the mains is switched off until all other equipment is connected).
3. Connect laptop/desktop, mechanical stimulation box, and electrical stimulation output box to the control module.
4. Connect neutral, anode (+ve), and cathode (−ve) from the electrical stimulation module to the electrical stimulation output box.
5. Connect the anode/cathode splitter to the electrical stimulation output box.
6. Feed electrodes through bioreactor chamber(s) and connect to the anode/cathode splitter (*see* **Notes 39** and **40**) (Fig. 6).
7. Attach bioreactor chamber(s) to mechanical stimulation unit, ensuring the resting length (12 mm) is maintained (*see* **Note 41**).
8. Once assembled, loosen the tension of the chamber (s) mechanical arm to permit movement when initiating the mechanical stretch regime.

3.9 Electrical and Mechanical Stimulation

1. Open the "Ebers TC-3 with Electrical Stimulation" software from the desktop/application menu.
2. From the "port settings" dropdown menu, ensure the control module is connected to the device containing the controlling software in order to proceed ("COM3").

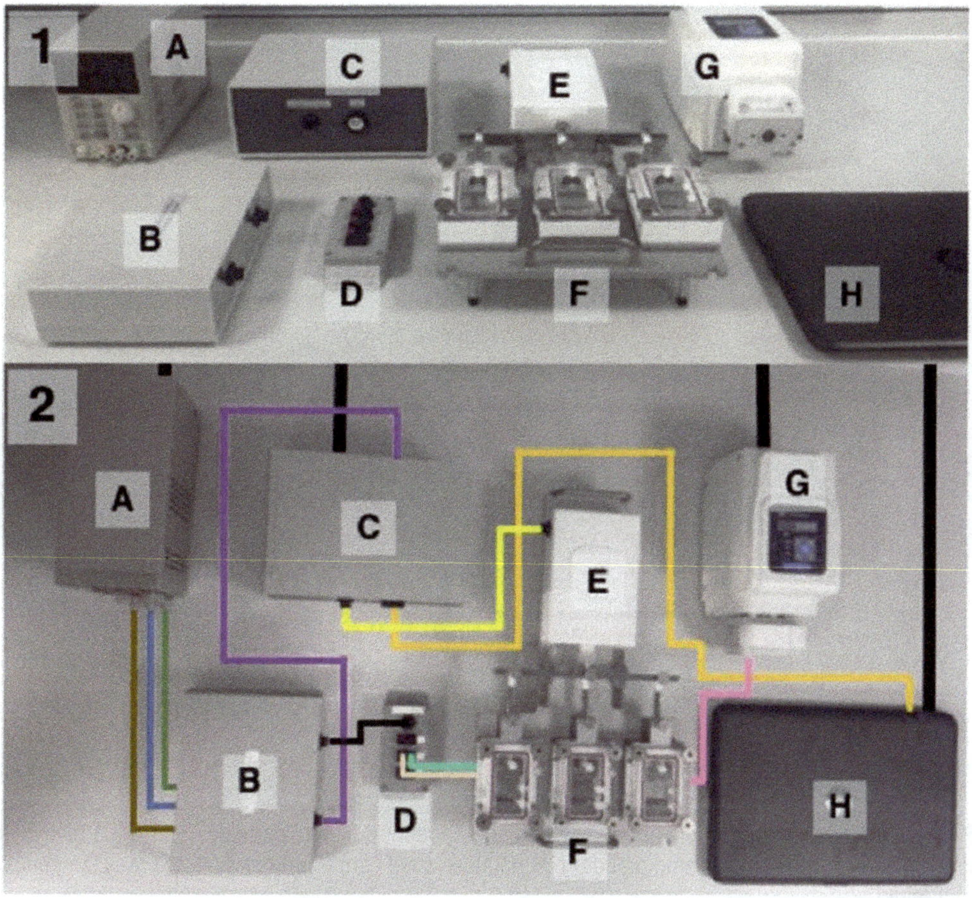

Fig. 5 (**1**) Frontal and (**2**) bird's-eye view of the TC-3 Bioreactor System, (EBERS Medical Technology, ESP) including the (**A**) electrical stimulation module, (**B**) electrical stimulation output box, (**C**) control module, (**D**) electrode anode/cathode splitter, (**E**) bioreactor mechanical stimulation box, (**F**) bioreactor 3-D cell culture chamber, (**G**) peristaltic pump/perfusion box, (**H**) laptop and controlling software. Image and figure legend adapted from [28] with permission from Wiley

3. Click "start" from the "manual mode" panel until the mechanical stretch arm reaches the desired resting length.
4. Click "stop" and "Set to 0" to confirm the starting length. It is at this point you may loosen the tension of the mechanical stretch arm as the chamber should be securely attached to the mechanical stimulation unit (*see* **Note 41**).
5. From the "program mode" panel, select the desired "Run (seconds; s)" and "Pause (seconds; s)" duration, the "ramp mode" and whether you would like the regime to "cycle" (Fig. 7).
6. Navigate to the "record number" to alter "velocity mode" and "frequency mode" settings (Fig. 7).
7. From the "velocity mode" function, the desired stretch limit/ "travel" (millimeters; mm), direction (backward/forward),

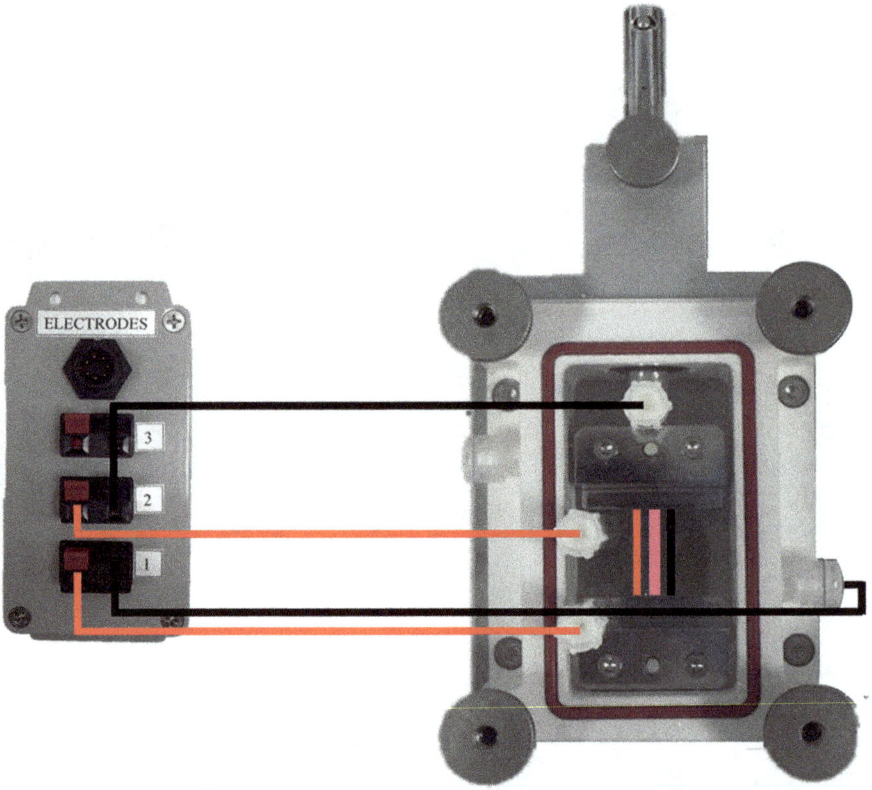

Fig. 6 Electrode placement for electrical stimulation. Anode (red) and cathode (black) pathways to the electrical anode/cathode splitter using platinum or stainless steel wire running parallel to the clamped muscle construct (pink)

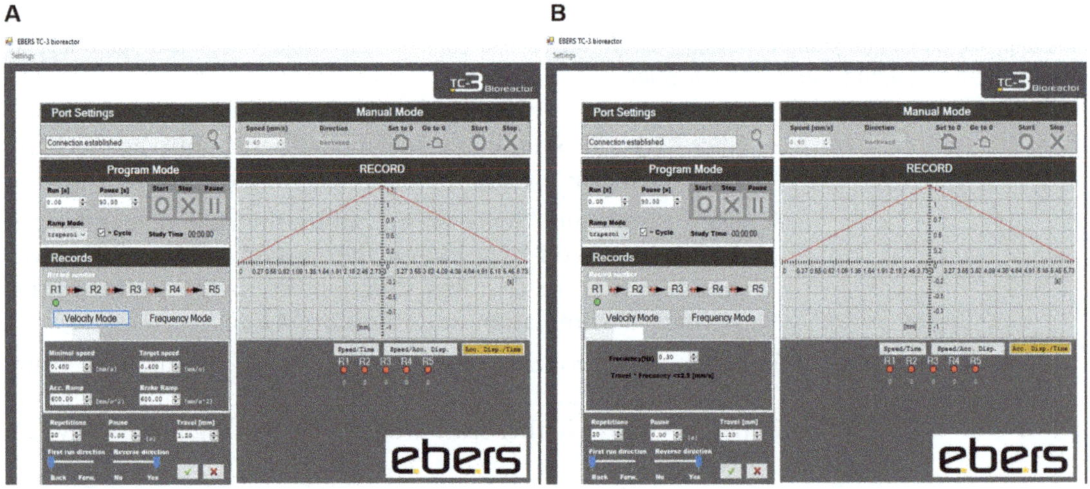

Fig. 7 Ebers TC-3 Software. Mechanical stretch navigation menu for adjusting (**a**) velocity and (**b**) frequency mode settings

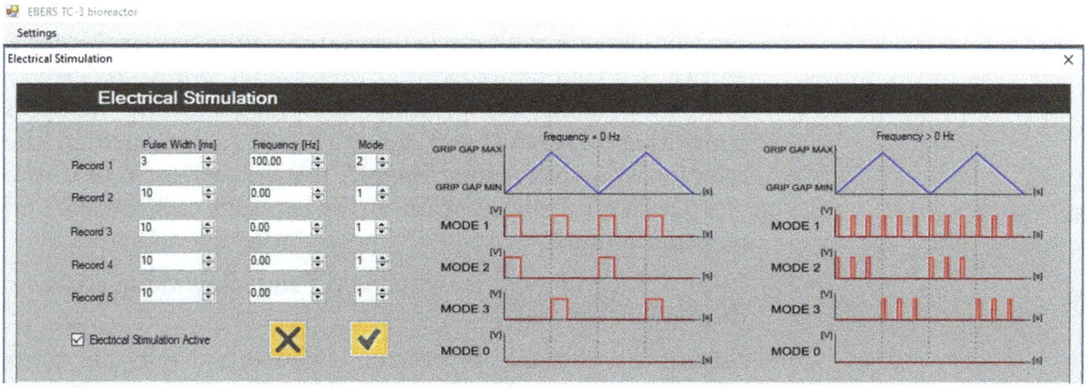

Fig. 8 Ebers TC-3 Software. Electrical stimulation navigation menu for adjusting pulse width (ms), frequency (Hz) and "MODE"

"minimal/target speed" (millimeters per second; mm/s) and "acceleration/brake ramp" (millimeters per second squared; mm/s^2) can be altered (Fig. 7; *see* **Note 42**).

8. From the "frequency mode" function, the stretch frequency (hertz; Hz) can be altered. Click the "green tick" to confirm the desired mechanical stretch regime (*see* **Note 42**).
9. To adjust the electrical stimulation protocol, navigate to the "electrical stimulation" function from the "settings" drop-down menu. Here, single impulses or pulse trains can be applied during the "forward" (when the muscle returns to its' resting length; MODE 3), "backward" (when the muscle is lengthening to reach the desired stretch limit; MODE 2), forward and backward movements combined (MODE 1) or at "rest" (MODE 0) (*see* **Notes 43** and **44**; Fig. 8).
10. Set the desired pulse width (milliseconds; ms), frequency (Hertz; Hz), amplitude (voltage; V), pulse train (seconds; s) and "MODE" number (*see* **Note 44**).
11. Confirm the desired electrical stimulatory parameters via clicking the "Electrical Stimulation Active" option followed by the tick option on screen (Fig. 8).
12. Following cessation of the mechanical and/or electrical stimulatory regime, click "stop" and lock the bioreactor chamber (s) stretch arm.
13. Ensure the electrical stimulation module is switched off at the mains and disconnect the electrodes from the anode/cathode splitter.
14. Place the chamber(s) in the incubator at 37 °C and incubate for the desired postexercise period.
15. Remove chamber(s) from the incubator and place in a sterile Class 2 cell culture hood.

16. Remove the clamps using sterile tools (*see* **Note 1**) and transfer the stimulated fibrin muscle using sterile tweezers (*see* **Notes 1** and **16**) for your desired sample preparation. For RNA/DNA/protein isolation we recommend homogenization in tubes containing MagNA Lyser Green Beads (submerged in the relevant lysis buffer) in a MagNA tissue Lyser for 6,000 rpm × 40 s (with 5 min rest on ice in between each cycle), and repeated 3 times.
17. Ensure all equipment is switched off and all working areas are thoroughly cleaned.
18. Ensure to package bioreactor chamber(s), relevant parts (clamps and screws) and tools and sterilise for future use (*see* **Note 1**).

4 Notes

1. Sterilize via autoclaving (e.g., 122 °C for 30 min).
2. hiFBS and hiNBCS are purchased from South America (EU approved) and New Zealand. It is also important to use serum from the same origin due to disparities in hormonal milieu (hiFBS from US vs. EU origin) which may alter the engineered muscle phenotype [29].
3. Use gelatin from porcine skin A (and not B) given that gelatin B may negatively affect skeletal muscle cell growth.
4. Pour curing agent into Sylgard, mixing thoroughly from side-to-side as well as top-to-bottom to guarantee curing when coating the culture dishes (it is normal for small bubbles to appear when mixing). It is also important to syringe (without a needle) Sylgard onto the culture dishes as soon as possible followed by agitation to ensure even distribution.
5. Cut pins in half to enable replacement of culture dish lids. There is no required length of pins, as long as the lid can be replaced.
6. Cut silk suture thread to 6 mm lengths. Using a disposable scalpel is preferable as using scissors causes greater incidence of fraying.
7. Dissolve 1 g of fibrinogen in 50 ml of Ham's F12-K medium to ensure enough reagent for approximately 250 fibrin gels given 200 μl is required per fibrin muscle construct.
8. It is suggested to filter fibrinogen through a 0.22 μm large vacuum filter as fibrinogen blocks the filter membranes when using 0.22 μm syringe filters.
9. Do not freeze-thaw.

10. Thrombin can be purchased as 1KU containing 27 mg. Therefore, there are approximately 37 units per mg, totalling 999 units. Each 1KU should therefore provide enough reagent for 200 fibrin gels at 10 U/ml when diluted with growth medium given 25 μl is required per fibrin muscle construct.
11. 6-Aminocaproic acid inhibits plasmin activity and subsequent fibrinolysis [30] induced by high levels of plasminogen by C2C12 myoblast cells. Overall use and specific concentration should therefore depend on the occurrence and rate of fibrinolysis since plasminogen is also required for muscle cell differentiation [31, 32]. It is suggested to culture without at first and then add 0.5% (final conc. 0.25 mg/ml) to growth medium, 1% (final conc. 0.5 mg/ml) to differentiation medium and 1% (final conc. 0.5 mg/ml) to maintenance medium if the fibrin matrix is degrading. If these concentrations do not reduce fibrinolysis, attempt to increase the concentrations to those described herein.
12. The amino acid L-proline and cofactor L-ascorbic acid (vitamin C) may encourage the production of ECM proteins in primary cultures, such as collagen and improve engineered musculoskeletal mechanics [33]. Use should therefore depend on the occurrence and rate of fibrin degradation. They are not always required if the gels are not degrading.
13. Genipin slows fibrin matrix degradation rates however at the expense of reduced force production [18]. Genipin also autofluoresces, negatively affecting any fluorescent/confocal microscopy performed and subsequent morphological analysis [18]. Therefore, its use should be determined by the most important dependant variables of interest within individual experiments.
14. Depending on the desired experimental measures, perhaps refrain from using streptomycin given the resultant alterations in force production and muscle phenotype following electrical stimulation [34, 35].
15. The TC-3 bioreactor system (Ebers Medical Technology, ESP) described herein is a commercially available, easy-to-use bioreactor system that permits simultaneous mechanical and electrical stimulation according to software amendments (*see* **Note 41**). Although the protocol described herein is specific to using the TC-3 bioreactor system, the principles remain the same if using customized/other commercially available bioreactor systems that enable similar stimulatory regimes.
16. Holding both sets of angled tweezers in one hand for simultaneous opening and closing permits greater control when transferring the muscle construct from the culture wells into the bioreactor chamber(s).

17. Local anesthetic, particularly bupivacaine are considered to induce myotoxic effects [36] which may be attributable to increased intracellular calcium release with simultaneous inhibition of calcium reuptake into the sarcoplasmic reticulum [37]. Inclusion and/or selection of local anesthetic should therefore be determined by the most important dependant variables of interest within individual experiments.
18. A needle biopsy yields 20–30 mg whereas a conchotome biopsy can yield 100–250 mg of skeletal muscle tissue. If conducting a needle biopsy, you may require multiple passes. Therefore, use the sterile wound dressing kit to wipe and clear the biopsy area before proceeding with the next pass of a needle biopsy in order to reduce risk of infection.
19. Hand pipette as opposed to vacuum aspirate to ensure greater control.
20. If the enzymatic digestion technique is successful, there are typically a low number of cells that grow from the remaining explanted tissue over the ensuing days. If so, dispose of the remaining explanted tissue according to the relevant legislations, (e.g. in the UK, the Human Tissue Act, 2004).
21. Primary human skeletal muscle derived cells start to senesce after several passages. Some may senesce even after 5–8 passages in-vitro depending on the age of the donor. This is because their drive in-vivo is to undergo a few rounds of proliferation only before fusing to fibers to repair/regenerate the fiber. Cells will still fuse once senescent but this may affect their initial growth when expansion/proliferation is required for the initial stage of making a self-assembling fibrin bioengineered muscle. Therefore, this should be a consideration when determining participant number, expected cell retrieval and the number of fibrin bioengineered constructs required to suit the experimental design of the study.
22. Homogenize suspension via pipetting up and down prior to resuspending.
23. Pipette DMSO down the side of the specimen container/tube in a dropwise manner while simultaneously swirling/mixing the solution in the tube.
24. It is best to allow sylgard spillages on bench surfaces to completely dry before removing using a gel scraper (or similar) and wiping.
25. Usually, there are lined marks on either side of the dish indicating the center line. Therefore, holding a ruler in line with these points when marking ensures accurate suture alignment when pinning in place.

26. Plates may only require 30 mins of sterilization to remove visible ethanol; however, it is likely that the sutures will still be saturated in ethanol. Therefore, air drying overnight is preferable.
27. Concentration of the protease inhibitor aprotinin should be optimized according to the rate of fibrinolysis (degradation of fibrin). Since the C2C12 myoblast cell line produces high levels of plasminogen and matrix metalloproteinases, we suggest adding 8 μl/ml of aprotinin to the thrombin solution. However, when using human skeletal muscle derived cells that contain a proportion of fibroblasts within the myoblast population that lay down additional matrix, we suggest reducing concentrations to 4 μl/ml.
28. Prepare sufficient excess thrombin gel solution for an additional gel (every 10 fibrin gels) to compensate for any potential minor pipetting error or loss of residual solution from the outside of pipette tips.
29. To ensure consistency between gels, pipette the thrombin solution along the sutures (outside to inside) and then across the center to join both sutures. Next, tilt the dish horizontally along the line of the sutures. Once the solution makes contact with both edges of the dish, rotate the dish in a circular motion so the solution now covers the entire edge of the dish. Resting the base of the dish on the culture hood surface, gently agitate to ensure even distribution across the dish surface (Fig. 2). It is important any bubbles appearing during this process are carefully removed, popping with a sterile syringe needle if required.
30. It is essential that cells are spread evenly across the dish to ensure consistent growth and attachment of the cells in a uniform fashion. We recommend adding 2 ml of cell suspension at 200,000 cells/ml for primary human gels [10] and 90,000 cells/ml for C2C12 gels.
31. Avoid pipetting PBS or media directly onto the fibrin gel in attempt to reduce cell disruption and potential tearing of the gel.
32. Gels are changed to lower serum to enable fusion/differentiation of single cells to form multinucleated myotubes.
33. Cells are switched from low serum (2% HS) differentiation medium to higher serum (7% hiFBS/hiNBCS) maintenance medium to reduce likelihood of gel degradation. Cells will continue to fuse and myotubes mature when under tension within the bioengineered system while in the presence of 7% serum in the medium.
34. Cells must be provided with sufficient nutrients to promote maturation and reduce the risk of gel degradation. Maintenance medium is changed following 2× PBS washes every

48 h. However, 0.5 ml of fresh medium is added on days where medium is not changed to help maintain any secreted factors by the cells that may promote differentiation and myotube hypertrophy.

35. Loosen the tension on the bioreactor chamber(s) arm to enable adjustment. Using a ruler, ensure a 12 mm gap between the medial edges of both clamps and tighten the bioreactor chamber(s) arm to avoid any unwanted changes in resting length.
36. Using the Ebers TC-3 bioreactor (Ebers Medical Technology, ESP), it is possible to mechanically load × 3 muscle constructs per chamber, totalling 9 muscle constructs per single regime (across 3 × chambers). However, the electrical stimulation setup described herein permits stimulation of just one muscle construct per chamber due to the space required for electrode wiring.
37. If the sutures are removed from the muscle construct during the transferring process, it is important to ensure the amount of muscle clamped at either side is the same. Ideally all other constructs in the same experiment would then be treated in this way, or that gel removed from the experiment.
38. If stimulating engineered muscle constructs in the absence or presence of specific nutrients (e.g. amino acids) according to the desired experimental measures, immersing constructs in Krebs Ringer HEPES (KRH) buffer rather than maintenance medium may be preferable. For Krebs-Ringer-HEPES (KRH) buffer at pH 7.4 (10×): 10 mM HEPES, 138 mM NaCI, 4.7 mM KCl, 1.25 mM $CaCl_2$, 1.25 mM MgSO, 5 mM Glucose and 5 mg/ml bovine serum albumin (BSA) diluted in dH_20.
39. Use rounded plyers to bend electrodes and feed through the culture chamber(s) to ensure electrodes run parallel with the muscle construct (Fig. 6).
40. For health and safety reasons, purchase electrode wire surrounded with plastic coating or self-coat with only the portion assembled through the bioreactor chamber(s) being exposed.
41. Attach the bottom of the bioreactor chamber(s) to the mechanical stretch unit, ensuring each chamber is slightly tilted (approximately 45°). Manually adjust the mechanical stretch unit arm onscreen using the "start" function until the mechanical arm attached to the bioreactor chamber(s) clicks in place. Click "stop" and "set to 0" to set the starting position.
42. Example Mechanical Stretch Regimes:
    - Cyclic stretch (1 cycle = 3 sets × 5 reps, each set interspersed with ~30 s rest). Undertake these cycles with 5% strain for 2 days, 10% strain for 2 days, and 15% strain for

4 days in an attempt to model progressive overload that has been demonstrated to evoke myotube hypertrophy [23].

- Static stretch, where the construct is stretched to 10% quickly (e.g., 0.4 mm/s resulting in 1.2 mm stretch of a 1.2 cm fibrin muscle after 3 s) and held at 10% stretch for 1 h to provide a model that would be most similar to synergistic ablation [22, 38].
- Continuous ramp load at 0.35 mm/h, which is the same rate at which stretch stimulates in-vivo bone elongation during development [9].
- 10% intermittent stretch – (1 interval = 4 sets × 10 reps at 0.4 mm/s during stretch and return to resting length, resulting in 1.2 mm stretch of a 1.2 cm fibrin muscle in 3 s and a return to resting length after 3 s, with each set interspersed with 90 s rest). Each interval is separated by 3.5 min rest and repeated, e.g., 3–5 times in an attempt to replicate a similar number of repetitions performed during a typical in-vivo bout of resistance type exercise that evokes hypertrophy (e.g., 4 sets × 10 reps × 3–5 exercises per muscle group).

43. Initially, only mechanical stretch or electrical stimulation could be applied at a given time. Specifically, a single pulse could only be applied when the desired stretch limit was attained or when at the initial resting position. Therefore, we liaised with the manufacturer (Ebers Medical Technology, ESP) regarding potential software (TC-3) modifications that would enable simultaneous electrical and mechanical stimulation to the 3D fibrin muscle constructs Ebers proceeded with the requested amendments, with the TC-3 bioreactor now permitting the application of pulse trains (rather than a single pulse at maximal stretch or at resting length) for continuous synchronized impulses during the whole mechanical cycle (Fig. 8). Furthermore, the bioreactor system now also permits the use of an external electrical stimulator for synchronization of mechanical stretch and electrical stimulation, providing a fail-safe if the internal stimulation box was to stop working but also enabling the use of more powerful or in-house stimulators in the future for higher sustained/repetitive pulse frequencies.
44. Example Electrical Stimulation Regimes:
    - Chronic low-frequency stimulation (CLFS) to evoke slow fiber formation (pulse frequency = 10 Hz; contraction duration = 60–600 s, adjusting rest intervals to ensure 60% active time, changing medium every 24 h and continue for a 14 day period) [39].
    - Higher frequency stimulatory regime to more closely mimic a more forceful contraction (pulse frequency = 100 Hz;

amplitude = 2.5 V/mm; 4 × 0.1 ms pulses delivered in a 400 ms pulse train followed by 3.6 s recovery) and repeated the desired number of times to mimic, e.g., resistance exercise in-vivo [5].

References

1. Brevet A, Pinto E, Peacock J, Stockdale FE (1976) Myosin synthesis increased by electrical stimulation of skeletal muscle cell cultures. Science 193(4258):1152–1154
2. Wehrle U, Dusterhoft S, Pette D (1994) Effects of chronic electrical stimulation on myosin heavy chain expression in satellite cell cultures derived from rat muscles of different fiber-type composition. Differentiation 58 (1):37–46. https://doi.org/10.1046/j.1432-0436.1994.5810037.x
3. Vandenburgh H, Kaufman S (1979) In vitro model for stretch-induced hypertrophy of skeletal muscle. Science 203(4377):265–268
4. Baar K, Torgan CE, Kraus WE, Esser K (2000) Autocrine phosphorylation of p70(S6k) in response to acute stretch in myotubes. Mol Cell Biol Res Commun 4(2):76–80. https://doi.org/10.1006/mcbr.2000.0257
5. Donnelly K, Khodabukus A, Philp A, Deldicque L, Dennis RG, Baar K (2010) A novel bioreactor for stimulating skeletal muscle in vitro. Tissue Eng Part C Methods 16 (4):711–718. https://doi.org/10.1089/ten.TEC.2009.0125
6. Nedachi T, Fujita H, Kanzaki M (2008) Contractile C2C12 myotube model for studying exercise-inducible responses in skeletal muscle. Am J Physiol Endocrinol Metab 295(5): E1191–E1204. https://doi.org/10.1152/ajpendo.90280.2008
7. Nikolic N, Bakke SS, Kase ET, Rudberg I, Flo Halle I, Rustan AC, Thoresen GH, Aas V (2012) Electrical pulse stimulation of cultured human skeletal muscle cells as an in vitro model of exercise. PLoS One 7(3):e33203. https://doi.org/10.1371/journal.pone.0033203
8. Vandenburgh HH (1988) A computerized mechanical cell stimulator for tissue culture: effects on skeletal muscle organogenesis. In Vitro Cell Dev Biol 24(7):609–619
9. Vandenburgh HH, Karlisch P (1989) Longitudinal growth of skeletal myotubes in vitro in a new horizontal mechanical cell stimulator. In Vitro Cell Dev Biol 25(7):607–616
10. Martin NR, Passey SL, Player DJ, Khodabukus A, Ferguson RA, Sharples AP, Mudera V, Baar K, Lewis MP (2013) Factors affecting the structure and maturation of human tissue engineered skeletal muscle. Biomaterials 34(23):5759–5765. https://doi.org/10.1016/j.biomaterials.2013.04.002
11. Sharples AP, Player DJ, Martin NR, Mudera V, Stewart CE, Lewis MP (2012) Modelling in vivo skeletal muscle ageing in vitro using three-dimensional bioengineered constructs. Aging Cell 11(6):986–995. https://doi.org/10.1111/j.1474-9726.2012.00869.x
12. Dennis RG, Kosnik PE 2nd (2000) Excitability and isometric contractile properties of mammalian skeletal muscle constructs engineered in vitro. In Vitro Cell Dev Biol Anim 36 (5):327–335. https://doi.org/10.1290/1071-2690(2000)036<0327:eaicpo>2.0.co;2
13. Dennis RG, Kosnik PE 2nd, Gilbert ME, Faulkner JA (2001) Excitability and contractility of skeletal muscle engineered from primary cultures and cell lines. Am J Physiol Cell Physiol 280(2):C288–C295
14. Huang YC, Dennis RG, Baar K (2006) Cultured slow vs. fast skeletal muscle cells differ in physiology and responsiveness to stimulation. Am J Physiol Cell Physiol 291(1): C11–C17. https://doi.org/10.1152/ajpcell.00366.2005
15. Huang YC, Dennis RG, Larkin L, Baar K (2005) Rapid formation of functional muscle in vitro using fibrin gels. J Appl Physiol 98 (2):706–713. https://doi.org/10.1152/japplphysiol.00273.2004
16. Martin NR, Passey SL, Player DJ, Mudera V, Baar K, Greensmith L, Lewis MP (2015) Neuromuscular junction formation in tissue-engineered skeletal muscle augments contractile function and improves cytoskeletal organization. Tissue Eng Part A 21 (19–20):2595–2604. https://doi.org/10.1089/ten.TEA.2015.0146
17. Khodabukus A, Baar K (2012) Defined electrical stimulation emphasizing excitability for the development and testing of engineered skeletal muscle. Tissue Eng Part C Methods 18 (5):349–357. https://doi.org/10.1089/ten.TEC.2011.0364
18. Khodabukus A, Baar K (2009) Regulating fibrinolysis to engineer skeletal muscle from the C2C12 cell line. Tissue Eng Part C

Methods 15(3):501–511. https://doi.org/10.1089/ten.TEC.2008.0286

19. Cheng CS, Ran L, Bursac N, Kraus WE, Truskey GA (2016) Cell density and joint microRNA-133a and microRNA-696 inhibition enhance differentiation and contractile function of engineered human skeletal muscle tissues. Tissue Eng Part A 22(7–8):573–583. https://doi.org/10.1089/ten.TEA.2015.0359
20. Shahini A, Choudhury D, Asmani M, Zhao R, Lei P, Andreadis ST (2017) NANOG restores the impaired myogenic differentiation potential of skeletal myoblasts after multiple population doublings. Stem Cell Res 26:55–66. https://doi.org/10.1016/j.scr.2017.11.018
21. Khodabukus A, Baar K (2015) Contractile and metabolic properties of engineered skeletal muscle derived from slow and fast phenotype mouse muscle. J Cell Physiol 230(8):1750–1757. https://doi.org/10.1002/jcp.24848
22. Player DJ, Martin NR, Passey SL, Sharples AP, Mudera V, Lewis MP (2014) Acute mechanical overload increases IGF-I and MMP-9 mRNA in 3D tissue-engineered skeletal muscle. Biotechnol Lett 36(5):1113–1124. https://doi.org/10.1007/s10529-014-1464-y
23. Powell CA, Smiley BL, Mills J, Vandenburgh HH (2002) Mechanical stimulation improves tissue-engineered human skeletal muscle. Am J Physiol Cell Physiol 283(5):C1557–C1565. https://doi.org/10.1152/ajpcell.00595.2001
24. Yaffe D, Saxel O (1977) Serial passaging and differentiation of myogenic cells isolated from dystrophic mouse muscle. Nature 270(5639):725–727
25. Blau HM, Pavlath GK, Hardeman EC, Chiu CP, Silberstein L, Webster SG, Miller SC, Webster C (1985) Plasticity of the differentiated state. Science 230(4727):758–766
26. Patel HP, Syddall HE, Martin HJ, Stewart CE, Cooper C, Sayer AA (2010) Hertfordshire sarcopenia study: design and methods. BMC Geriatr 10:43. https://doi.org/10.1186/1471-2318-10-43
27. Crown AL, He XL, Holly JM, Lightman SL, Stewart CE (2000) Characterisation of the IGF system in a primary adult human skeletal muscle cell model, and comparison of the effects of insulin and IGF-I on protein metabolism. J Endocrinol 167(3):403–415
28. Kasper AM, Turner DC, Martin NRW, Sharples AP (2018) Mimicking exercise in three-dimensional bioengineered skeletal muscle to investigate cellular and molecular mechanisms of physiological adaptation. J Cell Physiol 233(3):1985–1998. https://doi.org/10.1002/jcp.25840
29. Khodabukus A, Baar K (2014) The effect of serum origin on tissue engineered skeletal muscle function. J Cell Biochem 115(12):2198–2207. https://doi.org/10.1002/jcb.24938
30. Adelman B, Rizk A, Hanners E (1988) Plasminogen interactions with platelets in plasma. Blood 72(5):1530–1535
31. Lopez-Alemany R, Suelves M, Munoz-Canoves P (2003) Plasmin generation dependent on alpha-enolase-type plasminogen receptor is required for myogenesis. Thromb Haemost 90(4):724–733. https://doi.org/10.1160/th03-04-0291
32. Suelves M, Lopez-Alemany R, Lluis F, Aniorte G, Serrano E, Parra M, Carmeliet P, Munoz-Canoves P (2002) Plasmin activity is required for myogenesis in vitro and skeletal muscle regeneration in vivo. Blood 99(8):2835–2844
33. Paxton JZ, Grover LM, Baar K (2010) Engineering an in vitro model of a functional ligament from bone to bone. Tissue Eng Part A 16(11):3515–3525. https://doi.org/10.1089/ten.TEA.2010.0039
34. Khodabukus A, Baar K (2015) Glucose concentration and streptomycin alter in vitro muscle function and metabolism. J Cell Physiol 230(6):1226–1234. https://doi.org/10.1002/jcp.24857
35. Khodabukus A, Baar K (2015) Streptomycin decreases the functional shift to a slow phenotype induced by electrical stimulation in engineered muscle. Tissue Eng Part A 21(5–6):1003–1012. https://doi.org/10.1089/ten.TEA.2014.0462
36. Zink W, Sinner B, Zausig Y, Graf BM (2007) Myotoxicity of local anaesthetics: experimental myth or clinical truth? Anaesthesist 56(2):118–127. https://doi.org/10.1007/s00101-006-1121-5
37. Zink W, Graf BM, Sinner B, Martin E, Fink RH, Kunst G (2002) Differential effects of bupivacaine on intracellular Ca2+ regulation: potential mechanisms of its myotoxicity. Anesthesiology 97(3):710–716
38. Goldberg AL (1967) Work-induced growth of skeletal muscle in normal and hypophysectomized rats. Am J Phys 213(5):1193–1198. https://doi.org/10.1152/ajplegacy.1967.213.5.1193
39. Khodabukus A, Baehr LM, Bodine SC, Baar K (2015) Role of contraction duration in inducing fast-to-slow contractile and metabolic protein and functional changes in engineered muscle. J Cell Physiol 230(10):2489–2497. https://doi.org/10.1002/jcp.24985

Chapter 6

Isolation and Purification of Satellite Cells from Young Rats by Percoll Density Gradient Centrifugation

Yuji Matsuyoshi, Mariko Akahoshi, Mako Nakamura, Ryuichi Tatsumi, and Wataru Mizunoya

Abstract

Satellite cells (SCs) are myogenic stem cells that play an important role in skeletal muscle regeneration and hypertrophy. Primary cultures of SCs are useful to analyze cell functions; however, it is difficult to obtain highly pure SCs from young rats with the conventional procedures. The purpose of this study is to establish a purification method for SC isolation from young rats and quantitatively evaluate the purification procedure employing Percoll, a common research tool to purify cells. We elucidated the purity of SCs collected by Percoll density gradient centrifugation using real-time RT-qPCR and immunocytochemistry for desmin. Percoll treatment increased the purity of SCs isolated from young rats to nearly 90%, which was comparable to that achieved with the conventional method using middle-aged rats.

Key words Skeletal muscle, Cell isolation, Cell purification, Satellite cells, Percoll density gradient centrifugation

1 Introduction

Skeletal muscle is the largest organ essential for locomotion in animals. Noteworthy properties of skeletal muscle include its high regenerative ability and great adaptive capability to the external environment. However, these abilities may decline with aging, denervation, or muscle disuse.

Satellite cells (SCs), skeletal muscle-specific progenitor cells, are located between the basal lamina and sarcolemma of the postnatal muscle fiber [1]. SCs play an important role in maintaining the various functions of the muscle. SCs are thought to be indispensable for muscle regeneration [2, 3]. Skeletal muscle structure and function may be recovered in only a few weeks after injury [4], owing to the contribution of SCs.

The SCs usually reside in the G0 phase known as the quiescent state, but are activated upon muscle injury. Following their stimulation, SCs migrate toward the injured site and proliferate and

Sissel Beate Rønning (ed.), *Myogenesis: Methods and Protocols*, Methods in Molecular Biology, vol. 1889,
https://doi.org/10.1007/978-1-4939-8897-6_6,

differentiate to form multinucleated myofibers or may fuse to the damaged myofibers. At this time, not all SCs enter the differentiated phase and some proliferated SCs may return to the quiescent state to maintain reserve SC pool. This phenomenon is known as self-renewal [5]. While repeated injury experiments have shown that the number of SCs (reserve pool) remains constant even after multiple traumas, the population of SCs declines with age. Thus, the regenerative capacity of skeletal muscle decreases gradually with aging [6, 7]. The mechanisms that regulate these biological aspects of SCs are yet questionable.

To elucidate the characteristics of SCs, refined and efficient isolation method in primary culture is required. The established cell lines such as C2C12 and L6 are used as models of myoblasts. Many researchers still rely on SCs isolated from animals probably because primary cultures offer several advantages over cloned cell lines, especially by representing in vivo conditions.

The protocol to isolate SCs is already developed [8, 9]. However, to obtain high-purity SCs from young animals in sufficient quantity is difficult. To overcome this issue, fluorescence-activated cell sorting (FACS) is utilized as an alternative method for SC collection. However, FACS demands expensive apparatus and identification of SCs may be ambiguous, as distinct cell populations may share identical molecular markers [10]. Furthermore, FACS laser exposure tends to induce cell death during separation, which may decrease the yield of quiescent SCs [11].

Middle-aged rats (8–10 months old) are needed to get highly pure population of SCs, although the proliferative and differentiation activities of these SCs are diminished when compared with those obtained from young animals [9, 12, 13]. The reason young animals do not serve as a source of pure SCs is yet unknown. In our results, the proportion of SCs was about 50% of all adherent cells isolated from young rat muscles, although the procedures were similar to those for middle-aged rats. We assumed that the contamination of fibroblasts may reduce SC population in young animals.

In this protocol, we adopt density gradient centrifugation with Percoll to increase the purity of SCs from young rats. Percoll comprises colloidal silica particles of 15–30-nm diameter and is one of the commonly used materials for density gradient centrifugation of cells, viruses, and subcellular particles. This method is effective not only for obtaining highly pure SCs but also for removing contaminants (dust, yeast, blood, etc.). Percoll density gradient centrifugation has been already applied for the purification of SCs, but no quantitative data are available on the ratio of SCs after using Percoll methods [14, 15]. Here, we demonstrate the protocol to obtain highly purified SCs from young rats using Percoll with quantitative data.

We prepared SCs from systemic muscles of male Sprague-Dawley (SD) rats according to the method of Tatsumi et al. [9]

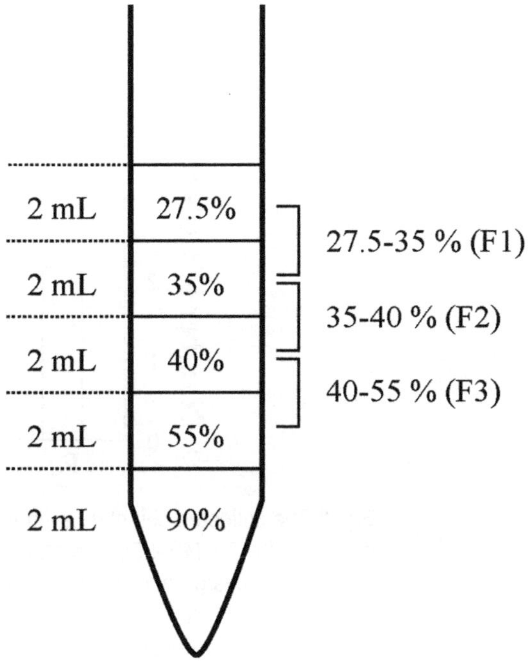

Fig. 1 Schematic representation of Percoll density gradient fractionation. Percoll solutions of different densities (27.5%, 35%, 40%, 55%, and 90%) are layered in the order from high to low density. The cell suspension is loaded on 27.5% Percoll solution (on top of the five-step Percoll gradient). After centrifugation (1680 × *g* for 20 min at room temperature), 2 mL cell suspensions were collected from each fraction (F1, F2, and F3) with a Pasteur pipet, as indicated at the right in parentheses

and performed Percoll density gradient centrifugation. In this study, we used a five-step Percoll gradient (Fig. 1). The collected cells from 27.5–35%, 35–40%, and 40–55% Percoll fractions were defined as F1, F2, and F3, respectively. The collected cell numbers were different among fractions, and F3 contained two times more cells than F1 and F2 (Fig. 2). We used real-time RT-qPCR to analyze the expression levels of two SC marker genes (*MyoD* and *c-met*) and two fibroblast marker genes (*FGF7* and *col1a1*) [16–19]. As a result, F3 fraction showed higher level of SC markers and lower level of fibroblast markers than F1 and F2 (Fig. 3), suggesting the presence of purified SCs in F3. Moreover, we examined the differentiated state of these cell fractions. F3 showed the most abundant myotubes than other fractions (Fig. 4), suggesting the presence of myogenic cells in F3.

To determine the degree of purity of SCs present in the F3 fraction, we performed immunostaining of desmin, a muscle-specific intermediate-filament protein and a hallmark of the population of replicating myoblasts that give rise to terminally differentiating myoblasts [20], with cultures of SCs at 40 h post-plating.

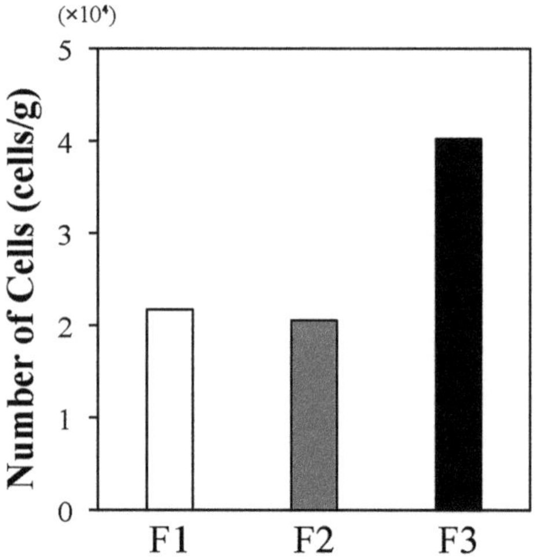

Fig. 2 The collected cell number in each fraction standardized by 1 g of minced muscle. The F3 (40–55%) fraction contained the maximum number of cells (4.0 $\times$ 10^4 cells/g), about twice as many in the F1 (27.5–35%) or F2 (35–40%) fraction. In fact, a layer of cells is observed macroscopically only in the F3 fraction in this protocol

Moreover, cell cultures isolated from a 10-month-old male SD rat or a 4-week-old male SD rat without Percoll treatment were prepared and immunostained with desmin under the same conditions to compare the effect of purification by Percoll. As a result, the proportion of desmin-positive cells from the young rat increased to 88.8% by Percoll density gradient centrifugation as compared with around 50.3% achieved without Percoll treatment (Fig. 5).

Thus, we demonstrate that Percoll density gradient centrifugation protocol is useful to purify SCs isolated from young rats without any expensive equipment or special techniques.

2 Materials

2.1 Experimental Animals

We purchased 10-month-old and 4-week-old male SD rats from a commercial supplier (Kyudo, Tosu, Japan) and housed them at 22 $\pm$ 2 °C and 55 $\pm$ 10% humidity on a 12:12 light/dark cycle (light on at 8 a.m.) with free access to regular food (CRF-1, Oriental Yeast, Tokyo, Japan) and water. All animal experiments were conducted in strict accordance with the recommendations in the Guidelines for Proper Conduct of Animal Experiments published by the Science Council of Japan and ethics approval from the Kyushu University Institutional Review Board (approval no. A21-117, A22-148, A24-138 and A28-092).

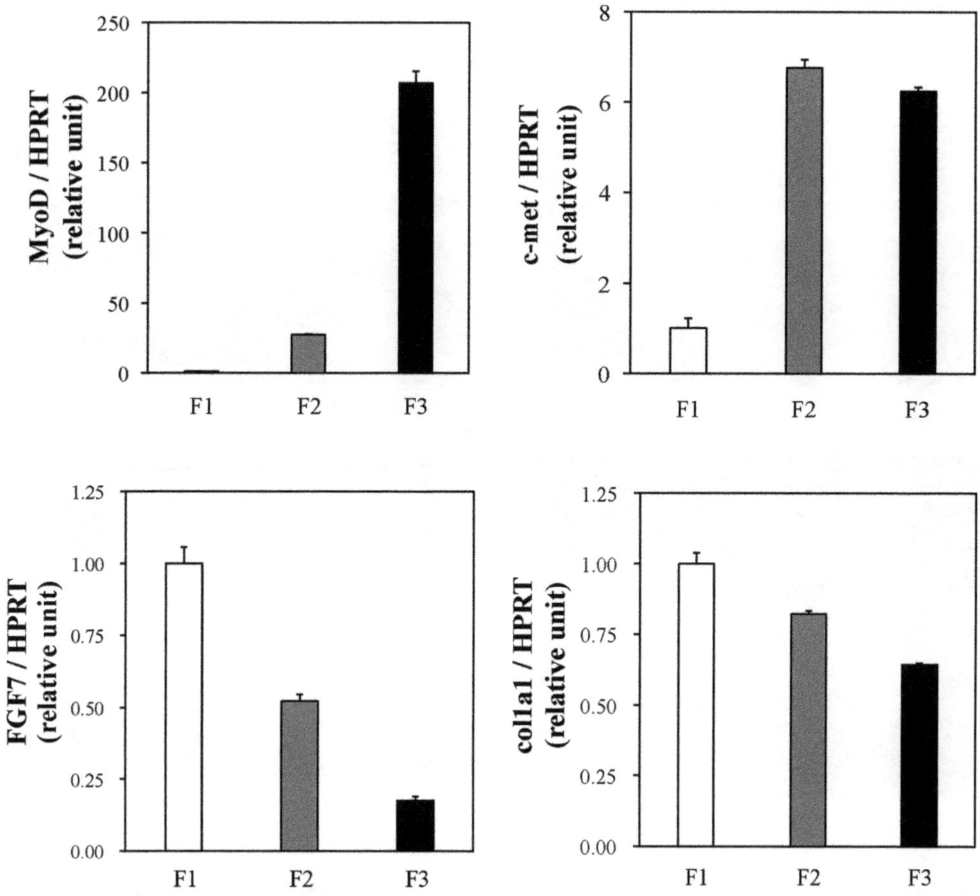

Fig. 3 The mRNA expression of *MyoD*, *c-met*, *FGF7*, and *col1a1* in the F1, F2, and F3 fractions measured by real-time RT-qPCR. The expression of SC markers (*MyoD* and *c-met*) is higher and that of the fibroblast markers (*FGF7* and *col1a1*) is lower in cells from the F3 fraction than those from F1 or F2. Data are shown as means $\pm$ SEM ($n = 3$ independent wells)

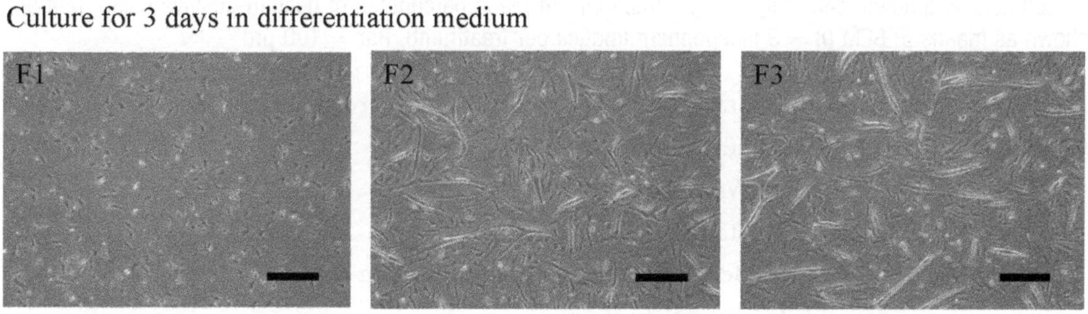

Fig. 4 Representative images of differentiated cells from F1, F2, and F3 cultured for 3 days in differentiation medium. Note that the F3 fraction showed the most abundant myotubes. Bar = 250 μm

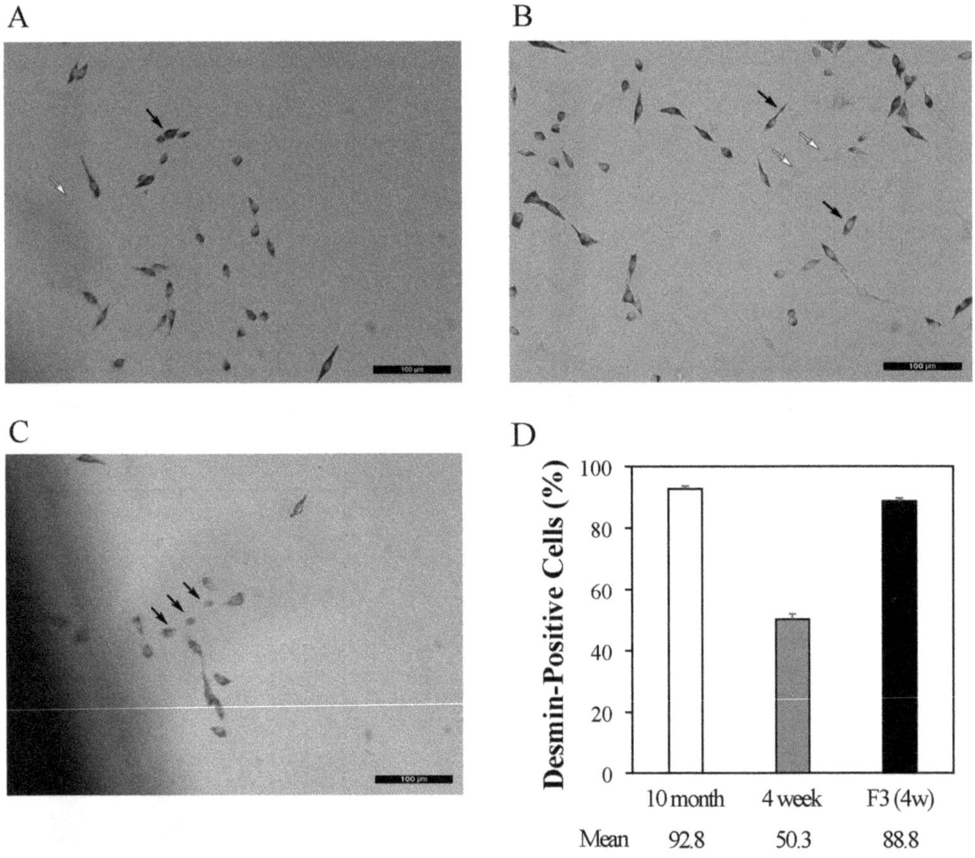

Fig. 5 Percentage of desmin-positive cells in the suspension of F3 (isolated from a 4-week-old male SD rat). Black arrows indicate desmin-positive cells and white arrows indicate desmin-negative cells. In addition, cells isolated from a 10-month-old male SD rat (10 month) and 4-week-old male SD rat (4 week) without Percoll treatment were assayed. By Percoll treatment, the proportion of SCs isolated from young rats increased from about 50% to nearly 90%. (**a**) Cells isolated from a 10-month-old male SD rat. (**b**) Cells isolated from a 4-week-old male SD rat without Percoll treatment. (**c**) Cells isolated from a 4-week-old male SD rat with Percoll density gradient centrifugation (F3 fraction). (**d**) Mean percentage of desmin-positive cells. Data are shown as means ± SEM ($n = 3$ independent images per treatment). Bar = 100 μm

2.2 General Equipment

1. Standard humidified tissue culture incubator (37 °C, 5% CO_2).
2. Standard clean bench.
3. Water bath (37 °C).
4. Standard auto pipette.
5. Bench-top centrifuge with swing rotor.
6. 70% ethanol.
7. Disposable serological pipettes, sterile, 2 and 10 mL.
8. Polypropylene centrifuge tubes, sterile, 15 and 50 mL.

2.3 Satellite Cell Isolation and Culture

All solutions are prepared using ultrapure grade water (referred to as water). The pH is adjusted with regular hydrochloric acid (HCl) and sodium hydroxide (NaOH) solution, unless otherwise indicated.

1. Phosphate-buffered saline (PBS, pH 6.5) containing 140 mM sodium chloride (NaCl), 10 mM potassium chloride (KCl), 8 mM disodium phosphate (Na2HPO4), and 2 mM potassium dihydrogen phosphate (KH2PO4) in water. Adjust pH to 6.5. Sterilize by autoclave. Warm the PBS to 37 °C just before use.
2. Culture medium: Minimum Essential Medium Alpha medium (αMEM) (Gibco) containing 10% normal horse serum (Gibco), 1% antibiotic–antimycotic mixture (Gibco), and 0.5% gentamicin (Gibco). At least 100 mL of this medium is required to prepare cells for isolation and culture. Adjust pH to 7.2. Culture medium is stored at 4 °C and used within 3 weeks after preparation. Warm the medium to 37 °C just before use.
3. Differentiation medium: Opti-MEM (Gibco) containing 2% normal horse serum, 1% antibiotic–antimycotic mixture, and 0.5% gentamicin is used. Adjust pH to 7.2. Differentiation medium is stored at 4 °C and used within 3 weeks after preparation. Warm the medium to 37 °C just before use.
4. 100 μg/mL Poly-L-lysine solution: Dissolve poly-L-lysine hydrobromide (Sigma) in sterile water. The solution is stored at 4 °C.
5. 20 μg/mL Fibronectin solution: Dissolve fibronectin from bovine plasma (Sigma) with PBS (pH 7.2) immediately before use.
6. PROTEASE TYPE XIV from *Streptomyces griseus* (Sigma).
7. Cell strainers, 100 and 40 μm.
8. Standard syringes and 0.22 μm syringe filters.
9. Dissecting instruments: Prepare and autoclave beakers and two pairs of scissors and forceps.

2.4 Percoll Density Gradient Centrifugation

1. Percoll (Sigma). Density: 1.13 ± 0.005 g/mL.
2. 10× Concentrated αMEM: Dissolve 10.2 g αMEM powder in 100 mL of water and filter-sterilize with a 0.22 μm bottle top filter.
3. Sterile Pasteur pipettes.

2.5 Real-Time RT-qPCR

1. RNeasy Micro kit (Qiagen).
2. SuperScript III (Invitrogen).
3. oligo(dT) primer (Invitrogen).

Table 1
Real-time RT-PCR primer for target genes

	Primer sequence (5′–3′)	
Primer	Forward	Reverse
MyoD	ccctgttgtttgtggagaca	ctgtgggaaagagtgggtgt
c-met	tcctgacgtgaacacatttga	gccagcattttagcatcactt
FGF7	cgagcgacacacgagaagt	actgggtgcgacagaacagt
Col1a1	aggcaacagtcgattcacct	gactgtcttgccccaagttc
HPRT	gaccggttctgtcatgtcg	acctggttcatcatcactaatcac

4. LightCycler TaqMan Master (Roche).
5. The primer sets (*see* Table 1).
6. LightCycler 1.5 Instrument (Roche).

2.6 Immuno-cytochemistry

1. PBS (pH 7.2).
2. Goat serum (Gibco).
3. Monoclonal anti-desmin antibody (Developmental Studies Hybridoma Bank, D3).
4. N-Histofine Simple Stain Rat MAX-PO (M) (Nichirei).

3 Methods

All procedures including centrifugation are performed at room temperature, unless otherwise indicated.

3.1 Satellite Cell Isolation [8, 9]

1. Prepare wells (24-well plates) for cell culture a day before dissection. Cover the surface of each well with 0.6 mL poly-L-lysine solution for 40 min and rinse three times with sterile water. Leave the plates to dry for at least 3 h.
2. Warm 1.5 L PBS (pH 6.5) to 37 °C and maintain the temperature throughout the whole procedure.
3. Coat each well with 0.6 mL fibronectin solution. This solution is removed immediately before plating cells. This step is performed after Percoll density gradient centrifugation.
4. Euthanize a rat following proper anesthesia treatment and place the rat body in a beaker containing 70% ethanol for 15 min.
5. Remove skin and excise the muscle tissues from the body, mainly the hind limbs and back muscles, with two pairs of scissors and forceps. Trim the fat and connective tissues and place the trimmed muscles in sterile beaker with warm PBS (pH 6.5) (*see* **Note 1**).

6. Mince the wiped muscles with scissors in sterile pre-weight Petri dish. Weigh them and record.
7. Weigh the protease (PROTEASE TYPE XIV). The required protease amount is 2.84 mg/g muscle.
8. The protease is dissolved in PBS (pH 6.5) at a final concentration of 1.25 mg/mL and vortexed until completely dissolved. Filter-sterilize protease solution with a 0.22 μm syringe filter.
9. Transfer 10–12 g of minced muscles into a 50 mL tube containing 25 mL filtered protease solution. Multiply the tubes according to the weight of obtained muscles (*see* **Note 2**).
10. Digest the tissue for 1 h in a water bath at 37 °C. During this step, mix the tubes by inverting gently every 15 min.
11. Centrifuge tubes at 1680 × *g* for 4 min.
12. Discard the excess supernatant containing protease.
13. Add warm PBS (pH 6.5) to the tube until the 40-mL mark and vortex for 20 s.
14. Centrifuge tubes at 186 × *g* for 10 min (*see* **Note 3**). Begin to warm the culture medium in a water bath at this time.
15. Transfer the supernatant into a new sterile 50 mL tube and incubate it in a water bath until **step 19**.
16. Repeat the same procedure as **step 13** using the precipitate in **step 15**. Centrifuge at 186 × *g* for 8 min and transfer the supernatant into a new 50 mL tube.
17. Repeat the same procedure as **step 13** using the precipitate in **step 16**. Centrifuge at 186 × *g* for 5 min and transfer the supernatant into a new 50 mL tube.
18. Centrifuge all the transferred supernatant (**steps 15–17**) at 1680 × *g* for 4 min.
19. Discard the excess supernatant and suspend the pellet in warmed culture medium. Collect all cell suspensions into one 50 mL tube.
20. To remove the residual fat and connective tissues, filter the cell suspension with 100- and 40 μm cell strainers in this order.
21. Centrifuge the suspension at 1680 × *g* for 3 min and discard the supernatant.
22. Add culture medium to the pellet until the 5 mL mark.
23. Count the cells in the suspension. During cell counting, keep the cell suspension in a water bath.

3.2 Percoll Density Gradient Centrifugation and Plating

1. Prepare 27.5%, 35%, 40%, 55%, and 90% Percoll solution before the isolation of SCs. To prepare 90% Percoll solution, one part of 10× αMEM is added to nine parts of undiluted 100% Percoll solution. Other Percoll solutions are prepared by

diluting 90% Percoll with 1× αMEM. Prepare Percoll solutions a day before dissection and store these at 4 °C. The solutions are left at room temperature and vortexed for 5 s before use.

2. Layer 2 mL of Percoll solutions one after the other into a 15 mL tube from higher to lower concentrations using 2 mL pipettes (*see* **Note 4** and Fig. 1).
3. Put 2 mL of the cell suspension on the top of Percoll solutions (*see* **Note 5**).
4. Centrifuge tubes at 1680 × *g* for 20 min.
5. After aspirating about 3 mL of the surface fraction, collect each 2 mL Percoll fraction (F1, F2, and F3) in different new tubes using different clean Pasteur pipettes (*see* **Note 6**).
6. Add five volumes of PBS (pH 6.5) to the collected suspension and mix by inverting. Centrifuge the tubes at 1680 × *g* for 4 min and discard the supernatant.
7. Repeat **step 6** one more time.
8. Add 5 mL culture medium to the pellet and count cells in each suspension.
9. Dilute the suspension with culture medium to adjust the cell concentration and plate cells on a 24-well plate coated with fibronectin and poly-L-lysine (*see* **Note 7**).
10. Differentiation may be induced by culturing cells in the presence of differentiation medium in a humidified tissue culture incubator (37 °C, 5% CO_2).

3.3 Real-Time RT-qPCR

1. Isolate SCs from a 4-week-old male SD rat according to Subheading 3.1.
2. Fractionate cells according to Subheading 3.2, and plate cells (F1, F2, and F3) on a 24-well plate coated with fibronectin and poly-L-lysine.
3. After 24 h from plating, isolate total RNA by RNeasy Micro Kit according to the manufacturer's instructions. Synthesize The cDNA from total RNA with reverse-transcriptase SuperScript III using oligo(dT) primer and other standard reagents such as dNTP, dithiothreitol (DTT), and reaction buffer.
4. Amplify specific genes by real-time RT-qPCR apparatus under the TaqMan probe-detection protocol using proper primers and probes (Table 1).
5. Determine threshold cycles (Ct) as the PCR cycle at which an increase in fluorescence above the baseline signal is first detected.
6. Calculate the level of mRNA expression using a standard curve constructed by serial dilution of aliquots of cDNA pooled from one randomly chosen sample.

7. Express values as fold change compared with the reference (F1) group, which are standardized by hypoxanthine guanine phosphoribosyl transferase (*HPRT*; NM_012583.2).

3.4 Immunocytochemistry (See Note 8, [21])

1. Isolate and purify SCs from rats according to Subheadings 3.1 and 3.2.
2. Plate cells on a 48-well plate coated with fibronectin and poly-L-lysine.
3. After 40 h from plating, rinse plated cells three times with warm PBS (pH 7.2).
4. Fix cells with cold methanol-0.1% H_2O_2 for 10 min at 4 °C.
5. Incubate cells with PBS (pH 7.2)-0.1% TritonX-100 for 10 min at room temperature.
6. Wash cells with PBS (pH 7.2) for 15 min, repeat twice more.
7. Block cells with 5% goat serum in PBS (pH 7.2) for 1 h.
8. Incubate cells with anti-desmin antibody solution (1:100 dilution in 0.1% BSA-PBS (pH 7.2)) overnight at 4 °C.
9. Wash cells with PBS (pH 7.2) for 15 min, repeat twice more.
10. Detect primary antibody with N-Histofine Simple Stain Rat MAX-PO (M), in which secondary antibody reaction time is 2 h at room temperature. Check the validity of positive reaction by negative control cells without primary antibody treatment.
11. Count positive and negative cells in randomly chosen fields containing 15–30 cells per field.

4 Notes

1. Blood was removed from the muscle tissues as much as possible. Blood (blood cells) causes collapse of Percoll fraction during centrifugation.
2. Here, the obtained muscle mass from the 10-month-old male SD rat (body weight was 737.7 g) was 46.3 g, while that from the 4-week-old male SD rat (body weight was 95.5 g) was 8.7 g.
3. Note that the cell fraction contains SC precipitates during centrifugation at 1680 × *g*, but SCs get suspended in the supernatant during centrifugation at 186 × *g*.
4. Carefully approach the tip of the Pasteur pipette onto the surface of top phase and gently add Percoll solutions along the tube wall. Do not blow out from the pipette. This step may be performed during SC isolation with centrifugation.
5. The upper limit of the loaded cell number is 3.0×10^7 cells in 2 mL of the suspension. It is recommended to use additional

tubes when the cell number exceeds the permissible limit. Conversely, it may be difficult to collect fractionated cells when the cell number is below 5.0×10^6 (the cells at F3 may not be observed by the naked eye).

6. To compare the fractions, we collected all three fractions (F1, F2, and F3). However, just F3 fraction is needed for purification purpose. The 55–90% fraction predominantly contains red blood cells and almost no cells adhere to the coated wells in this fraction.
7. We routinely collect about 1.2×10^6 cells in F3 from about 18 g muscles isolated from two 4-week-old male SD rats using this protocol. This cell number is equivalent to about 50 wells of a 24-well plate at 2.0×10^4 cells/well as our standard initial cell density.
8. This immunocytochemistry protocol to detect desmin protein in individual SCs was performed as previously reported [21], with slight modifications.

Acknowledgments

This work was funded by the Japan Society for the Promotion of Science (JSPS) KAKENHI #26712023 to WM. Many thanks to Ms. Akiko Sato and Mr. Shuichi Kitaura (Kyushu University, Japan) for animal care and for help with handling chemical reagents.

References

1. Mauro A (1961) Satellite cell of skeletal muscle fibers. J Biophys Biochem Cytol 9:493–495
2. Sambasivan R, Yao R, Kissenpfennig A, Van Wittenberghe L, Paldi A, Gayraud-Morel B, Guenou H, Malissen B, Tajbakhsh S, Galy A (2011) Pax7-expressing satellite cells are indispensable for adult skeletal muscle regeneration. Development 138(17):3647–3656. https://doi.org/10.1242/dev.067587
3. Fry CS, Lee JD, Mula J, Kirby TJ, Jackson JR, Liu F, Yang L, Mendias CL, Dupont-Versteegden EE, McCarthy JJ, Peterson CA (2015) Inducible depletion of satellite cells in adult, sedentary mice impairs muscle regenerative capacity without affecting sarcopenia. Nat Med 21(1):76–80. https://doi.org/10.1038/nm.3710
4. Bentzinger CF, Wang YX, Dumont NA, Rudnicki MA (2013) Cellular dynamics in the muscle satellite cell niche. EMBO Rep 14 (12):1062–1072. https://doi.org/10.1038/embor.2013.182
5. Dumont NA, Wang YX, Rudnicki MA (2015) Intrinsic and extrinsic mechanisms regulating satellite cell function. Development 142 (9):1572–1581. https://doi.org/10.1242/dev.114223
6. Shi X, Garry DJ (2006) Muscle stem cells in development, regeneration, and disease. Genes Dev 20(13):1692–1708. https://doi.org/10.1101/gad.1419406
7. Shefer G, Van de Mark DP, Richardson JB, Yablonka-Reuveni Z (2006) Satellite-cell pool size does matter: defining the myogenic potency of aging skeletal muscle. Dev Biol 294(1):50–66. https://doi.org/10.1016/j.ydbio.2006.02.022
8. Allen RE, TemmGrove CJ, Sheehan SM, Rice G (1997) Skeletal muscle satellite cell cultures. In: Emerson CP, Sweeney HL (eds) Methods in cell biology, Methods in muscle biology. methods in cell biology, vol 52. Elsevier Academic Press Inc, San Diego, pp 155–176. https://doi.org/10.1016/s0091-679x(08)60378-7

9. Tatsumi R, Yamada M, Katsuki Y, Okamoto S, Ishizaki J, Mizunoya W, Ikeuchi Y, Hattori A, Shimokawa H, Allen RE (2006) Low-pH preparation of skeletal muscle satellite cells can be used to study activation in vitro. Int J Biochem Cell Biol 38(10):1678–1685. https://doi.org/10.1016/j.biocel.2006.04.003
10. Kallestad KM, McLoon LK (2010) Defining the heterogeneity of skeletal muscle-derived side and main population cells isolated immediately ex vivo. J Cell Physiol 222(3):676–684. https://doi.org/10.1002/jcp.21989
11. Mollet M, Godoy-Silva R, Berdugo C, Chalmers JJ (2007) Acute hydrodynamic forces and apoptosis: a complex question. Biotechnol Bioeng 98(4):772–788. https://doi.org/10.1002/bit.21476
12. Barani AE, Durieux AC, Sabido O (1985) Freyssenet D (2003) Age-related changes in the mitotic and metabolic characteristics of muscle-derived cells. J Appl Physiol 95 (5):2089–2098. https://doi.org/10.1152/japplphysiol.00437.2003
13. Charge SB, Brack AS, Hughes SM (2002) Aging-related satellite cell differentiation defect occurs prematurely after Ski-induced muscle hypertrophy. Am J Physiol Cell Physiol 283(4):C1228–C1241. https://doi.org/10.1152/ajpcell.00206.2002
14. Kastner S, Elias MC, Rivera AJ, Yablonka-Reuveni Z (2000) Gene expression patterns of the fibroblast growth factors and their receptors during myogenesis of rat satellite cells. J Histochem Cytochem 48(8):1079–1096. https://doi.org/10.1177/002215540004800805
15. Che X, Guo J, Wang B, Bai Y (2011) Rapid isolation of muscle-derived stem cells by discontinuous Percoll density gradient centrifugation. In Vitro Cell Dev Biol Anim 47 (7):454–458. https://doi.org/10.1007/s11626-011-9433-4
16. Yablonkareuveni Z, Rivera AJ (1994) Temporal expression of regulatory and structural muscle proteins during myogenesis of satellite cells on isolated adult-rat fibers. Dev Biol 164 (2):588–603. https://doi.org/10.1006/dbio.1994.1226
17. Cornelison DD, Wold BJ (1997) Single-cell analysis of regulatory gene expression in quiescent and activated mouse skeletal muscle satellite cells. Dev Biol 191(2):270–283. https://doi.org/10.1006/dbio.1997.8721
18. Powers CJ, McLeskey SW, Wellstein A (2000) Fibroblast growth factors, their receptors and signaling. Endocr Relat Cancer 7(3):165–197
19. Wang Q, Peng ZH, Xiao SX, Geng SM, Yuan JY, Li ZX (2007) RNAi-mediated inhibition of COL1A1 and COL3A1 in human skin fibroblasts. Exp Dermatol 16(7):611–617. https://doi.org/10.1111/j.1600-0625.2007.00574.x
20. Kaufman SJ, Foster RF (1988) Replicating myoblasts express a muscle-specific phenotype. Proc Natl Acad Sci U S A 85(24):9606–9610. https://doi.org/10.1073/pnas.85.24.9606
21. Do MK, Sato Y, Shimizu N, Suzuki T, Shono J, Mizunoya W, Nakamura M, Ikeuchi Y, Anderson JE, Tatsumi R (2011) Growth factor regulation of neural chemorepellent Sema3A expression in satellite cell cultures. Am J Physiol Cell Physiol 301(5):C1270–C1279. https://doi.org/10.1152/ajpcell.00257.2011

Chapter 7

Lentivirus-Mediated RNAi in Skeletal Myogenesis

Dongwook Kim, Adriana Reyes-Ordoñez, and Jie Chen

Abstract

RNA interference (RNAi) has greatly facilitated investigation of gene functions in vitro as well as in vivo. Recombinant lentivirus is widely used to deliver small hairpin RNA (shRNA) because of its high transduction capacity into diverse cell types and tissues. Here, we describe methods of lentivirus-mediated delivery of shRNA for the study of skeletal muscle cell differentiation in vitro and injury-induced muscle regeneration in mice.

Key words RNAi, shRNA, Lentivirus, Gene knockdown, Skeletal myogenesis, Muscle regeneration

1 Introduction

RNA interference (RNAi) has been widely used as a tool to study specific genes by knockdown of their expression *in vitro* [1]. In addition, RNAi has become an essential technique for genetic screening in various animal models such as C. *elegans* [2], *Drosophila* [3], and mice [4]. RNAi is achieved mainly by direct delivery of chemically-synthesized double-stranded small interfering RNA (siRNA) or vector-based short hairpin RNA (shRNA) into cells. Adenovirus, adeno-associated virus, and retrovirus are all commonly used vectors for the delivery of shRNA. Lentivirus, a subtype of retrovirus, has the unique capacity to infect cells independent of cell proliferation, which allows both dividing and nondividing cell types to be transduced efficiently by lentiviral vectors [5, 6]. Therefore, transduction of cells or tissues with lentiviruses encoding shRNAs is an effective method to aid loss-of-function studies. This approach has been made especially feasible by the availability of lentiviral shRNA libraries for the human and mouse genomes (e.g., [7]).

Adult skeletal muscle has the unique capacity for robust regeneration upon injury, owing to the presence of resident stem cells called satellite cells [8]. However, the molecular circuitry that regulates skeletal muscle regeneration is still incompletely

Sissel Beate Rønning (ed.), *Myogenesis: Methods and Protocols*, Methods in Molecular Biology, vol. 1889,
https://doi.org/10.1007/978-1-4939-8897-6_7,

understood. The RNAi approach was indispensable for our deciphering of the mammalian target of rapamycin (mTOR) signaling network in the regulation of myogenesis [9]. We also performed lentivirus-mediated RNAi screen in C2C12 murine myoblast cells to identify muscle-expressed cytokines that regulate myogenic differentiation [10], and some of those cytokines have been shown to function in vivo using a $BaCl_2$ injury-induced muscle regeneration model in mice coupled with lentiviral shRNA injection [11, 12]. Here, we describe procedures for gene knockdown using lentivirus-mediated delivery of shRNA to study myogenesis. Lentivirus packaging of shRNA constructs, concentration of the viruses, and validation of target gene knockdown in myoblasts are described, followed by protocols of phenotypic assessment of myotube formation *in vitro* with both C2C12 and mouse primary myoblasts (Fig. 1). We also illustrate procedures for delivery of lentiviruses expressing shRNAs into mouse tibialis anterior (TA) muscles in vivo and for characterization of knockdown effect on injury-induced muscle regeneration (Fig. 2).

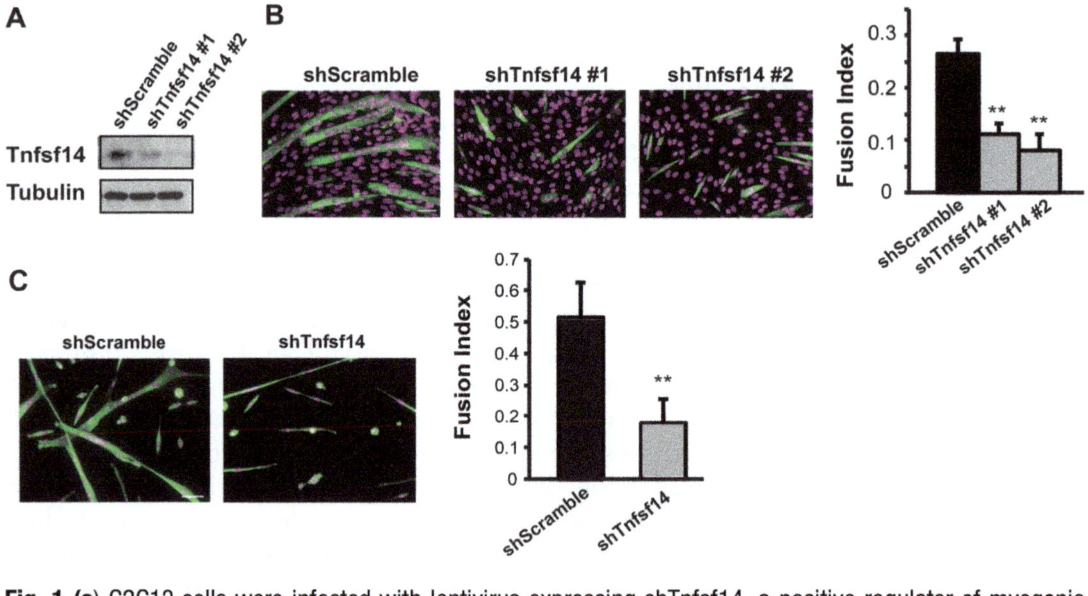

Fig. 1 (**a**) C2C12 cells were infected with lentivirus expressing shTnfsf14, a positive regulator of myogenic differentiation, or shScramble (negative control), selected for 2 days with puromycin (3 μg/mL), lysed and followed by western blot analysis. (**b**) Tnfsf14 knockdown cells, treated as in (A) were induced to differentiate for 72 h, followed by staining for MHC (green) and DAPI (magenta) and quantification of fusion index. (**c**) Mouse primary myoblasts were infected with lentivirus expressing shTnfsf14 or shScramble and induced to differentiate for 48 h, followed by staining for MHC (green) and DAPI (magenta) and quantification of fusion index. Scale bars: 50 μm. Paired two-tailed *t*-test was performed. $^{*}P < 0.05$, $^{**}P < 0.001$. All error bars represent s.d. of independent replicates. (Reproduced from ref. 11, Copyright © 2015, Springer Nature, an open access article distributed under the terms of the Creative Commons CC BY license)

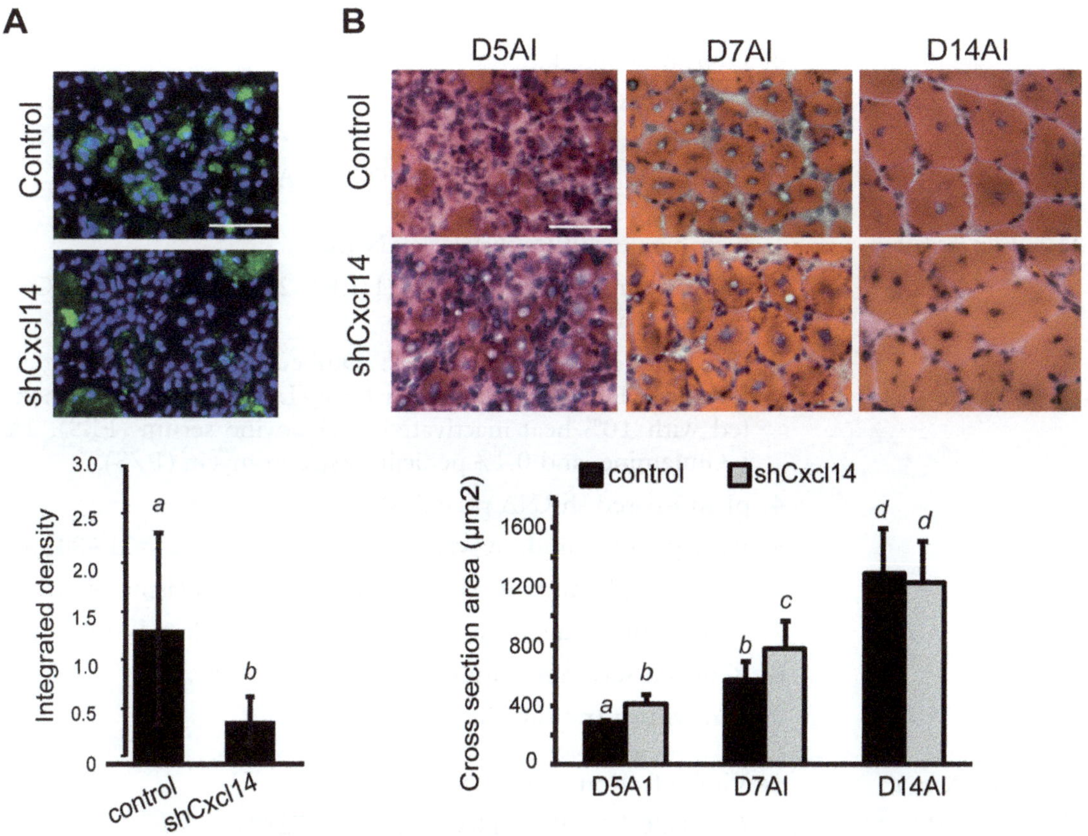

Fig. 2 (**a**) TA muscles were co-injected with $BaCl_2$ and shRNA viruses and isolated on day 3 after injury (AI), cryosectioned and immunostained for the target protein Cxcl14 (green) along with DAPI (blue). Fluorescence intensity was quantified using ImageJ software. (**b**) TA muscles were co-injected with $BaCl_2$ and shRNA viruses, and isolated on days 5, 7, and 14 AI. Upon cryosection, H&E staining was performed and regenerating myofiber cross-sectional area (CSA) was quantified. Scale bars: 50 μm. Paired two-tailed *t*-test was performed. The data denoted by different letters (a–d) are significantly different from each other ($P < 0.05$). All error bars represent s.d. of independent replicates. (Reproduced from ref. 12, Copyright © 2017, Springer Nature, an open access article distributed under the terms of the Creative Commons CC BY license)

2 Materials

2.1 In Vitro Analysis

2.1.1 Preparation and Identification of shRNA Plasmids

1. MISSION® shRNA Bacterial Glycerol Stock (Sigma-Aldrich), containing the pLKO.1-puro or TRC2-pLKO-puro plasmid of shRNA for the gene of interest. At least 2 pre-validated plasmids with distinct target sequences for the gene should be obtained. Three to five plasmids of distinct target sequences are recommended if they are not pre-validated.
2. A negative control shRNA plasmid containing a scrambled sequence in lieu of a target sequence (Addgene, plasmid #1864).

3. DNA miniprep kit.
4. DNA midiprep kit.
5. Restriction enzyme PvuII.
6. Sequencing primer:
   LKO.1 5′: 5′-GACTATCATATGCTTACCGT-3′.

2.1.2 Virus Production

1. 6 cm tissue culture dishes (*see* **Note 1**).
2. Human embryonic kidney (HEK) 293 T cells (ATCC®, CRL-3216™).
3. Medium for culturing 293T: Dulbecco's Modified Eagle's Medium (DMEM) containing 4.5 g/L of glucose supplemented with 10% heat-inactivated fetal bovine serum (FBS), 1% L-Glutamine, and 0.1% penicillin/streptomycin (P/S).
4. pLKO-based shRNA plasmids.
5. Packaging plasmid: pCMV-dR8.2 (Addgene, plasmid #8455).
6. Envelope plasmid: pCMV-VSV-G (Addgene, plasmid #8454).
7. Transfection reagent, e.g., TransIT®-LT1 (Mirus Bio LLC).
8. Reduced-serum medium, e.g., Opti-MEM™.
9. Lentivirus titration Kit.

2.1.3 Knockdown In Vitro—C2C12 Cells

1. Viral infection and selection.
   (a) C2C12 cells (ATCC®, CRL-1772™).
   (b) C2C12 growth medium: DMEM supplemented with 10% FBS and 1% P/S.
   (c) Puromycin selection medium: C2C12 growth medium with 3 μg/mL puromycin.
   (d) Polybrene: 8 mg/mL in water, filter-sterilized (store at −20 °C).
2. Western blot analysis.
   (a) Cell lysis buffer (e.g., 1× standard SDS sample buffer).
   (b) SDS-PAGE gel running system.
   (c) Western blot transfer system.
   (d) Polyacrylamide gels prepared with a 37.5:1 acrylamide/bis-acrylamide solution (2.7% crosslinker), at acrylamide concentrations appropriate for the molecule weights of target proteins.
   (e) SDS-PAGE running buffer: 0.025 M Tris–HCl, pH 8.3, 0.25 M Glycine, 0.1% SDS.
   (f) PVDF membrane.
   (g) Transfer buffer: 10 mM CAPS (3-[Cyclohexylamino]-1-propanesulfonic acid), pH 11, 20% methanol.

   (h) PBST (PBS containing 0.05% Tween-20).
   (i) Blocking buffer: 5% non-fat dry milk in PBST. Store at 4 °C.
   (j) Enhanced Chemiluminescence (ECL) solution.
3. Differentiation.
   (a) Differentiation medium: DMEM supplemented with 2% horse serum.
   (b) Gelatin: 0.2% solution in 1× PBS.
4. Immunofluorescence microscopy and quantification of differentiation.
   (a) Fixation buffer: 3.7% Paraformaldehyde in 1× PBS.
   (b) Cell permeabilization buffer: 0.1% Triton in 1× PBS.
   (c) Blocking buffer: 3% bovine serum albumin (BSA) in 1× PBS.
   (d) MF-20 anti-myosin heavy chain (MHC) antibody.
   (e) Goat anti-mouse IgG (H + L) antibody, fluorescein isothiocyanate (FITC) conjugated.
   (f) DAPI: 2 μg/mL in 1× PBS.
   (g) Mounting medium.
   (h) Round coverslips (18–20 mm).

2.1.4 Knockdown In Vitro—Mouse Primary Myoblasts (In Addition to Materials Listed Above)

1. Mouse primary myoblasts isolation.
   (a) Dissecting tools.
   (b) 70% ethanol.
   (c) Neonatal mice of any desirable strain at 5–7 days of age.
   (d) Collagen-coated cell culture plates: Prepare a working stock of 0.1 mg/mL Collagen using sterile water. Pipette 5 mL onto a 10 cm plate or 500 μL per well for a 6-well plate. Incubate for 2–3 h at room temperature. Aspirate all excess solution and allow to dry. Store plates at 4 °C. Rinse once with sterile water or PBS before plating cells.
   (e) 70 μm and 40 μm cell strainers.
   (f) Antibiotic solution: 100 mL sterile PBS containing 20.5 μg/mL Amphotericin, 100 U/mL penicillin and 100 μg/mL streptomycin. (This solution can be stored at 4 °C for a few weeks.)
   (g) Enzyme solution (made fresh before each use): 1.5 U/mL collagenase D, 2.4 U/mL dispase II, 2.5 mM $CaCl_2$.
   (h) Trypsin solution (made fresh before each use): mix 3 mL of 0.05% trypsin and 2 mL of calcium- and magnesium-free HBSS.

(i) Inhibition solution: 100 mL HBSS containing 20% FBS, 100 U/mL penicillin, and 100 μg/mL streptomycin.

(j) F10-based primary culture medium: Ham's F10 medium, containing 20% FBS, 5 ng/mL bFGF, 100 U/mL penicillin, and 100 μg/mL streptomycin.

(k) Freezing medium: 50% FBS, 40% F10 culture medium, 10% DMSO.

2. Differentiation.
Differentiation medium: DMEM supplemented with 2% horse serum.

2.2 In Vivo Analysis

2.2.1 Concentrating Lentivirus

1. Refrigerated Ultracentrifuge.
2. Ultra centrifugation tubes.
3. Lentivirus titration Kit.

2.2.2 Tibialis Anterior (TA) Muscle Injury

1. Insulin syringes with 28 g × 1/2″ needles.
2. Barium chloride ($BaCl_2$) Solution: 1.2% $BaCl_2$ prepared in 0.9% NaCl solution, filter-sterilized and stored at room temperature.
3. Concentrated shRNA-containing lentivirus for both your target gene and scrambled sequence to be used as a negative control.
4. Isoflurane or other IACUC authorized anesthetic drugs.
5. Ten- to twelve-week-old C57BL/6 or FVB mice.

2.2.3 TA Muscle Isolation

1. Isoflurane or other IACUC authorized anesthetic drugs.
2. Dissecting tools.
3. Liquid nitrogen-cooled 2-methyl butane.
4. Dry ice.
5. Optimum Cutting Temperature (O.C.T) freezing medium.
6. Cryomolds.
7. Aluminum foil, precut in 5 × 5 cm pieces.

3 Methods

3.1 In Vitro Analysis

3.1.1 Preparation and Identification of shRNA Plasmids

1. Streak out the bacteria containing shRNA plasmids on an LB agar plate with 100 μg/mL ampicillin. Incubate overnight at 37 °C.
2. Inoculate a single colony from each plate to 3 mL LB medium supplemented with 100 μg/mL ampicillin, and incubate at 37 °C overnight on a shaker.

3. Isolate plasmid DNA from cultures using a miniprep kit according to the manufacturer's instructions. Save the residual cultures at 4 °C for **step 6** below.
4. Digest DNA with PvuII to confirm the plasmids. The expected three fragments are 0.8, 2.5, and 3.8 kb.
5. If necessary, sequence with LKO.1 5′ primer to confirm the identity of the shRNA.
6. Upon validation of the miniprep DNA, inoculate culture from **step 3** into 50 mL LB containing 100 μg/mL ampicillin, and incubate at 37 °C overnight on a shaker.
7. Isolate plasmid DNA using a midiprep kit, according to the manufacturer's instructions.

3.1.2 Virus Production

1. HEK293T cells are cultured at 37 °C with 5% CO_2.
2. Seed cells in 6 cm tissue culture dishes using 293 T medium, aiming for 60–80% confluency on the day of transfection (*see* **Note 2**).
3. Mix plasmids: 0.9 μg pCMV-dR8.2, 0.1 μg pCMV-VSV-G, and 1 μg pLKO-shRNA with 20 μL Opti-MEM™ in a microcentrifuge tube. This mixture is referred to as Solution I (*see* **Note 3**).
4. Mix 6 μL transfection reagent (e.g., TransIT®-LT1) and 90 μL Opti-MEM™ in a separate microcentrifuge tube (referred to as Solution II), incubate at room temperature for 5 min.
5. Combine Solution I and Solution II, pipet up and down gently to mix.
6. Incubate at room temperature for 20–30 min.
7. During the incubation, remove medium from cells and replenish with 2 mL fresh 293T medium.
8. Add the transfection mixture drop-wise into the plate of cells. Gently rock the plates back and forth to mix well.
9. Incubate cells for 12–15 h at 37 °C with 5% CO_2.
10. Remove medium from cells and replenish with 2 mL fresh 293T medium. Care should be taken to avoid detachment of the cells.
11. Twenty-four hours later, collect the medium, which contain viruses, and store at 4 °C. Add 2 mL fresh 293T medium to cells and collect the medium again at 48 h.
12. The virus-containing medium harvested at 24 h and 48 h can be processed separately or combined. Centrifuge the collected medium at 300 × *g* for 5 min at 4 °C to remove cell debris.
13. Transfer the supernatant to a new tube, make aliquots and store at −80 °C.

14. Measure the virus titer using qPCR Lentivirus Titration Kit according to the manufacturer's instructions. The titer should be $\geq 4 \times 10^6$ IU/mL.

3.1.3 Knockdown In Vitro—C2C12 Cells

1. Viral infection and selection.
   (a) Seed 3–5 × 10^5 C2C12 cells in a 6 cm tissue culture plate, aiming for 20–30% density on the following day (*see* **Note 4**).
   (b) Add to each plate of cells 2 mL of fresh growth medium containing 0.2–0.5 mL lentivirus (target shRNA or negative control with scrambled hairpin sequence) and 8 μg/mL polybrene (*see* **Note 5**).
   (c) Incubate overnight (16–20 h) at 37 °C with 7.5% CO_2 (*see* **Note 6**).
   (d) Replace the medium with 2 mL of puromycin selection medium and incubate for 2–4 days. During this selection period, if cells become confluent they should be split (*see* **Note 7**).
2. Western blot analysis.
   (a) To confirm knockdown, seed 2–3 × 10^5 puromycin-selected cells per well (see above) in a 12-well plate.
   (b) On the following day, lyse the cells with 80–100 μL of 1× SDS sample buffer.
   (c) Sonicate the lysate 1–3 times for 5 s each time to shear DNA and reduce sample viscosity.
   (d) Centrifuge the samples at 16,000 × *g* for 5 min, transfer supernatant to a new tube, and incubate at 95 °C for 5 min.
   (e) Separate equal volume of each sample (5–10 μL) on an SDS-PAGE gel of the appropriate concentration.
   (f) While the gel is running, prepare the PVDF membrane: cut the membrane to size of gel, wet membranes in 100% methanol, then soak in transfer buffer until use (at least 5 min).
   (g) Assemble the transfer cassette following the manufacturer's instructions, insert the cassette into transfer system, fill with transfer buffer. Transfer at 250 mA for 70 min, using an ice block to avoid overheating.
   (h) Upon completion of transfer, incubate the membrane in blocking buffer either at 4 °C overnight or at room temperature for at least 10 min.
   (i) Incubate the membrane with primary antibody diluted in blocking buffer to an appropriate concentration at 4 °C overnight or at room temperature for 60 min.

(j) Wash the membrane with PBST by rinsing several times and then incubate, at least once, for 5 min at room temperature.

(k) Incubate the membrane with secondary antibody diluted in blocking buffer at room temperature for 30–60 min.

(l) Wash the membrane five times with PBST, 4 min each time.

(m) Incubate the membrane in an ECL solution following the manufacturer's instructions.

(n) Capture the chemiluminescence signals with either X-ray films or a digital imager (*see* Fig. 1a for an example of western blot results for knockdown of the target protein Tnfsf14).

(o) If necessary, quantify the western blot band intensities using an image processing software (such as ImageJ) and compare the knockdown level of target proteins between samples.

3. Differentiation.

(a) To examine the effect of knockdown on myoblast differentiation, seed the puromycin-selected cells at 2–3 × 10^5/well in a 12-well plate coated with 0.2% gelatin (*see* **Note 8**).

(b) When the cells reach 90–100% confluency, remove the growth medium, wash cells with 1× PBS, and add 0.5 mL differentiation medium.

(c) Replenish cells with fresh differentiation medium daily for 3 days. By day 3 myotubes should be fully formed (*see* **Note 9**).

4. Immunofluorescence microscopy and quantification of differentiation.

(a) For immunofluorescent staining of cells, at day 3 of differentiation, remove medium, wash the cells twice with 500 μL of 1× PBS, and fix the cells using 500 μL of 3.7% formaldehyde for 15 min.

(b) Permeabilize the fixed cells with 500 μL of 0.1% Triton in 1× PBS for 5 min.

(c) Block with 500 μL of blocking buffer.

(d) Add to each well 500 μL of 0.5 μg/mL MF-20 antibody in blocking buffer, incubate for 1 h at room temperature.

(e) Wash five times with 1× PBS.

(f) Add to each well 500 μL of blocking buffer containing 6.7 μg/mL of FITC-conjugated anti-mouse antibody and

0.2 ng/mL of DAPI. Incubate for 30 min at room temperature.

(g) Wash five times with 1× PBS.

(h) Add 1 drop of mounting medium to each well and carefully place a round coverslip on top, making sure no air bubbles are trapped. Aspirate excess liquid and allow to dry overnight at 4 °C.

(i) Samples are examined under a fluorescence microscope, and images are captured. The fusion index is calculated as the percentage of total nuclei in myotubes (cells with 2 or more nuclei). Each sample (i.e., one well of stained cells) should be analyzed by quantifying all cells in randomly chosen microscopic fields, totaling ≥2000 nuclei (*see* Fig. 1b for the effect of Tnfsf14 knockdown on differentiation).

3.1.4 Knockdown In Vitro—Mouse Primary Myoblasts

1. Mouse primary myoblasts isolation [13]:

(a) Euthanize a neonatal mouse (5 to 7-day-old) following IACUC approved protocol. Cut away tails and paws and wash the body by immersion in 70% ethanol followed by 3 min immersion in antibiotic solution.

(b) Dissect and collect skeletal muscle tissues from fore and hind limbs, and mince the tissues finely with scissors or razor blades. Add 50–100 μL of PBS to keep the tissue moist during dissection.

(c) Add 1 mL of pre-warmed enzyme solution per dissected limb and transfer the suspension into a 15 mL sterile conical tube and incubate in a 37 °C water bath for 1 h. Several times during the incubation, gently pipet solution up and down using a 5 mL pipette to break up the tissue clamps and mix the solution.

(d) Allow the undigested tissue to settle and transfer the supernatant containing the digested tissues into a new tube. Add an equal volume of inhibition solution to the supernatant to stop the enzymatic activity.

(e) Add 2 mL of trypsin solution to the remaining undigested tissue fragments and incubate for 30 min to 1 h at 37 °C (*see* **Note 10**).

(f) Transfer the dislodged cells to a new tube and add an equal volume of inhibition solution. Combine the cells and filter using a 70 μm cell strainer followed by a passage through a 40 μm cell strainer to remove cell clumps.

(g) The filtered solution is then centrifuged at 350 × *g* for 10 min, and the cell pellet is resuspended in 5 mL of F10-based primary cell culture medium. Plate the cells in a culture dish and incubate at 37 °C with 7.5% CO_2 for 30 min (*see* **Note 6**).

(h) Collect cells that are not adhered to the plate and transfer to a new cell culture plate. Repeat **steps g–h** three times to remove fibroblasts before finally plating the cells on a collagen-coated plate. Incubate as above overnight.

(i) The next day, transfer non-adherent cells to a new collagen-coated plate, and add fresh medium into the overnight plate for expansion. Incubate both plates as above.

(j) The next day, transfer again non-adherent cells into a new collagen-coated plate. Maintain cells in culture, replenishing medium every other day (*see* **Note 11**).

(k) To continue the enrichment of myoblast in culture, once there is a significant number of myoblasts on the plate (may take 2–3 days) split cells by incubating them in 1 mL PBS for 5–10 min at room temperature, then tap the plate on a surface to release loosened myoblasts. The more adherent fibroblasts will not be knocked off as easily. Transfer the non-adherent cells to a new collagen-coated dish and continue incubation at 37 °C, 7.5% CO_2.

(l) To achieve 100% myoblasts purity, repeat the last step 2–3 times until no fibroblasts are remaining. The now pure primary myoblasts can be expanded and frozen down (*see* **Note 12**).

(m) To freeze down myoblasts, split a 60–70% confluent 10 cm plate of cells into eight 10 cm plates. The next day when they are at a 10–20% confluency freeze the cells using 1 mL of freezing medium for each plate. It is very important that the cells are frozen down when they are at low density, in order to preserve viability and differentiation potential.

2. Differentiation.

(a) Seed the primary myoblasts on collagen-coated 12-well tissue culture plates and culture overnight, aiming at 20–30% density on the following day (*see* **Note 13**).

(b) Replace the cell medium with 1 mL fresh growth medium and 5–10 μL of concentrated lentivirus (*see* Subheading 3.2.1 below for virus concentration procedure).

(c) After overnight incubation (16–20 h), replace the medium with 1 mL of fresh growth medium (without puromycin). Continue culturing the cells until they reach 60–80% confluence (~24 h).

(d) Remove the medium, wash cells with PBS, and add 0.5 mL differentiation medium. Replenish with fresh differentiation medium daily for 2 days, at the end of which myotubes should be fully formed.

(e) To visualize and quantify the effect of knockdown on differentiation, immunofluorescent staining can be conducted as described in Subheading 3.1.3, with ≥300 nuclei quantified for each sample (*see* Fig. 1c for the effect of Tnfsf14 knockdown on differentiation).

3.2 In Vivo Analysis

3.2.1 Concentrating Lentivirus

1. Sterilize the ultracentrifuge tubes to be used by soaking them in 70% ethanol overnight, and then exposing to UV light inside a biosafety cabinet for 15 min. Allow the alcohol to drain completely from the tube.
2. Load 10 mL of viral supernatant (*see* Subheading 3.1.2) in each ultracentrifuge tube, and centrifuge at 68,000 × *g* for 2 h at 4 °C.
3. Following centrifugation, carefully transfer medium into a bleach-filled container (*see* **Note 14**).
4. Resuspend the pellet in 100 μL of 293T growth medium.
5. Store the concentrated virus in small aliquots (15–20 μL) at −80 °C until use.
6. Using a commercial kit (e.g., qPCR Lentivirus Titration Kit, Abm Inc.), determine the titer of the concentrated virus. Typically, $\geq 1 \times 10^8$ IU/mL is expected.
7. Use at least 1×10^6 IU of virus in each in vivo knockdown experiment.

3.2.2 Tibialis Anterior (TA) Muscle Injury

1. Mix $BaCl_2$ with the concentrated lentivirus expressing shRNA of interest or negative control to a final concentration of ~0.9% $BaCl_2$ and at least 1×10^6 IU of virus in a total volume of 50–70 μL. Fill the insulin syringes with the mixtures above, one with the target shRNA and one with negative control (scramble shRNA).
2. Apply IACUC approved protocol to anesthetize the mouse, e.g., 2–3% Isoflurane aspiration delivered by a vaporizer.
3. Confirm adequate level of anesthesia by monitoring respiration and withdrawal reflex (toe pinch), applying supplementary anesthetic if necessary. When using isoflurane, ensure constant anesthesia by placing a facial mask on the mouse to allow constant aspiration of 2% isoflurane throughout the procedure.
4. Place the mouse lying in a supine position and soak the hind limb with 70% alcohol to facilitate visualization of leg anatomy (*see* **Note 15**).

5. Locate TA distal tendon on the distolateral portion of the hind leg, visible as a white structure under the skin.
6. Position the needle bevel up and insert at a 45° angle just above the TA tendon.
7. Slowly, slide the needle intramuscularly until it reaches the proximal portion of the TA muscle. Inject the solution slowly while withdrawing the needle, aiming to distribute the solution throughout the entire TA.
8. Repeat the procedure on the other leg.
9. Allow the mouse to recover from anesthesia and return to cage. Monitor throughout the day for any signs of distress.
10. Evaluate muscle regeneration by recovering injured muscles at different days post-injury (e.g., days 3, 5, 7, 14, 21, and 28).

3.2.3 TA Muscle Isolation

1. Anesthetize mouse following IACUC approved procedures and confirm adequate level of anesthesia by monitoring respiration and withdrawal reflex (toe pinch), apply supplementary anesthetic if necessary (*see* Subheading 3.2.2).
2. With the mouse lying in a supine position expose the TA muscle by dissecting the skin toward the foot.
3. Gently remove the fascia covering the muscle.
4. Cut the distal TA tendon and use it to hold the muscle as you cut the proximal attachment.
5. Place the muscle on a cryomold on top of OCT medium, orient the muscle flat on the OCT, and then cover the entire muscle with the freezing medium (*see* **Note 16**).
6. Snap-freeze by immersing cryomold in precooled 2-methyl butane for 30–60 s.
7. Quickly, wrap the specimen in aluminum foil, label it, and store at −80 °C until use. Dry ice can be used during transport of the frozen samples.
8. After both TA muscles have been removed, euthanize the mouse following IACUC approved protocols.

3.2.4 Knockdown Confirmation—-Immunohistochemistry

1. Using a cryostat microtome, section 10 μm slices of the muscle at −20 °C and place on glass slides, which should be stored at −80 °C until use.
2. Place slides in a Coplin staining jar and hydrate sections by submerging in PBS for 10 min. Proceed to immunofluorescent staining of sections following specific protocols for antibodies of interest.
3. After staining, visualize and capture images on a fluorescent microscope, using a 20× dry objective.

4. Using an image processing software (such as ImageJ) quantify the presence of the protein of interest by measuring fluorescence intensity on both the knockdown and control muscle sections (*see* Fig. 2a for an example).

3.2.5 Muscle Regeneration Analysis—Cross-Sectional Area (CSA) Measurement

1. Cryosection the muscle as described in Subheading 3.2.4.
2. Stain sections with hematoxylin-eosin and visualize samples under a bright-field microscope. Capture multiple images at 20× magnification to cover as much regenerative area as possible. Regenerating myofibers are easily identified by their centrally localized nuclei (as opposed to nuclei located at the periphery in non-injured myofibers). See images in Fig. 2b.
3. Perform blind analysis of the digital images using ImageJ:
   (a) Set the measurement scale before starting to analyze the images (*see* **Note 17**).
   (b) Using the free-hand selection tool, outline the circumference of each regenerating fiber and measure each outlined area (CSA). Include all centrally nucleated (i.e., regenerating) myofibers within an area of 614,400 μm^2. Disregard myofibers that are distorted or folded.
   (c) Generate an average CSA value per TA muscle and compare values obtained from knockdown vs control TA muscles. An example of the analysis is shown with the knockdown of Cxcl14, which leads to increased regenerating myofiber CSA compared to the control (Fig. 2b).

4 Notes

1. Various sizes of tissue culture plates can be used depending on the amount of virus required for experiments. For virus concentration, we typically use 10 cm tissue culture plates.
2. Plate cells 18–24 h before transfection for the cells to be actively dividing at the time of transfection.
3. If a different size of plates is used, adjust the amounts of components in mixture proportionally.
4. Seed an extra plate as a control for puromycin selection. The non-transduced cells should be killed completely by puromycin during the selection period.
5. The volume of virus required for knockdown experiments depends on factors such as the viral titer and the stability and/or abundance of the protein to be knocked down.
6. All myoblasts are cultured at 37 °C with 7.5% CO_2 unless indicated otherwise. We find 7.5% CO_2 to be optimal for the

maintenance and differentiation of both C2C12 cells and mouse primary myoblasts.

7. Changing medium is not necessary during this period, unless there is a massive amount of dead cells that need to be removed.
8. Make a 1% gelatin solution using gelatin powder in 1× PBS and autoclave. Add 0.2% gelatin in 1× PBS to plates, enough to cover the bottom—0.5 mL per well of 12-well plate, let it stand at room temperature at least 15 min, remove the gelatin and seed cell immediately. It is important that the plates do not dry out after gelatin treatment.
9. Some C2C12 sublines may take up to 6 days to fully differentiate. Optimization of differentiation time course is required before any functional study.
10. This step of trypsin digestion is optional. It is only necessary if clumps of tissues remain after initial incubation with enzymes. When fresh dispase and collagenase are used, there is usually little undigested tissue left at **step d**.
11. There may be a low number of cells left at this point, but the myoblasts are expected to start growing rapidly in about a week. If the myoblasts are growing in colonies or if there are very few myoblasts, trypsinize the cells very briefly (less than a minute). After resuspending in medium, pre-plate for 20–30 min on a noncoated plate, and incubate all non-adherent cells on a new collagen-coated plate.
12. Depending on the purity of myoblasts upon isolation, the number of cycles of myoblast enrichment needed (cycles of differential plating on non-coated and coated-plates) can vary.
13. While maintaining mouse primary myoblasts, it is critical to replenish growth medium every other day and not let myoblasts grow on the same plate more than 5 days or to higher than 50–60% confluency.
14. A small pellet should be visible. Take great care to avoid disturbing the pellet when removing medium.
15. Online tutorials are available to help visualize, inject, and recover the TA muscle (e.g., [14]).
16. If analysis of proteins or RNAs is desired, divide the isolated muscle into 2 portions. Snap-freeze one portion to be used for protein or RNA isolation, and embed and freeze the portion to be used for histological analysis.
17. Adjust the scale bar to match the image acquisition conditions before beginning to measure the CSA: take an image of a stage micrometer using the same equipment and settings that are used to capture the images, and use this to set the scale bar in ImageJ. Be sure to select micron as the output unit.

Acknowledgment

We thank members of the Chen laboratory for developing and refining the protocols described here. This work was supported by NIH grant R01 AR048914.

References

1. McManus MT, Sharp PA (2002) Gene silencing in mammals by small interfering RNAs. Nat Rev Genet 3(10):737–747. https://doi.org/10.1038/nrg908
2. Kamath RS, Fraser AG, Dong Y, Poulin G, Durbin R, Gotta M, Kanapin A, Le Bot N, Moreno S, Sohrmann M, Welchman DP, Zipperlen P, Ahringer J (2003) Systematic functional analysis of the Caenorhabditis elegans genome using RNAi. Nature 421 (6920):231–237. https://doi.org/10.1038/nature01278
3. Boutros M, Kiger AA, Armknecht S, Kerr K, Hild M, Koch B, Haas SA, Paro R, Perrimon N, Heidelberg Fly Array C (2004) Genome-wide RNAi analysis of growth and viability in Drosophila cells. Science 303 (5659):832–835. https://doi.org/10.1126/science.1091266
4. Zuber J, Shi J, Wang E, Rappaport AR, Herrmann H, Sison EA, Magoon D, Qi J, Blatt K, Wunderlich M, Taylor MJ, Johns C, Chicas A, Mulloy JC, Kogan SC, Brown P, Valent P, Bradner JE, Lowe SW, Vakoc CR (2011) RNAi screen identifies Brd4 as a therapeutic target in acute myeloid leukaemia. Nature 478(7370):524–528. https://doi.org/10.1038/nature10334
5. Kafri T, Blomer U, Peterson DA, Gage FH, Verma IM (1997) Sustained expression of genes delivered directly into liver and muscle by lentiviral vectors. Nat Genet 17 (3):314–317. https://doi.org/10.1038/ng1197-314
6. Naldini L, Blomer U, Gage FH, Trono D, Verma IM (1996) Efficient transfer, integration, and sustained long-term expression of the transgene in adult rat brains injected with a lentiviral vector. Proc Natl Acad Sci U S A 93 (21):11382–11388
7. Moffat J, Grueneberg DA, Yang X, Kim SY, Kloepfer AM, Hinkle G, Piqani B, Eisenhaure TM, Luo B, Grenier JK, Carpenter AE, Foo SY, Stewart SA, Stockwell BR, Hacohen N, Hahn WC, Lander ES, Sabatini DM, Root DE (2006) A lentiviral RNAi library for human and mouse genes applied to an arrayed viral high-content screen. Cell 124 (6):1283–1298. https://doi.org/10.1016/j.cell.2006.01.040
8. Tidball JG (2011) Mechanisms of muscle injury, repair, and regeneration. Compr Physiol 1(4):2029–2062. https://doi.org/10.1002/cphy.c100092
9. Ge Y, Chen J (2012) Mammalian target of rapamycin (mTOR) signaling network in skeletal myogenesis. J Biol Chem 287 (52):43928–43935. https://doi.org/10.1074/jbc.R112.406942
10. Ge Y, Waldemer RJ, Nalluri R, Nuzzi PD, Chen J (2013) RNAi screen reveals potentially novel roles of cytokines in myoblast differentiation. PLoS One 8(7):e68068. https://doi.org/10.1371/journal.pone.0068068
11. Waldemer-Streyer RJ, Chen J (2015) Myocyte-derived Tnfsf14 is a survival factor necessary for myoblast differentiation and skeletal muscle regeneration. Cell Death Dis 6:e2026. https://doi.org/10.1038/cddis.2015.375
12. Waldemer-Streyer RJ, Reyes-Ordonez A, Kim D, Zhang R, Singh N, Chen J (2017) Cxcl14 depletion accelerates skeletal myogenesis by promoting cell cycle withdrawal. NPJ Regen Med 2. https://doi.org/10.1038/npjregenmed.2016.17
13. Ge Y, Sun Y, Chen J (2011) IGF-II is regulated by microRNA-125b in skeletal myogenesis. J Cell Biol 192(1):69–81. https://doi.org/10.1083/jcb.201007165
14. Guardiola O, Andolfi G, Tirone M, Iavarone F, Brunelli S, Minchiotti G (2017) Induction of acute skeletal muscle regeneration by cardiotoxin injection. J Vis Exp 119. https://doi.org/10.3791/54515

Chapter 8

Adipogenesis from Bovine Precursors

Frea Mehta, Ruud Theunissen, and Mark J. Post

Abstract

Cultured meat, or tissue engineered meat, is a promising alternative to conventional meat production. In order to realistically mimic the multiple tissue types found in beef, food-compatible methods for bovine fat tissue engineering must be developed. We present a protocol for the isolation of adipose tissue-derived preadipocytes and subsequent adipogenic differentiation through free fatty acid stimulation. Differentiating preadipocytes can be either grown in 2D culture conditions or seeded in 3D alginate scaffolds. Differentiation is visually confirmed through lipophilic staining.

Key words Cultured meat, Bovine, Adipose, Preadipocyte, Adipogenic differentiation, Lipid, Free fatty acid, Monounsaturated fatty acid, Branched chain fatty acid, Alginate

1 Introduction

Conventional livestock production is a resource and environmental burden, the detrimental effects of which are exacerbated by increasing population growth and therefore protein demand [1]. Beef production is responsible for 41% of global greenhouse gas emissions associated with livestock [2], 33% of global water consumption associated with livestock [3], and is a driver of biodiversity loss [4]. Engineered skeletal muscle tissue, or cultured meat, is an alternative to traditional livestock rearing, and cultured beef represents a particularly high-impact application of the technology [5, 6]. Meat is predominantly skeletal muscle but contains other tissue types, most prominently fat. Cultured fat contributes to the taste and texture of cultured meat, and is therefore critical to its consumer acceptance.

Adipose tissue is a heterogeneous tissue, containing monolocular lipid-filled "white" adipocytes as well as a stromal vascular fraction consisting of mesenchymal stem cells, endothelial progenitors, and immune-related cells [7]. Mesenchymal stem cells with

Sissel Beate Rønning (ed.), *Myogenesis: Methods and Protocols*, Methods in Molecular Biology, vol. 1889,
https://doi.org/10.1007/978-1-4939-8897-6_8, © Springer Science+Business Media, LLC, part of Springer Nature 2019

adipogenic potential are termed preadipocytes, and are a promising source of replicative cells for cultured fat. In vitro adipogenesis has been extensively described in mammalian model systems, in both primary preadipocytes as well as preadipocyte cell lines such as murine 3T3-L1. Differentiation is typically induced by a chemical cocktail consisting of 3-isobutyl-1-methylxanthine (IBMX), dexamethasone, rosaglitizone, and insulin [8]. Collectively, this cocktail promotes adipogenesis by activation of transcription factors peroxisome proliferator-activated receptor-gamma (PPAR-γ) and CCAAT/enhancer binding protein-alpha (C/EBP-α) [9–13]. However, this chemical cocktail is not food-compatible and therefore not suitable for differentiation of cultured fat intended for human consumption. A potential solution to this problem is induction of adipogenesis by free fatty acids. Unsaturated and branched-chain fatty acids are agonists of PPAR-family transcription factors [14–17] and are typically found in dietary sources. In this protocol, we propose the use of monounsaturated ω-9, ω-7, and ω-5 and branched chain fatty acids for adipogenic differentiation. Selected fatty acids are tabulated below.

Fatty acid	Type	Dietary source
Erucic acid	Monounsaturated ω-9	Vegetable oils
Elaidic acid	Monounsaturated ω-9	Vegetable oils
Oleic acid	Monounsaturated ω-9	Vegetable oils
Palmitoleic acid	Monounsaturated ω-7	Dairy, animal fat, vegetable and fish oils
Myristoleic acid	Monounsaturated ω-5	Animal fat
Phytanic acid	Branched chain	Dairy, ruminant fat
Pristanic acid	Branched chain	Dairy, ruminant fat

Adipogenic differentiation is typically conducted in 2D cell culture environments, but 3D culturing is essential for structure in the case of cultured fat. We propose the use of food-compatible alginate hydrogel scaffolds for 3D culturing of differentiating adipocytes. Alginate is a well-established biocompatible material [18, 19] as well as a widely-used thickening agent in the food industry.

This chapter is divided into the following three sections: isolation of adipogenic precursors from primary bovine adipose tissue, food-compatible strategies for adipogenic differentiation in 2D culture, and adipogenic differentiation in 3D culture. The first section describes methods for obtaining and maintaining primary cells with preadipogenic potential. Adipogenic differentiation can be induced in these preadipocytes in either 2D or 3D culture systems, both of which are described in the subsequent two sections. Adipogenic differentiation is visually confirmed by staining with Oil Red O, a lipophilic dye.

2 Materials

2.1 Bovine Stromal Vascular Cell Isolation and Cell Culture

1. Primary bovine adipose tissue (*see* **Note 1**).
2. Ice, plastic bag, and insulated box for transport.
3. Biological safety cabinet.
4. 70% ethanol.
5. 1 L glass beaker.
6. Sterile phosphate buffered saline (PBS), pH 7.2.
7. Sterile forceps, scalpel, and scissors.
8. 15 cm cell culture dish or other work surface.
9. 15 and 50 mL conical centrifuge tubes.
10. Tube racks.
11. Weigh scale.
12. 5, 10, 25, and 50 mL sterile serological pipettes.
13. Pipette controller.
14. 37 °C water bath.
15. Dulbecco's Modified Eagle Medium (DMEM) + 4.5 g/L D-glucose + pyruvate and GlutaMAX™ (*see* **Note 2**).
16. 100× penicillin/streptomycin/amphotericin B (P/S/A) (*see* **Note 3**).
17. Type II collagenase (collagenase, type 2; Worthington Biochemical Corporation).
18. Collagenase solution: 200 units type 2 collagenase per 1 mL DMEM with 1% P/S/A.
19. Shaking incubator.
20. Heat inactivated fetal bovine serum (HI FBS) (*see* **Note 4**).
21. 440 μm nylon cell strainer.
22. Benchtop centrifuge.
23. Culture medium solution: DMEM with 10% HI FBS and 1% P/S/A (*see* **Note 5**).
24. 6-well cell culture plates.
25. Cell culture incubator with temperature, humidity, and CO_2 control.
26. Aspiration pump and pipette.
27. Inverted light microscope.
28. 0.05% trypsin-EDTA.

29. 0.4% trypan blue vital stain.
30. 10–1000 μL micropipettes and sterile tips.
31. Automated cell counter and cell counting chamber slides or hemocytometer and cover slip.
32. Cell culture flasks or plates.
33. Dimethyl sulfoxide (DMSO).
34. Freezing medium: 10% DMSO in HI FBS.
35. 2 mL cryotubes.
36. Controlled-rate cell freezing container.
37. −80 °C freezer.
38. Liquid nitrogen cell storage container.

2.2 Free Fatty Acid-Mediated In Vitro Bovine Adipogenesis

1. Isolated stromal vascular cells from bovine adipose tissue.
2. Cell culture incubator with temperature, humidity, and CO_2 control.
3. Biological safety cabinet.
4. 37 °C water bath.
5. 24-well cell culture plate.
6. Culture medium (*see* Subheading 2.1, **item 23**).
7. 10–1000 μL micropipettes and sterile tips.
8. Individual free fatty acids (Sigma-Aldrich): (*see* **Note 6**).
   (a) Phytanic acid.
   (b) Pristanic acid.
   (c) Erucic acid.
   (d) Oleic acid.
   (e) Palmitoleic acid.
   (f) Elaidic acid.
   (g) Myristoleic acid.
9. Aspiration pump and pipette.

2.2.1 Lipid Accumulation Visualization with Oil Red O

1. 5, 10, 25, and 50 mL sterile serological pipettes.
2. Pipette controller.
3. PBS.
4. Paraformaldehyde (PFA).
5. Fixation solution: 3% PFA in PBS.
6. 0.5% Oil Red O solution in isopropanol.
7. Distilled water.

8. Filter paper.
9. Glass funnel.
10. Inverted light microscope.

2.3 Alginate Scaffolds for 3D Bovine Adipose Tissue Culture

1. Isolated stromal vascular cells from bovine adipose tissue.
2. Cell culture incubator with temperature, humidity, and CO_2 control.
3. 3-layer tissue culture multi flask.
4. 5, 10, 25, and 50 mL sterile serological pipettes.
5. Pipette controller.
6. 10–1000 μL micropipettes and sterile tips.
7. Aspiration pump and pipette.
8. 37 °C water bath.
9. PBS.
10. Trypsin-EDTA.
11. Culture medium (*see* Subheading 2.1, **item 23**)
12. Benchtop centrifuge.
13. 0.4% trypan blue vital stain.
14. Cell counter and slides or hemocytometer and cover slip.
15. Low-viscosity guluronic acid (LVG) ultrapure alginate (PRONOVA™ UP LVG Alginate; NovaMatrix).
16. DMEM.
17. Shaker table.
18. Calcium Chloride.
19. 10 mM HEPES (4-(2-hydroxyethyl)-1-piperazineethanesulfonic acid) buffer solution in distilled water.
20. 50 mL glass beaker.
21. Stirring hotplate with stir bar.
22. 2 mL sterile disposable syringe.
23. 6-well cell culture plate.
24. Free fatty acid solutions (*see* Subheading 2.2, **item 8**).

2.3.1 Lipid Accumulation Visualization Within 3D Scaffolds

1. 100–1000 μL micropipettes and sterile tips.
2. 1.5 mL microcentrifuge tubes.
3. Benchtop centrifuge.
4. Plastic disposable cryomolds.
5. Optimal Cutting Temperature (OCT) embedding compound.
6. Liquid nitrogen in a dewar or −80 °C isopentane.
7. −80 °C freezer.
8. Cryomicrotome.

9. Clean glass slides.
10. −20 °C freezer.
11. 0.5% Oil Red O solution in isopropanol.
12. Ethanol.
13. Distilled water.
14. Filter paper.
15. Glass funnel.
16. Humid chamber with staining tray.
17. Aqueous mounting medium.
18. Clean glass coverslips.
19. Inverted light microscope.

3 Methods

3.1 Bovine Stromal Vascular Cell Isolation

1. Obtain 50–100 g of bovine subcutaneous fat tissue (*see* **Note 7**) from an abattoir (*see* **Note 8**) and store in a clean plastic bag on ice for transport. Transport to the laboratory within 1 h.
2. Transfer the tissue to a biological safety cabinet (*see* **Note 9**). Submerge tissue in 70% ethanol for 5 min to disinfect, then briefly submerge in sterile PBS.
3. Using sterile forceps, scalpel, and scissors, mince the tissue into small, 2–3 mm diameter pieces. Avoid the outer 1–3 mm of tissue (*see* **Note 10**), and remove visible connective tissue and blood vessels (*see* **Note 11**). Use a sterile 15 cm cell culture dish as a sterile work surface.
4. Transfer the tissue to 50 mL centrifuge tubes and weigh.
5. Distribute the minced tissue between 50 mL centrifuge tubes with no more than 10 g of tissue per tube. Add 5 mL of 37 °C collagenase solution per 1 g of tissue to each tube.
6. Cap the lids tightly, mix well by inversion, and incubate the tubes in a shaking incubator for 1 h at 37 °C with 150 rpm shaking.
7. Add heat inactivated fetal bovine serum (HI FBS) to each tube to attain a 10% FBS solution.
8. Filter through a 440 μm sieve into a sterile 50 mL centrifuge tube.
9. Centrifuge the filtrate for 5 min at 450 × *g*.
10. Figure 1 shows the contents of the centrifuged tube. The floating fraction contains buoyant mature adipocytes, ruptured adipocytes, and lipids. This floating fraction can be discarded

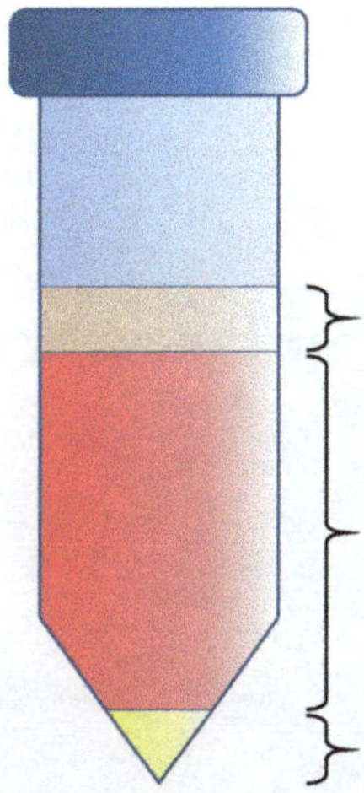

Fig. 1 Adipose tissue following collagenase digestion and separation by centrifugation. Preadipocytes are found in the stromal vascular fraction, while mature adipocytes and lipids float to the upper fraction

along with the aqueous supernatant so that only the stromal vascular fraction (SVF) remains.

11. Resuspend the pelleted SVF in 5 mL of warm culture medium.
12. Centrifuge for 5 min at 450 × *g*.
13. Resuspend pellet in 6 mL of warm culture medium. Transfer to a 6-well cell culture plate, plating 2 mL of cell suspension per well.
14. Incubate for 6 h or overnight at 37 °C and 5% CO_2.
15. Aspirate the supernatant from the wells and rinse twice with PBS.
16. Add fresh medium and return to the incubator.
17. Monitor every day (*see* **Note 12**) with an inverted light microscope and refresh the medium every 2–3 days (*see* **Note 13**).

3.1.1 Maintenance of Stromal Vascular Cells in Culture

1. Once cells reach at least 50% confluency, they can be passaged.
2. Aspirate the wells and wash with PBS. Add half a well volume's worth of warm trypsin per well to dissociate. Incubate for 2–3 min, monitoring for detachment. Gently tap the plate against the table to aid dissociation if necessary.
3. Add an equal volume of warm culture medium to each well to neutralize the trypsin.
4. Collect the cell suspension in a 15 mL centrifuge tube and centrifuge 5 min at 450 × *g*. Aspirate the supernatant and resuspend the cells in 2–5 mL culture medium.

Fig. 2 Stromal vascular cells one passage after isolation. Scale bar: 1000 μm

5. Mix 10 μL of trypan blue stain with 10 μL of cell suspension and load 10 μL into either a counting chamber slide or hemocytometer. Count live cells (*see* **Note 14**) and calculate their viability (*see* **Note 15**).
6. Transfer cells to cell culture flasks or plates, depending on yield and downstream use. Seed cells at 5000 cells/cm^2.
7. Passage cells at 75–80% confluency, typically every 3 days. Cells display a typical SVF fibroblast-like cell morphology (Fig. 2).
8. If cells are not needed, they can be frozen and stored in liquid nitrogen.
   (a) To freeze back cultures, dissociate cells from the culture dishes, collect, and quantify as described in **steps 2–5**.
   (b) Resuspend in freezing medium and distribute the cell suspension into cryotubes. The number of cells frozen back per tube depends on individual needs, but one million cells in 1 mL of freezing medium per tube is sufficient.
   (c) Transfer the sealed, labeled cryotubes to a cooling box and store at −80 °C between 2 and 48 h (*see* **Note 16**).
   (d) Transfer the tubes to liquid nitrogen for long-term storage (*see* **Note 17**).
   (e) To recover frozen cells, remove cryotubes from liquid nitrogen storage and thaw at room temperature. As soon as the samples are liquid, add 1 mL of culture medium, remove all the liquid, and transfer to a sterile 15 mL

centrifuge tube. Centrifuge for 5 min at 450 × *g*, aspirate the supernatant, and resuspend the pellet in 15 mL of culture medium. Seed on a T-75 flask.

3.2 Free Fatty Acid-Mediated In Vitro Bovine Adipogenesis

1. At least one passage after isolation (*see* **Note 18**), cells can be treated with free fatty acids (FFAs) to induce lipid accumulation. Seed cells in a 24-well culture plate (*see* **Note 19**). If using recovered frozen cells, passage at least once before seeding.
2. Maintain cells for 3 days in culture medium.
3. In a biological safety cabinet, prepare FFA solutions by diluting stock FFA solutions in warm (*see* **Note 20**) culture medium. The FFAs can be administered as a mix or as single fatty acids (*see* **Note 21**). Regardless, the final concentration of each fatty acid in the medium solution should be 50 μm.
4. Aspirate the culture medium from the wells and replace with the same volume of warm FFA solutions. Keep plain culture medium in at least one well as a negative control.
5. Replace the FFA-supplemented medium and control culture medium every 2–3 days. Cultures can be maintained for 1–2 weeks. Lipid accumulation can be assessed with Oil Red O (*see* **Note 22**) staining as early as 2–3 days after FFA introduction.

3.2.1 Lipid Accumulation Visualization with Oil Red O

1. To confirm lipid accumulation, fix and stain desired wells with Oil Red O (ORO).
2. Remove FFA medium from desired wells and rinse twice with PBS.
3. Add 1 mL of paraformaldehyde (PFA) fixation solution to each well to fix. Incubate for 15 min at room temperature.
4. Remove the PFA solution and dispose of in an appropriate waste container. Rinse the wells twice with PBS.
5. Dilute the ORO stock solution (0.5% ORO in isopropanol) to 0.3% in distilled water. Incubate for 15 min at room temperature.
6. Place a piece of filter paper in a glass funnel and filter the ORO solution into a clean new bottle (*see* **Note 23**).
7. Replace PBS in fixed wells with 0.5 mL ORO per well.
8. Aspirate the staining solution after 30 min and rinse the wells 3 to 5 times with distilled water, gently so as not to disturb the cell layer.
9. Add 1 mL PBS to the wells and photograph immediately. ORO-stained lipids will appear red. Three days following initial FFA treatment, FFA-treated SVF preadipocytes display lipid accumulation (Fig. 3).

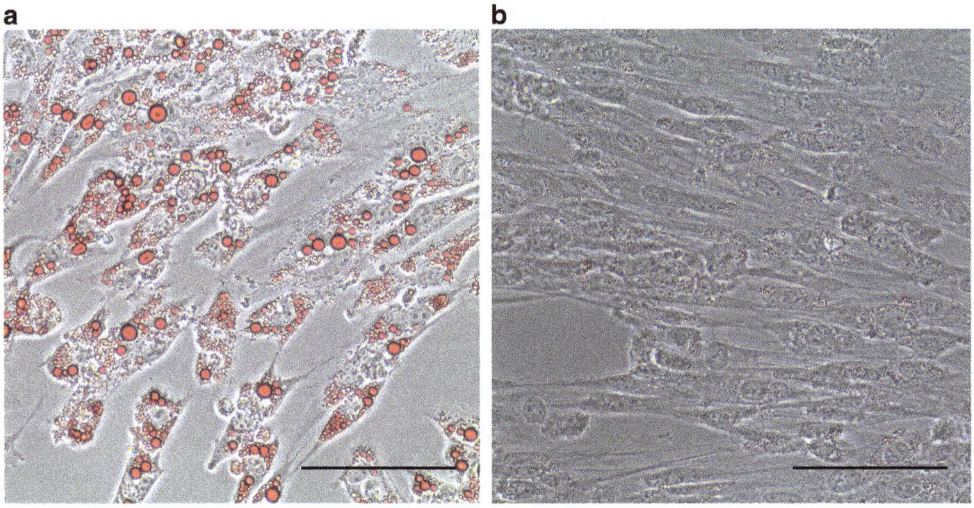

Fig. 3 (**a**) Oil Red O-stained preadipocytes after 3 days in free fatty acid-treated medium. (**b**) Oil Red O-stained preadipocytes after 3 days in normal culture medium. Scale bar = 100 μm

3.3 Alginate Scaffolds for 3D Bovine Adipose Tissue Culture

1. At least one passage after isolation or one passage after recovery of frozen cells, seed the stromal vascular (SV) cells in a 3-layer tissue culture multi flask (*see* **Note 24**) at 5000 cells/cm^2.
2. Harvest the cells when they have reached confluency. Dissociate and quantify cells from the flask as described in Subheading 3.1.1, **steps 2–5**.
3. Following quantification, centrifuge cells for 5 min at 450 × *g*.
4. Resuspend the pellet in warm culture medium with a concentration of 4×10^7 cells/mL.
5. Prepare the low-viscosity high guluronic acid (LVG) ultrapure alginate solution. Dissolve 10 mg LVG alginate in 1 mL DMEM by gently mixing for 3 h on a shaker table.
6. Prepare 100 mL of 66 mM calcium chloride in 10 mM HEPES buffer in a glass beaker and warm to 37 °C. Place on a hot plate with stirring and add a stir bar.
7. Warm the 1% LVG alginate solution to 37 °C and add to the cell suspension in a 1:1 ratio with the culture medium. Mix well by pipetting.
8. Transfer the LVG alginate and cell suspension to a sterile 2 mL syringe without a needle.
9. Gently stir the buffered $CaCl_2$ solution. Submerge the tip of the syringe in the solution and extrude the alginate and cell suspension to produce a tube-shaped "spaghetti" scaffold (Fig. 4, *see* **Note 25**).

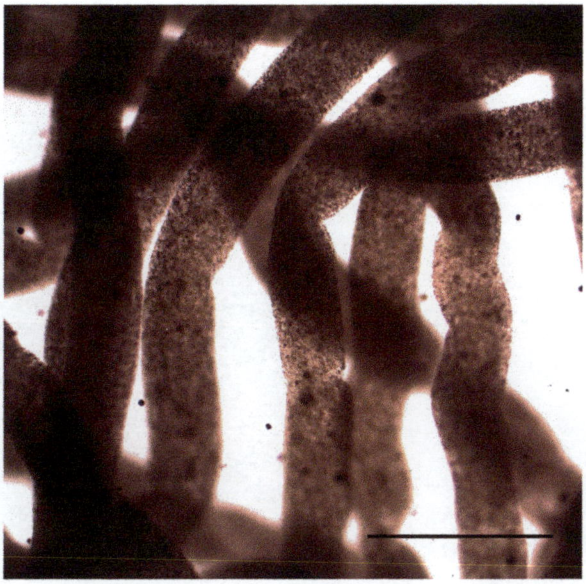

Fig. 4 Alginate "spaghetti" cylindrical scaffolds containing preadipocytes. Scale bar: 2000 μm

10. Drain the remaining buffer from the beaker and wash the scaffolds twice with PBS.
11. Transfer the scaffolds to a 6-well culture plate and submerge in 4 mL of culture medium containing free fatty acids (FFAs) as described in Subheading 3.2. Store in an incubator at 37 °C and 5% CO_2.
12. Change the FFA-supplemented medium every 2–3 days. The scaffolds can be maintained for 1–4 weeks in culture. Lipid accumulation can be assessed with Oil Red O staining as early as 2–3 days after FFA introduction.

3.3.1 Lipid Accumulation Visualization Within 3D Scaffolds

1. Transfer 1–2 cm of "spaghetti" scaffolds to a 1.5 mL microcentrifuge tube using a sterile P-1000 micropipette tip.
2. Centrifuge for 1 min at 400 × *g* and remove the supernatant.
3. Transfer the scaffolds to a plastic disposable cryomold and add Optimal Cutting Temperature (OCT) embedding compound to completely cover the scaffolds.
4. Snap freeze the mold in either liquid nitrogen or isopentane at −80 °C. Store the frozen scaffolds at −80 °C.
5. Cut 50 μm sections from the block in a cryomicrotome and mount the sections to a clean glass slide by thawing (*see* **Note 26**). Air dry at room temperature for 5–10 min.

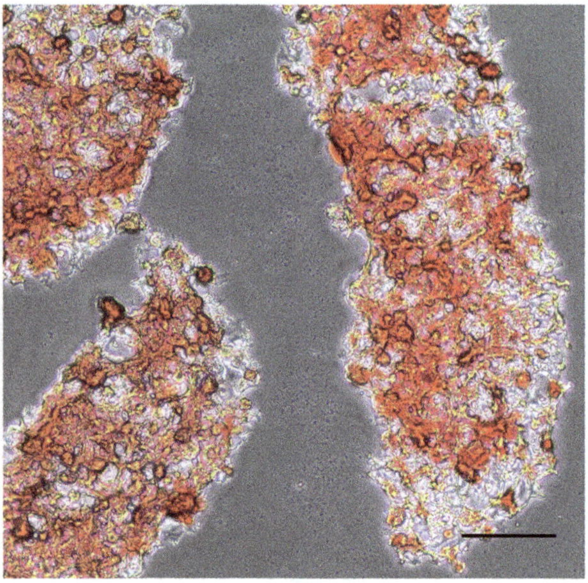

Fig. 5 Oil Red O-stained 50 μm sections of alginate "spaghetti" scaffolds containing preadipocytes after 3 days in free fatty acid-treated medium. Scale bar = 100 μm

6. The sections can either be stored at −20 °C for 1 week or stained immediately.
7. To stain, prepare a 0.3% working solution of Oil Red O (ORO) as described in Subheading 3.2.1, **steps 5–6**.
8. Place the slides in a staining tray with wet paper towels to maintain humidity. Add 50 μL ORO solution to each slide and close the humid chamber.
9. After 15 min, remove the ORO solution by gently rinsing twice with distilled water.
10. Air dry the slides, cover with 5–6 drops of aqueous mounting solution, and cover with a coverslip.
11. Image the slides immediately; ORO-stained lipids will appear red (Fig. 5). Stained slides can be stored for 2–3 days at −20 °C.

4 Notes

1. Retrieval, handling, and shipping of animal tissues is subject to national guidelines of animal tissue handling.
2. If GlutaMAX™-supplemented DMEM is unavailable, L-glutamine can be added to a final concentration of 2 mM. L-glutamine degrades rapidly in culture conditions, and should therefore be added immediately prior to medium use.

3. Antibiotic-free adipocyte culture is necessary for edible cultured fat applications. This protocol uses antibiotics and antimycotics to minimize infection risk from primary cell isolations under non-sterile conditions such as those found in an abbatoir.
4. In long-term cultured meat goals of global flock reduction, serum-free cell culture conditions are necessary.
5. Store culture medium at 4 °C for up to 2 weeks.
6. Make up 20 mM stock solutions of each individual free fatty acid (FFA) in ethanol. Store stock solutions at −20 °C, flushing the vial with nitrogen prior to closing.
7. Adipose tissue has depot-specific properties [20], so depot selection can be experimentally relevant. However, bovine tissue is often sourced from commercial slaughterhouses, limiting depot selection.
8. Samples should be as fresh as possible, ideally obtained within 30 min of slaughter.
9. All steps in which unfixed cells are exposed to the outside environment should take place in the biological safety cabinet.
10. Avoid tissue that was directly exposed to unsterile slaughterhouse and lab environments to minimize the risk of microbial contamination.
11. Adipose is a highly vascularized tissue and it is impossible to manually separate all blood vessels. Remove visible blood vessels to minimize endothelial cell contamination, as these fast-growing cells are likely to overgrow the culture.
12. Additionally, monitor for fungal or microbial infection.
13. Depending on yield, cells may not be visible for 3–5 days.
14. Cells stained with trypan blue are dead, as live cells with intact cell membranes do not take up the dye. Live cells can be identified by a blue ring around their perimeter.
15. Healthy cultures should have >90% viability. Cultures that look healthy but result in low viability counts might be suffering from the following common problems: incubating in trypsin for too long damages cell membrane integrity or crystals in old trypan blue solution can generate false negatives that look like stained dead cells. To avoid these issues, closely monitor cells for detachment during the trypsin dissociation step and filter old trypan blue with a 0.2 μm filter.
16. The rate of cooling depends on the cooling box used. Cryotubes should be transferred to long-term storage in liquid nitrogen as soon as they have reached −80 °C. Do not store cells at −80 °C for longer than 48 h.

17. Liquid nitrogen (LN_2) is a cryogen that can cause bodily harm if in contact with exposed skin or eye tissue. Always wear appropriate personal protective equipment when handling it. Additionally, store and handle LN_2 in well-ventilated spaces with oxygen monitors to minimize asphyxiation risk.
18. The stromal vascular (SV) cell isolates contain a heterogeneous population of cells including preadipocytes. Early passages are better suited for adipogenic treatment, as the preadipocyte population is less likely to have been overgrown by other competing cell types.
19. A 24-well plate is recommended to test the adipogenic effects of various FFAs, but the system can be scaled up to any desired culture size.
20. FFAs can precipitate in culture medium. To minimize precipitation, dilute in warm medium and vortex gently to mix.
21. The crude nature of this SV cell isolation coupled with adipose depot-dependent variations results in a heterogeneous cell population. It is recommended to initially screen both FFA mixes and single FFAs in order to optimize adipogenesis within a given isolation.
22. Oil Red O (ORO) is a lipophilic dye that can be visualized with transmitted light.
23. Use the working solution of ORO within 2 h.
24. For 3D culture, a large number of cells are required. If multi-flasks are unavailable, simply grow the desired number of SV cells in multiple flasks or dishes.
25. Alginate gelation occurs through ionic cross-linking in which Ca^{2+} binds to guluronate blocks of alginate. These guluronate blocks form junctions between alginate polymers, creating a cross-linked gel structure [18]. A chelator such as ethylenediaminetetraacetic acid (EDTA) can be used to dissociate divalent calcium ions from the alginate and dissolve the gel, if necessary.
26. Cryosectioning methods are beyond the scope of this chapter. An overview of methods is available in Methods in Molecular Biology, Volume 1018, pp. 301–311: Cryosectioning, Alexander I. Son, Katie Sokolowski, Renping Zhou. 26 April 2013.

Acknowledgments

This work was supported by the Brin-Wojcicki Foundation, the Fulbright U.S. Student Program, and the Netherland-America Foundation.

References

1. de Vries M, van Middelaar CE, de Boer IJM (2016) Comparing environmental impacts of beef production systems: a review of life cycle assessments. Livest Sci 178:279–288
2. Opio C, Gerber P, Mottet A et al (2013) Greenhouse gas emissions from ruminant supply chains–a global life cycle assessment. Food and Agriculture Organization of the United Nations (FAO), Rome
3. Mekonnen MM, Hoekstra AY (2012) A global assessment of the water footprint of farm animal products. Ecosystems 15:401–415
4. Gerber PJ, Mottet A, Opio CI et al (2015) Environmental impacts of beef production: review of challenges and perspectives for durability. Meat Sci 109:2–12
5. Tuomisto HL, Teixeira De Mattos MJ (2011) Environmental impacts of cultured meat production. Environ Sci Technol 45:6117–6123
6. Mattick CS, Landis AE, Allenby BR et al (2015) Anticipatory life cycle analysis of in vitro biomass cultivation for cultured meat production in the United States. Environ Sci Technol 49:11941–11949
7. Capurso C, Capurso A (2012) From excess adiposity to insulin resistance: the role of free fatty acids. Vasc Pharmacol 57:91–97
8. Scott MA, Nguyen VT, Levi B et al (2011) Current methods of Adipogenic differentiation of Mesenchymal stem cells. Stem Cells Dev 20:1793–1804
9. Wu Z, Bucher NL, Farmer SR (1996) Induction of peroxisome proliferator-activated receptor gamma during the conversion of 3T3 fibroblasts into adipocytes is mediated by C/EBPbeta, C/EBPdelta, and glucocorticoids. Mol Cell Biol 16:4128–4136
10. Kim SP, Ha JM, Yun SJ et al (2010) Transcriptional activation of peroxisome proliferator-activated receptor-gamma requires activation of both protein kinase A and Akt during adipocyte differentiation. Biochem Biophys Res Commun 399:55–59
11. Kim CY, Kim KH (2013) Dexamethasone-induced selenoprotein S degradation is required for adipogenesis. J Lipid Res 54:2069–2082
12. Farmer SR (2006) Transcriptional control of adipocyte formation. Cell Metab 4:263–273
13. Rosen ED, MacDougald OA (2006) Adipocyte differentiation from the inside out. Nat Rev Mol Cell Biol 7:885–896
14. Grygiel-Górniak B (2014) Peroxisome proliferator-activated receptors and their ligands: nutritional and clinical implications–a review. Nutr J 13:1–10
15. Hanhoff T, Benjamin S, Börchers T et al (2005) Branched-chain fatty acids as activators of peroxisome proliferator-activated receptors. Eur J Lipid Sci Technol 107:716–729
16. Ding ST, Wang JC, Mersmann HJ (2003) Effect of unsaturated fatty acids on porcine adipocyte differentiation. Nutr Res 23:1059–1069
17. Zomer AWM, van Der Burg B, Jansen GA et al (2000) Pristanic acid and phytanic acid: naturally occurring ligands for the nuclear receptor peroxisome proliferator-activated receptor alpha. J Lipid Res 41:1801–1807
18. Lee KY, Mooney DJ (2012) Alginate: properties and biomedical applications. Prog Polym Sci 37:106–126
19. Augst AD, Kong HJ, Mooney DJ (2006) Alginate hydrogels as biomaterials. Macromol Biosci 6:623–633
20. Dodson MV, Du M, Wang S et al (2014) Adipose depots differ in cellularity, adipokines produced, gene expression, and cell systems. Adipocytes 3:236–241

Chapter 9

Transcriptomic Profiling During Myogenesis

Alicja Majewska, Tomasz Domoradzki, and Katarzyna Grzelkowska-Kowalczyk

Abstract

Microarray-based transcriptomic profiling enables simultaneous measurement of expression of multiple genes from one biological sample. Here we describe a detailed protocol, which serves to examine global gene expression using whole genome oligonucleotide microarrays. We also provide examples of bioinformatics tools, which are helpful in analyses and interpretation of microarray data, and propose further biological assays, to warrant conclusions drawn from transcriptomic signature.

Key words Microarrays, RNA isolation and validation, cRNA labeling and hybridization, Gene ontology classification, Biological association networks

1 Introduction

Transcriptomic profiling based on microarray method enables simultaneous monitoring of expression of thousands of genes and identification of genes/proteins that contribute to regulation of cellular processes. It provides a powerful tool for understanding the critical transcriptional changes attributed to particular stages of myogenesis, which is a complex and highly regulated process involving crosstalk between numerous cellular pathways.

There are many types of microarrays used for gene expression analysis. They differ from each other with reference to length of probe ("probe" is used to describe the nucleotide sequence that is attached to the microarray surface). Complementary DNA (cDNA) microarrays are composed of long probes (nucleotide sequences) of hundreds to thousands nucleotides, whereas in oligo arrays the probes are short and their length is defined by the platform, e.g., very short 25 nucleotides in Affimetrix, or longer: 50–70 nucleotides (Agilent—60 nucleotides). Nowadays oligo microarrays are used more often, because they are more specific for the intended binding target ("target" in microarray experiments refers to what is

Sissel Beate Rønning (ed.), *Myogenesis: Methods and Protocols*, Methods in Molecular Biology, vol. 1889,
https://doi.org/10.1007/978-1-4939-8897-6_9,

hybridized to the probes). The probes are designed for the regions of genes which are dissimilar from other genes regions.

Before starting the microarray experiment the most important thing is to determine the aim of the specific experiment and to plan it well. The question is how many replicates should be done? Statisticians do not give a simple answer. It depends on biological variability of investigated material, whether samples come from cell culture, laboratory animals, or different types of specimens. It seems that for cell culture and laboratory animal experiments the lowest admissible number of replicates should be 3–4. In the case of experiments on biological material of higher variability, the number of replicates should be greater. When planning experiment you also need to make a decision whether you perform one- or two-color microarray. In most microarray platforms it is possible to do both options. The exception is Affimetrix platform, where only one-color microarrays are possible. In the case of one-color microarrays one labeled sample is hybridized on one microarray. Two-color microarrays use two different samples, which are labeled with two different fluorophores, and then they are hybridized together on one microarray, e.g., control sample or common reference sample is labeled with cyanine 3 (Cy3) and investigated sample is labeled with cyanine 5 (Cy5). The common reference sample could represent total RNA isolated from tissues from wild-type organism, or it could be a pool of all samples of interest, or a pool of samples which do not take part in the experiment. Common references are frequently used to provide easy means of comparing many samples against one another.

The analysis conducted by Oberthuer et al. [1] showed a largely equivalent overall performance of both methods: one- and two-color microarrays, thereby suggesting that the choice of a one-color or of a two-color platform does not need to be a primary factor in decisions regarding experimental microarray design [1]. Also Slonim and Yanail [2] stated a good overall agreement in the data produced by these two methods: one- and two-color platforms.

There are several microarray platforms available on the market, but in most of the published studies using the microarray analyses (which can be found in the NCBI base) the gene expression experiments were done using Affimetrix or Agilent microarrays. These two platforms adhere to strict procedures and conditions which have been standardized and optimized. They have dedicated reagents for samples preparation, labeling, hybridization, and washing. However, many other technologies may use different components and protocols for array production, sample labeling, hybridization, and image acquisition which introduce many sources of variation.

In our experiments we used Agilent Platform and two-color microarrays. Agilent has different formats of microarray slides

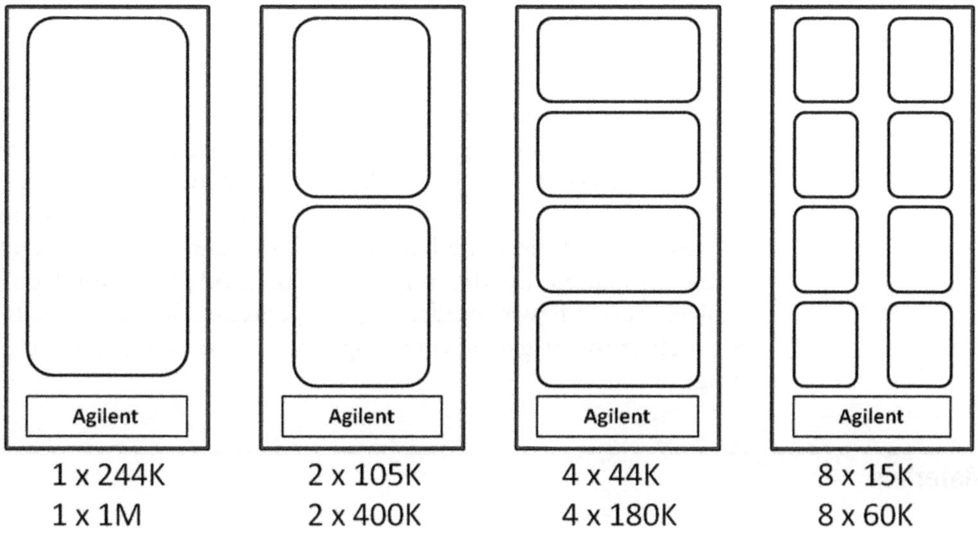

Fig. 1 Schematic illustration of the types of Agilent microarray slides

depending on the number of microarrays on a slide and density of printed probes (Fig. 1). For mouse genome two formats are available: 4×44K (Mouse Gene Expression v2 4×44K Microarray—four arrays per slide, each array comprises about 44,000 probes) and 8×60K (SurePrint G3 Mouse Gene Expression v2 8×60K Microarray—eight arrays per slide, each array comprises about 60,000 probes). Aside from the microarrays, this platform also offers the whole set of reagents for every preparation step of the microarrays.

Here, we present a detailed protocol for the analysis of global gene expression using whole genome oligonucleotide microarrays. Microarray profiling is multistep analysis that involves extraction of RNA from a biological sample, cDNA synthesis, cRNA synthesis and amplification, RNA labeling, and hybridization on microarray slide, followed by signal detection, quantification, and analysis. In the method described here two different samples, which are labeled with two different fluorophores, are hybridized together on one microarray, control sample is labeled with cyanine 3 (Cy3) and examined sample is labeled with cyanine 5 (Cy5).

There are several bioinformatics tools that enable statistical analyses of microarray-based gene expression changes and help to classify these genes according to different criteria. Here, we describe examples of ontological classification of genes in terms of their involvement in different biological processes or their molecular function, using the PANTHER (Protein Analysis Through Evolutionary Relationships) Classification System [3]. We present examples of gene network generation using Pathway Studio (Elsevier) software, which is useful for searching and examination of relevant relationships among genes, proteins, metabolites, based on information on molecular interactions present in databases. Another web-accessible program, the DAVID (Database for Annotation, Visualization and Integrated Discovery), allows identification of gene categories/subcategories and finding enriched functionally related gene groups [4], which is useful for interpretation of microarray-based profiling in the context of myogenesis. Exact roles of numerous genes included to transcriptomic profile are usually not described in skeletal muscle biology; therefore, results obtained by the microarray method and bioinformatics analyses reveal new candidate genes, proteins, or cellular pathways for further investigation on myogenesis in physiological and pathological states.

2 Materials

Work with RNA requires attentiveness to avoid RNase contamination (*see* **Note 1**). Use only DNase- and RNase-free reagents and plastic ware.

2.1 Equipment and Materials for RNA Extraction

1. Set of adjustable automatic micropipettes only for RNA work (P10, P100, P1000).
2. Sterile, DNase/RNase-free, filter pipette tips.
3. DNase/RNase-free 1.5 mL tubes.
4. Cell-scrapers.
5. Vortex mixer.
6. Microcentrifuge with rotor for 2 mL tubes (capable of reaching at least 14,000 $\times g$).
7. NanoDrop spectrophotometer.
8. Agilent Bioanalyzer 2100 (Agilent Technologies).
9. 96–100% ethanol, 70% ethanol (for dilution use DNase/RNase-free water).
10. Phosphate buffered saline (PBS), pH 7.4.
11. RNA isolation kit, e.g., RNeasy Mini Kit (Qiagen) includes all buffers: RLT, RW1, RPE, DNase/RNase-free water, and RNeasy spin column.
12. DNase I.
13. RNA 6000 Nano Chip kit (Agilent Technologies).
14. Ice bucket.

2.2 Equipment and Materials for Microarray

1. Set of adjustable automatic micropipettes only for RNA work (P10, P100, P1000).
2. Vortex mixer.

3. NanoDrop spectrophotometer.
4. Thermocycler.
5. Thermoblock or thermomixer with thermotop.
6. Microcentrifuge with rotor for 2 mL and 0.2 mL tubes.
7. Hybridization gasket slides.
8. Hybridization chambers (Agilent Technologies).
9. Agilent Hybridization Oven.
10. Magnetic stir plate.
11. Three slide-staining dishes, with slide rack.
12. Clean forcep.
13. Agilent Microarray Scanner.
14. Ozone-Barrier Slide Cover.
15. DNase/RNase–free 0.2, 0.5, and 1.5 mL microcentrifuge tubes.
16. Nuclease-free aerosol barrier pipette tips.
17. G3 Mouse GE 8×60K v2 Microarray or HD Mouse GE 4×44K v2 Microarray.
18. Low Input Quick Amp Labeling Kit, Two-Color Agilent includes: T7 Primer, 5× First Strand Buffer, 0.1 M DTT—Dithiothreitol, 10 mM dNTP Mix, Affinity Script RNase Block Mix, 5× Transcription Buffer, NTP Mix, T7 RNA Polymerase Blend, Nuclease-free Water Cyanine 3-CTP Cyanine 5-CTP.
19. RNA Spike-In Kit, Two-Color Agilent includes: Spike A Mix, Spike B Mix and Dilution Buffer.
20. RNeasy Mini Kit (Qiagen).
21. Gene Expression Hybridization Kit Agilent Includes: 10× Gene Expression Blocking Agent, 25× Fragmentation Buffer, 2× Hi-RPM Hybridization Buffer.
22. Gene Expression Wash Buffer Kit Agilent includes: Gene Expression Wash Buffer 1, Gene Expression Wash Buffer 2, 10% Triton X-102. When the buffers are opened for the first time add 2 mL of 10%Triton X-102 to each of them. The final dilution of Triton X-102 in Gene Expression Wash Buffers is 0.005%. The addition of Triton X-102 reduces the possibility of array wash artifacts.
23. 95% to 100% ethanol (molecular biology grade).
24. DNase/RNase-free water.
25. Milli-Q water.
26. Ice bucket.

2.3 Software for Microarray Scanning, Extraction, and Date Analysis

1. Agilent Scan Control software.
2. Agilent Feature Extraction software.
3. GeneSpring GX Agilent (commercial software) or freely available limma R-package of Bioconductor (Bioconductor—open source open-development software project for the analysis and comprehension of high-throughput data in genomics and molecular biology [5]).

3 Methods

3.1 Cell Culturing and Sampling

Seed myoblasts and perform experiments according to the experimental design (*see* **Note 2**).

3.2 Cell Collection and Lyses

1. We usually culture cells on 60 mm diameter dishes in order to extract RNA (*see* **Note 3**). For RNA extraction we use $1–2.5 \times 10^6$ cells per sample (*see* **Note 4**).
2. Cells have to be processed immediately after removal from the incubator. Aspirate the cell culture medium completely, and then wash cells with PBS (*see* **Note 5**).
3. After removing PBS add lysis buffer (e.g., RLT, Fenozol, or TRI Reagent) which should cover the entire surface of the dish, scrape the cells using cell scraper, collect the lysate, and pipet into 1.5 or 2 mL microcentrifuge tube (*see* **Note 6**).
4. Pipet the lysate up and down 10 times and vortex for 20 s to mix it well, ensure that no cell clumps are visible. After this step you can continue the procedure of RNA isolation or freeze and store the samples at −70 °C (*see* **Note 7**).

3.3 RNA Extraction Protocol

In transcriptomic analyses (using microarrays or quantitative PCR methods) the starting biological material is total RNA extracted from cells or tissues. The RNA has to be of very good quality (*see* **Note 8**).

From this point the present description follows the protocol of the RNeasy Mini Kit (*see* **Note 9**).

1. Centrifuge lysate for 5 min at 3000–5000 × *g* at 20–25 °C and transfer the supernatant to a new RNase-free 1.5 or 2 mL polypropylene tube.
2. Add 1 volume of 70% ethanol to the supernatant and mix well by pipetting. Do not centrifuge.
3. Transfer up to 700 μL of the sample to an RNeasy spin column, close the lid gently, and centrifuge for 15 s at 10,000 × *g* 20–25 °C. Discard the flow-through. Reuse the collection tube (*see* **Note 10**).

4. Wash the column with 350 μL RW1 Buffer and centrifuge for 20 s at 10,000 × *g*. Discard the flow-through and reuse the collection tube (*see* **Note 11**).
5. We usually digest traces of genomic DNA on the column during extraction and washing RNA using RNase-Free DNase Set (*see* **Note 12**). Mix 10 μL DNase I stock solution with 70 μL Buffer RDD. (Prepare one mix for all samples) Mix the solution by gently inverting the tube. Do not vortex.
6. Pipet the DNase I incubation mix (80 μL) directly onto the RNeasy Mini Spin Column membrane and leave it on the bench (20–30 °C) for 15 min but not longer (*see* **Note 13**).
7. Wash the column with 350 μL RW1 Buffer and centrifuge for 20 s at 10,000 × *g*. Discard the flow-through.
8. Wash the column membrane adding 500 μL Buffer RPE to the column and centrifuge for 20 s at 10,000 × *g*. Discard the flow-through, repeat the wash, and centrifuge 2 min at 10,000 × *g*. Remove carefully the column from the collection tube and place it in a new collection tube.
9. For drying column membrane and removing the residual ethanol centrifuge the column at full speed for 1 min.
10. Place the column on the new 1.5 mL tube and add 40 μL of water directly on the membrane, centrifuge 1 min at 10,000 × *g*. Put the tubes with RNA on ice. RNA dissolved in RNase/DNase-free or DEPC-treated water should be stored at −80 °C.

3.4 RNA Quantity and Quality

To evaluate the quantity and quality of extracted RNA measure the quantity and check purity using NanoDrop Spectrophotometer (*see* **Note 14**) and check the quality of RNA using Agilent Bioanalyzer (*see* **Note 15**). The high quality of RNA is very important for microarrays and quantitative PCR analysis.

3.5 Workflow for Sample Preparation and Two-Color Microarray Processing

In this chapter we will go through every step of the Agilent platform microarray procedure and we will give some tips to help readers in their future work (*see* **Note 16**).

Each microarray experiment is a multistep procedure (Fig. 2) (*see* **Note 17**).

1. After removing the RNA samples from freezer and thawing them, put the samples on ice (*see* **Note 18**) and measure the concentration of RNA using NanoDrop Spectrophotometer (*see* **Note 19**).
2. Prepare adequate concentration of RNA in 0.2 mL microcentrifuge tube. The Agilent protocol expects quantity of total RNA between 10 ng and 200 ng in final volume of 1.5 μL. We usually use 100 ng of total RNA (*see* **Note 20**).

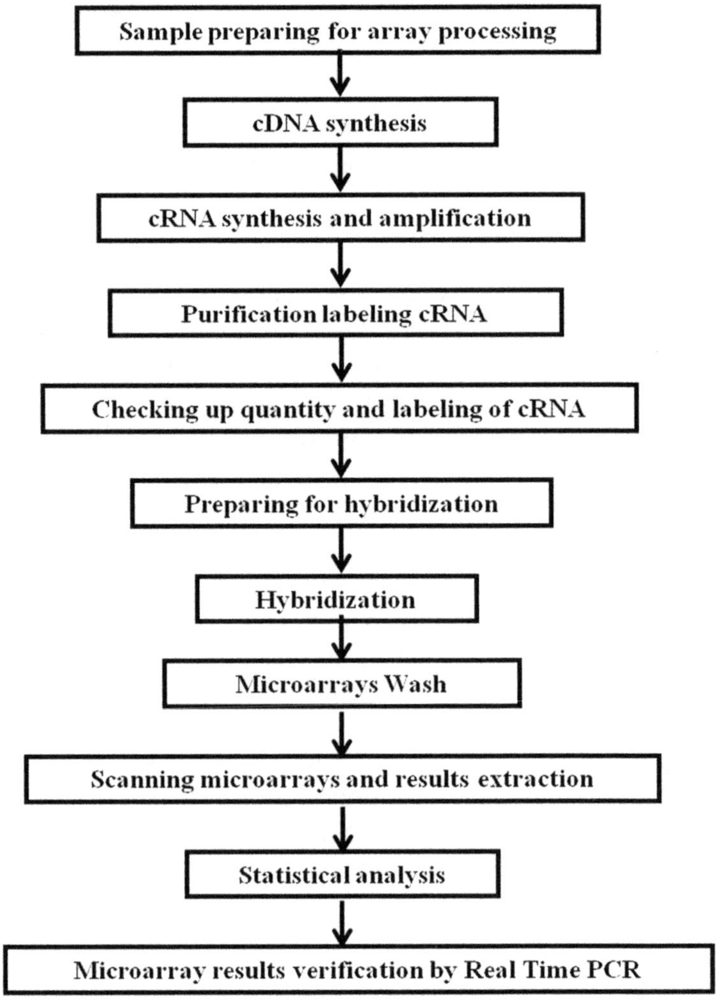

Fig. 2 Multistep procedure of microarray experiment

3. Prepare the Spike-in (positive control) dilution using RNA Spike-In Kit, Two-Color Agilent (*see* **Note 21**). There are two concentrated stock solutions of Spike Mixes (A and B). Spike A Mix is designated for control samples which will be labeled with Cy3 and Spike B Mix is used for investigated samples which will be labeled with Cy5. After thawing the stock solutions of Spike Mixes and Dilution Buffer vortex them vigorously and then heat the Spike Mixes at 37 °C for 5 min in thermoblock. Vortex again and briefly centrifuge (after finishing incubation set the thermoblock at 80 °C). You have to prepare four dilutions for Spike A and Spike B separately, the first three dilutions are the same for every amount of RNA (between 10 and 200 ng) taken to the experiment, and the fourth dilution depends on the taken amount of RNA.

First dilution (1:20): Pipet 38 μL of Dilution Buffer into a sterile 0.5 mL tube and add 2 μL of the Spike Mix stock. Do this dilution parallel for both Spike Mixes A and B in separate tubes. Vortex the solutions accurately, and spin down quickly in a microcentrifuge (to make sure that all of the liquid is at the bottom of the tube).

Second dilution (1:40): Pipet 78 μL of Dilution Buffer into sterile 0.5 mL tube and add 2 μL of the First dilution. Vortex solution accurately, and spin down quickly in a microcentrifuge.

Third Dilution (1:16): Pipet 30 μL of Dilution Buffer into a sterile 0.5 mL tube and add 2 μL of the Second dilution. Vortex solution accurately, and spin down quickly in a microcentrifuge.

Fourth Dilution has to be appropriate for the amount of RNA. The designated dilution for 100 ng of total RNA is 1:2. The volume of the prepared dilution depends on the number of samples. Two microliters of Spike Mix are added to each sample, e.g., for 4 samples 16 μL of the Spike Mix are needed, but it is always better to prepare more, in this example—20 μL. Pipet 10 μL of Dilution Buffer into the sterile 0.5 mL tube and add 10 μL of Third Dilution. Vortex solution accurately, and spin down quickly in a microcentrifuge (*see* **Note 22**).

4. Add 2 μL of the Fourth Dilution into 1.5 μL of prepared RNA. Add Spike Mix (A or B) to the adequate samples. Add Spike Mix A to the sample which will be labeled with Cy3 and spike B to sample labeled with Cy5. The total volume will be 3.5 μL.

5. *cDNA synthesis*: Prepare T7 Primer Mix composed of 0.8 μL T7 Primer Mix and 1 μL Nuclease-free Water for one sample (*see* **Note 23**). Add 1.8 μL T7 Primer Mix to each sample, put them into thermocycler and denature primer at 65 °C for 10 min. Then put the samples at 4 °C for 5 min, briefly spin each sample in a microcentrifuge to remove the content from the tubes' walls and lid. During primer denaturation prepare Master Mix for cDNA synthesis. Briefly spin 5× First Strand Buffer and put it in the thermoblock at 80 °C for 4–5 min to accurately resuspend the buffer components. Afterward briefly vortex and spin the contents of the tube (*see* **Note 24**). Thaw other reagents (0.1 M DTT, 10 mM dNTP Mix), gently vortex, spin and keep at room temperature. Do not vortex the Affinity Script RNase Block Mix (blend of enzymes). Just spin it and keep on ice prior to use. Prepare the cDNA Master Mix just before adding to the samples. Mix 2 μL of 5× First Strand Buffer (*see* **Note 25**), 1 μL DTT, 0.5 μL dNTP Mix, and 1.2 μL Affinity Script RNase Block Mix. Mix by pipetting up and down. Add 4.7 μL of cDNA Master Mix to each sample tube and mix by pipetting up and down. Place the samples in

thermocycler and perform the reaction using the following conditions: 40 °C for 2 h, 70 °C for 15 min, 4 °C for 5 min, hold at 4 °C (incubation at 70 °C inactivates the Affinity Script enzyme). Spin samples briefly after incubation.

6. *cRNA synthesis*: Prepare two Transcription Master Mixes, one with Cyanine 3-CTP and one with Cyanine 5-CTP, in separate microcentrifuge tubes. Do this just before adding the appropriate Transcription Master Mix to the newly synthesized cDNA samples. Thaw the reagents, slightly vortex, and spin down quickly in a microcentrifuge and keep at room temperature. Do not vortex the enzyme, just spin it, and keep on ice. Mix the reagents in proper quantity. The quantities per one sample are as follows: 0.75 μL of nuclease-free water, 3.2 μL of 5× Transcription Buffer, 0.6 μL of 0.1 M DTT, 1 μL of NTP Mix, 0.21 μL of T7 RNA Polymerase Blend, 0.24 μL of Cyanine 3-CTP or Cyanine 5-CTP. Pipet gently the Master Mix up and down. Add 6 μL of appropriate Master Mix with Cy3 or Cy5 to proper samples. Place the samples in thermocycler at 40 °C for 2 h. After incubation spin quickly the tubes with samples.
7. *Purification of labeled cRNA*: Agilent recommends using RNeasy Mini Kit for purification of labeled cRNA (the same kit that we used for RNA extraction). Cool down the microcentrifuge to 4 °C. Use 1.5 mL microcentrifuge tubes and pipet 84 μL of nuclease-free water. Transfer 16 μL of samples from 0.2 mL microcentrifuge tubes to tubes with water. Add 350 μL of RLT Buffer and mix well by pipetting. Add 250 μL of 96% to 100% purity ethanol, mix accurately by pipetting, and transfer the 700 μL of the cRNA sample to RNeasy Mini Spin Column. Centrifuge the sample at 4 °C for 30 s at 14,500 × *g*. Discard the flow-through and collection tube. Transfer the RNeasy column to a new Collection Tube (2 mL) and add 500 μL of RPE Buffer to the column and spin the sample at 4 °C for 30 s at 14,500 × *g*. Discard the flow-through. Reuse the collection tube (*see* **Note 26**). Add again 500 μL of RPE Buffer to the column. Centrifuge the sample at 4 °C for 1 min at 14,500 × *g*. Discard the flow-through and the collection tube. Transfer the RNeasy column to a new Collection Tube (1.5 mL) and spin the sample in the centrifuge at 4 °C for 30 s at 14,500 × *g* to remove any remaining traces of RPE Buffer. Discard this collection tube and use a fresh microcentrifuge tube (1.5 mL). Add 30 μL RNase-Free Water directly onto the RNeasy filter membrane. Wait 1 min, then centrifuge at 4 °C for 30 s at 14,500 × *g*. Put the eluted cRNA on ice. Discard the RNeasy column.
8. *Checking the quantity and labeling of cRNA*: Measure the quantity of cRNA using the Nanodrop. In Nanodrop

Spectrophotometer software choose the tab "Microarray" (this option allows you to measure the concentration of cRNA and cyanines). Blank Nanodrop using water which was used for cRNA elution. Use 1.5–2 μL (not 1 μL) for measurement of samples concentration and make sure that the baseline is a flat horizontal line on the graph (*see* **Note 27**). Clean the sample loading area well with a laboratory wipe after each measuring. Export data from Nanodrop to an Excel file and calculate the yield of cRNA (μg), specific activity (pmol Cy3 or Cy5 per μg cRNA) of each reaction (*see* **Note 28**) and how many μL of cRNA and water you need to prepare for samples hybridization. This amount depends on the type of microarrays slide. For microarray on 8-pack slide 300 ng of cRNA are needed for each sample. Two samples and water give total volume of 19 μL.

9. To prepare samples for hybridization use Gene Expression Hybridization Kit. Prepare the 10× Blocking Agent. This reagent is delivered in a lyophilized form. Add 500 μL of nuclease-free water to the vial, gently vortex. If the pellet does not dissolve completely in water, incubate it for 5 min at 37 °C, vortex again, and spin for a few seconds. (10× Blocking Agent minimizes nonspecific binding). Set the thermoblock at 60 °C (*see* **Note 29**). Prepare the samples for hybridization in 1.5 mL centrifuge tube. At this moment pool two samples (Fig. 3): Cy3-labeled (300 ng cRNA) and Cy5-labeled (300 ng cRNA). These samples are designed to compare the

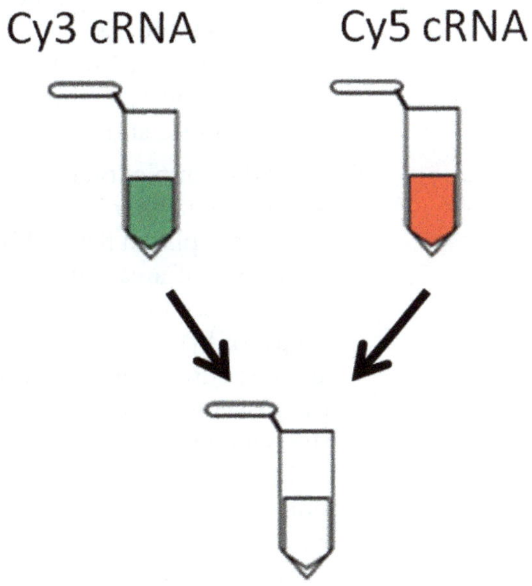

Fig. 3 Pooling samples labeled with Cy3 (control or common reference) and Cy5 (examined)

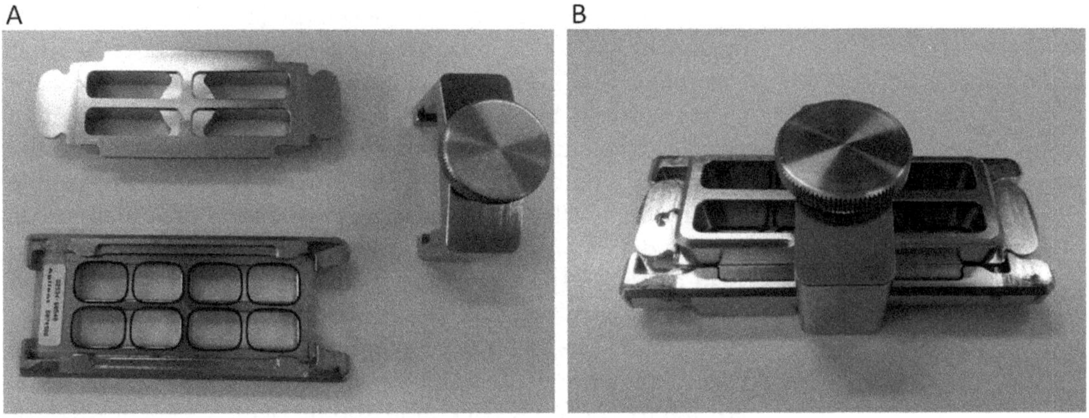

Fig. 4 Chamber for slides: (**a**)—chamber base with loaded gasket for slide with 8 micorarrays; (**b**)—closed chamber ready to put to the hybridization oven

gene expression (e.g., control sample and examined sample). Add adequate volume of water to obtain total volume of 19 μL. Add 5 μL of 10× Blocking Agent and 1 μL 25× Fragmentation Buffer to each probe (Fragmentation buffer reduces cRNA to optimum sized fragments, for hybridization to microarrays). Total volume in the tube is now 25 μL. Incubate samples at 60 °C for exactly 30 min in thermoblock or thermomixer with thermotop to fragment the RNA. Warm the hybridization oven (set the temperature at 65 °C). Immediately after incubation cool the samples on ice for 1 min (*see* **Note 30**). Add 25 μL of 2× Hi-RPM Hybridization Buffer to stop the fragmentation reaction. Mix well by pipetting up and down, but don't make bubbles. Don't mix on the vortex. Spin for 1 min at room temperature at 12,000 × *g* in a microcentrifuge to drive the sample off the walls and lid and to aid in bubble reduction. Put samples on ice and load onto the array as soon as possible.

10. Load the samples on gaskets. Load gasket slide into the Agilent SureHyb chamber base with the label facing up. Make sure that the slide is placed flat and is not ajar (Fig. 4a). Load the sample on center of area limited by gasket. The liquid cannot reach gasket, because this causes leakage of seal. Slowly dispense 40 μL of hybridization sample onto the gasket well in a "drag and dispense" manner. Avoid the introduction of air bubbles to the gasket wells (it is easy to make bubbles at the end of loading the sample when you press pipette bottom to the second resistance). Load microarray slide on gasket slide ("active side" facing the gasket slide) (*see* **Note 31**) Put the SureHyb chamber cover onto the sandwiched slides and slide the clamp assembly onto both pieces. Firmly hand-tighten the clamp onto the chamber (Fig. 4b). Vertically rotate the assembled chamber to wet the gasket and assess the mobility of the

bubbles. If necessary, tap the assembly on a hard surface to move stationary bubbles.

11. *Hybridization*: Load each assembled chamber into the oven rotator rack. Set hybridization rotator to rotate at 10 rpm and hybridize at 65 °C for 17 h.
12. For microarray wash we use Gene Expression Wash Buffer Kit (Agilent) consisting of two buffers: Gene Expression Wash Buffer 1 and Gene Expression Wash Buffer 2. Pre-warm the Wash Buffer 2 at 37 °C in a water bath on the night before washing the arrays. Prepare three slide-staining dishes. Fill two slide-staining dishes with Gene Expression Wash Buffer 1 at room temperature. Put slide rack into one of them. Remove hybridization chambers from hybridization oven. Put the hybridization chamber assembly on a flat surface and with gloved fingers, remove the array-gasket sandwich from the chamber. Transfer the sandwich to the slide-staining dishes with Wash Buffer 1. Keeping still between fingers submerge the array-gasket totally and slip one of the blunt ends of the forceps between the slides. Gently turn the forceps upward or downward to separate the slides. Let the gasket slide drop to the bottom of the staining dish. Transfer the microarray slide to the next slide-staining dish with Wash Buffer 1, put it into slide rack and using magnetic stirrer wash the slide 1 min (*see* **Note 32**). Fill the third slide-staining dish with Wash Buffer 2 warmed to 37 °C. After 1 min immediately transfer the slide rack to third slide-staining dish and wash for 1 min on magnetic stirrer. Remove very slowly the slide rack from the buffer. Slow movement in vertical direction should allow you to receive a dry microarray slide without droplets. If liquid remains on the bottom edge of the slide, dab it on a cleaning tissue. Load microarray slide into the slide holder, "active side" microarray surface facing upward, put an ozone-barrier slide cover on the top of the array and close the cover holder (*see* **Note 33**). After washing the microarrays remember to clean very carefully the staining dishes, rack, and stir bars to prepare the equipment for next experiments. It is necessary to remove any traces of contaminating material to avoid wash artifacts on your slides. Do not use any detergent to wash this equipment. Run copious amounts of Milli-Q water through the staining dish. Pour out completely and fill again with Milli-Q water. For next use dishes have to be dry and clean (*see* **Note 34**).
13. *Scanning microarrays*: Turn on Microarray Scanner before starting washing slides to warm up the lasers. We use Agilent DNA Microarray Scanner G2505C, so the next points of the following protocol will describe procedures devoted to this equipment. In Scan Control Agilent Software select appropriate protocol: AgilentG3_GX_2color for G3, format 8×60K,

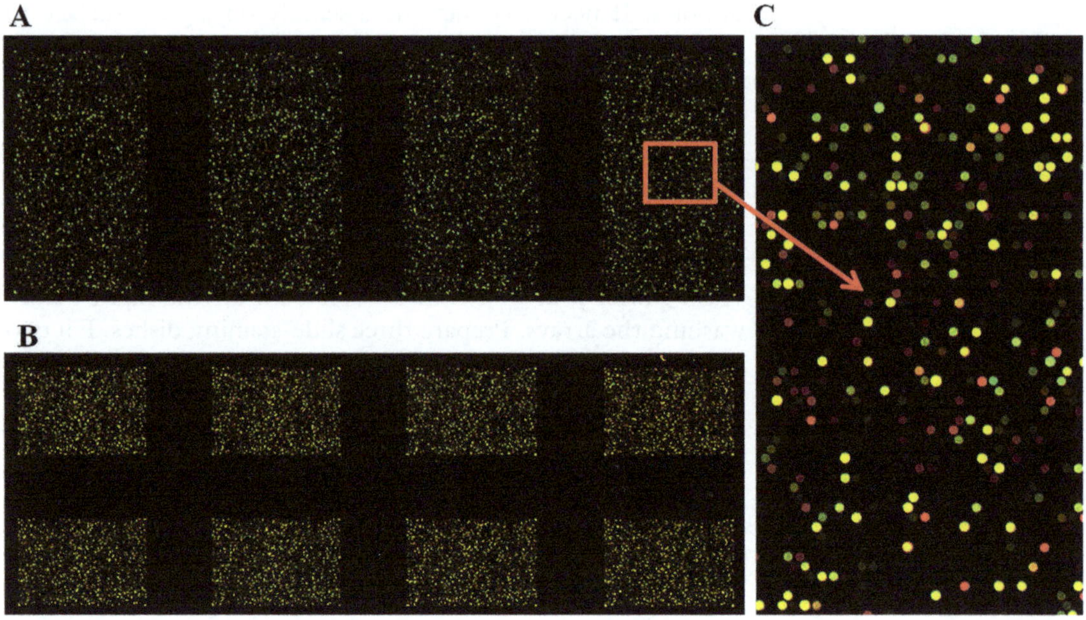

Fig. 5 Scan images (TIFF images) of two-color microarray slides generated by Scan Control Software after scanning: (**a**)—4×44K slide; (**b**)—8×60K slide; (**c**)—enlarged fragment of picture A

AgilentHD_GX_2color for HD format 4×44K and correct resolution for given format: for 8×60K is 3 μm and for 4×44K 5 μm (*see* **Note 35**). Scan the microarrays using Scan Agilent Software immediately after washing slides to avoid the degradation of cyanine dyes and signal intensity. Scanner reads out fluorescence intensity of hybridized RNA (*see* **Note 36)** and then visualizes it generating scan image—TIFF image (Fig. 5).

14. Data preprocessing and extraction is done by Agilent Feature Extraction (FE) Software. The FE program automatically assigns a default grid template and protocol for each extraction set. If the software does not assign the grid, it is possible to download it from the eArray website (*see* **Note 37**).

3.6 Statistical Analysis of Microarray Data

For further analyses of the microarray data we used Gene Spring Software, which is commercial software dedicated for microarray analyses.

1. First open a new project and create new experiment (select analysis type: Expression, Experiment type: Agilent Expression two color, Work flow type: Analysis Biological Significance or Data Import Wizard). Analysis option: Biological Significance is dedicated for new users and is designed to assist the user step by step through the creation and basic analysis of microarray experiment, but does not give a possibility to change the

default settings. We used the second option of analysis: Data Import Wizard, which offers the user a variety of options, and the user decides, which options will be chosen during microarray analysis (flag options can be changed, raw signal thresholding can be altered, there are options for baseline transformation of the data and for creating different interpretations).

2. Load the txt. files generated by the Feature Extraction.
3. Perform baseline to median of all samples. (For each probe the median of the log summarized values from all the samples is calculated and subtracted from each of the samples.)
4. Define the parameters of experiment (e.g., treatment, time, etc.) and group the samples.
5. Create interpretation, determine how the samples would be grouped into experimental conditions for display and used for analysis, e.g., if there are more parameters and groups of samples than one, it is possible to choose all of them or some of them. Any combination of these conditions can be chosen to form an interpretation. Create both a non-averaged and averaged interpretation at the same time by selecting the appropriate option in the dialog window. In following analysis the program will choose adequate option, e.g., while performing statistical analysis, the interpretation that is used is always the non-averaged interpretation.
6. Quality control (QC) allows you to decide which samples are ambiguous and which are passing the quality criteria. Based on the QC results, you can remove unreliable samples from any downstream analysis. The QC dialog displays three tiled windows: 3D PCA scores (Fig. 6), Quality Control Metrics—Report (the same tables are generated by Feature extraction shown in Table 7), Quality Control Metrics—Plot (graphical presentation of the previous Quality Control Metrics—Report table). Principal Component Analysis (PCA) is used to simplify the analysis and visualization of multidimensional data sets, calculates the PCA scores, and visually presents them in a 3D scatter plot. The scores are used to check data quality. This allows viewing of separations between groups of replicates. Ideally, replicates within a group should cluster together and separately from arrays in other groups (Fig. 6).
7. Filter probe sets by flags to remove poor quality probes (absent flags). Select Entity List: All Entities and Interpretation—Non-averaged. Choose flag values that an entity must satisfy to pass by defining the acceptable flag. Mark Detected (Present), do not mark Not Detected (Marginal), and Compromised (Absent). Select the minimum number of samples out of all samples which must pass the filter (e.g., at least 1 out of 8 samples has acceptable values).

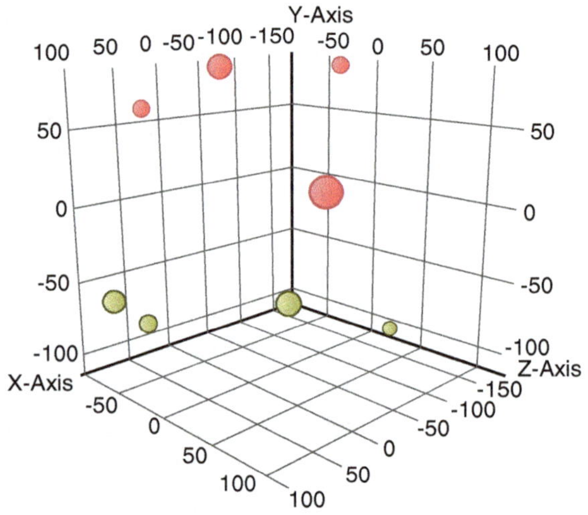

Fig. 6 3D PCA (Principal Component Analysis) score. This graph visualizes relations between variables and assesses the existence of groups of individuals and groups of variables. In that regard, it can be used to see if the sample groups separate according to prior knowledge. One dot corresponds to one microarray. Microarrays representing one investigated group are shown as dots of the same color. This graph clearly shows two investigated groups: red dots are grouped on the top of the plot and the yellow dots are gathered at the bottom of the plot. The groups are visibly separated

8. If statistical analysis does not indicate gene, which is significantly differentially expressed (DE), it is possible to limit entities and use filtering by error on the standard deviation or coefficient of variation (CV). We used filtering by CV $<50.0\%$ to remove genes in which case the samples showed high variability.
9. *Statistical analysis*: Choose the Entity list (the list after filtering, e.g., filtering by error) and interpretation (Non-averaged) and then select conditions and their order, as well as appropriate test (Moderated *t*-test, *t* test unpaired, *t*-test paired, Mann-Whitney unpaired, Mann-Whitney paired—depends on the type of experiment, for comparing more than two groups use Anova test) (*see* **Note 38**). Afterward apply multiple testing correction Benjamini Hochberg False Discovery Rate (FDR) (*see* **Note 39**).
10. After statistical analysis export the data as a text file or Excel file for further analysis. These files include statistically significant differentially expressed (DE) genes and other information: probe Name, level of *P* value, *P* after correction, fold change, log fold change, direction of gene expression regulation (up/down). It is also possible to add Gene symbol, Gene Description, Sequence of probe, Ensembl ID, Entrez Gene ID, Gene Bank Accession Gene Name raw signal values, Normalized signal values.

11. Preselect differentially expressed genes for further analyses, based on the *P* value ($P \leq 0.05$) and/or differential fold-change value (*see* **Note 40**).
12. Choose fold-change cutoff value for the results. Keep in mind that different fold-change cutoffs can affect interpretation of the data (*see* **Note 41**).

3.7 Bioinformatics Analyses—General Instructions

1. Choose a bioinformatics tool for data analysis (*see* below for examples).
2. Prepare microarray data in the form of gene list containing official gene symbols and upload the gene list according to guidelines provided by the program/software.
3. Follow a protocol/tool manual provided by the program to perform classification and analysis of results.
4. Examine the microarray data set or selected gene subset(s) in terms of potential gene network important in myogenesis. Identify highly connected nodes, which correspond to key genes/proteins involved in potential mechanisms of experimental factor(s) effects (see below for example).
5. Localize proteins encoded by experimentally modified genes in the cell, in order to establish their particular biological role in myogenesis.

Simplified examples of step-by-step bioinformatics analyses and data presentation, using different software and databases are given below.

3.8 Ontological Analysis Using PANTHER Classification System

1. Let's consider a hypothetical set of genes ($n = 2463$) with expression modified by an experimental factor. Gene Ontology (GO) analysis using the PANTHER Classification System assigns these genes to different GO categories in terms of their involvement in biological processes (*see* **Note 42**). Ten-top list of gene ontology terms (accession category—Biological Process) enriched in genes provided on the hypothetical gene list is presented in Table 1 (*see* **Note 43**).
2. Drill down to child categories of GO Cellular process (the most enriched gene term on the ten-top list in Table 1), according to the instruction provided by the PANTHER system. The result of this step of analysis is presented in Fig. 7a.
3. Repeat the previous step to obtain child categories of GO Cell communication (the most enriched gene term on the ten-top list in Fig. 7a). The result of this step of analysis is presented in Fig. 7b.
4. Choose more enriched ontological term. The selected gene set is annotated to "Signal transduction (GO:0007165, $n = 303$)

Table 1
Ten-top list of gene ontology terms (accession category—Biological Process) enriched in genes provided on the hypothetical gene list. The analysis was performed using PANTHER Classification system

No	Term and GO number	Gene count	Percentage
1.	Cellular process GO:0009987	1051	46.8
2.	Metabolic process GO:0008152	731	32.6
3.	Biological regulation GO:0065007	348	15.5
4.	Localization GO:0051179	277	12.3
5.	Response to stimulus GO:0050896	275	12.3
6.	Cellular component organization and biogenesis GO:0071840	261	11.6
7.	Developmental process GO:0032502	234	10.4
8.	Multicellular organismal process GO:0032501	196	8.7
9.	Immune system process GO:0002376	44	2.0
10.	Biological adhesion GO:0022610	43	1.9

Functional ontology analysis of the uploaded gene list ($n = 2463$) identifies 2244 genes annotated to indicated species (*Rattus norvegicus*) and reveals 13 ontology terms. The most enriched ten-top list is presented as an example. Gene count—number of genes from the hypothetical list classified to particular term, Percentage—percent of genes involved in the term against total genes

and it will serve as an example for illustrating further bioinformatics tools (*see* **Note 44**).

5. Perform further classification of genes annotated to "Signal transduction" in terms of their molecular function. The analysis reveals 7 categories (presented on Fig. 7c), and the gene category named "Binding" (molecular function GO:0005488) is the most enriched ($n = 181$).
6. Continue functional classification and find child categories of gene subset annotated to "Binding." The identified subcategories (7) are presented in Fig. 8a.
7. Choose the most enriched gene category ("Protein Binding", GO:0005515, $n = 154$) for further identification of its child categories (Fig. 8b).
8. Select the most interesting (or the most enriched) group ("Receptor binding," GO:0005102, $n = 64$) and, again, drill down child categories. In this step the categories of the lowest level of classification are identified (Fig. 8c).
9. Retrieve a gene list for each gene group (*see* **Note 45**). An example of a gene list is presented in Table 2. Each individual gene name/symbol is linked to the Panther Gene Information page, which provides summary information and enables

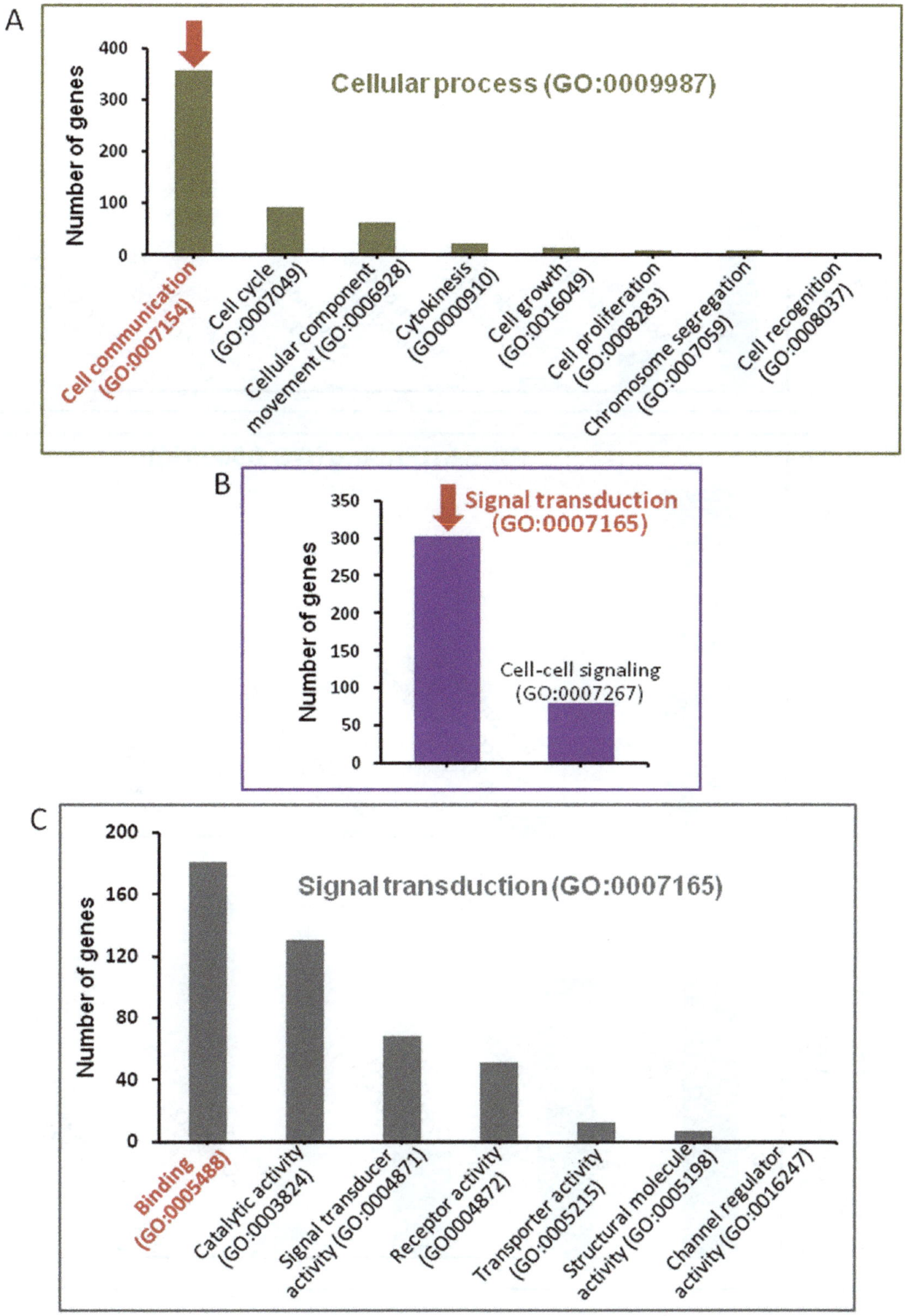

Fig. 7 An example of step-by-step gene ontology analysis in terms of involvement in biological processes and molecular functions, performed using Panther Classification System. (**a**–**c**)—Subsequent levels of analysis present gene ontology groups enriched in genes from the hypothetical list. Gene category indicated by red block arrows and described in red serve to drill down to child gene categories

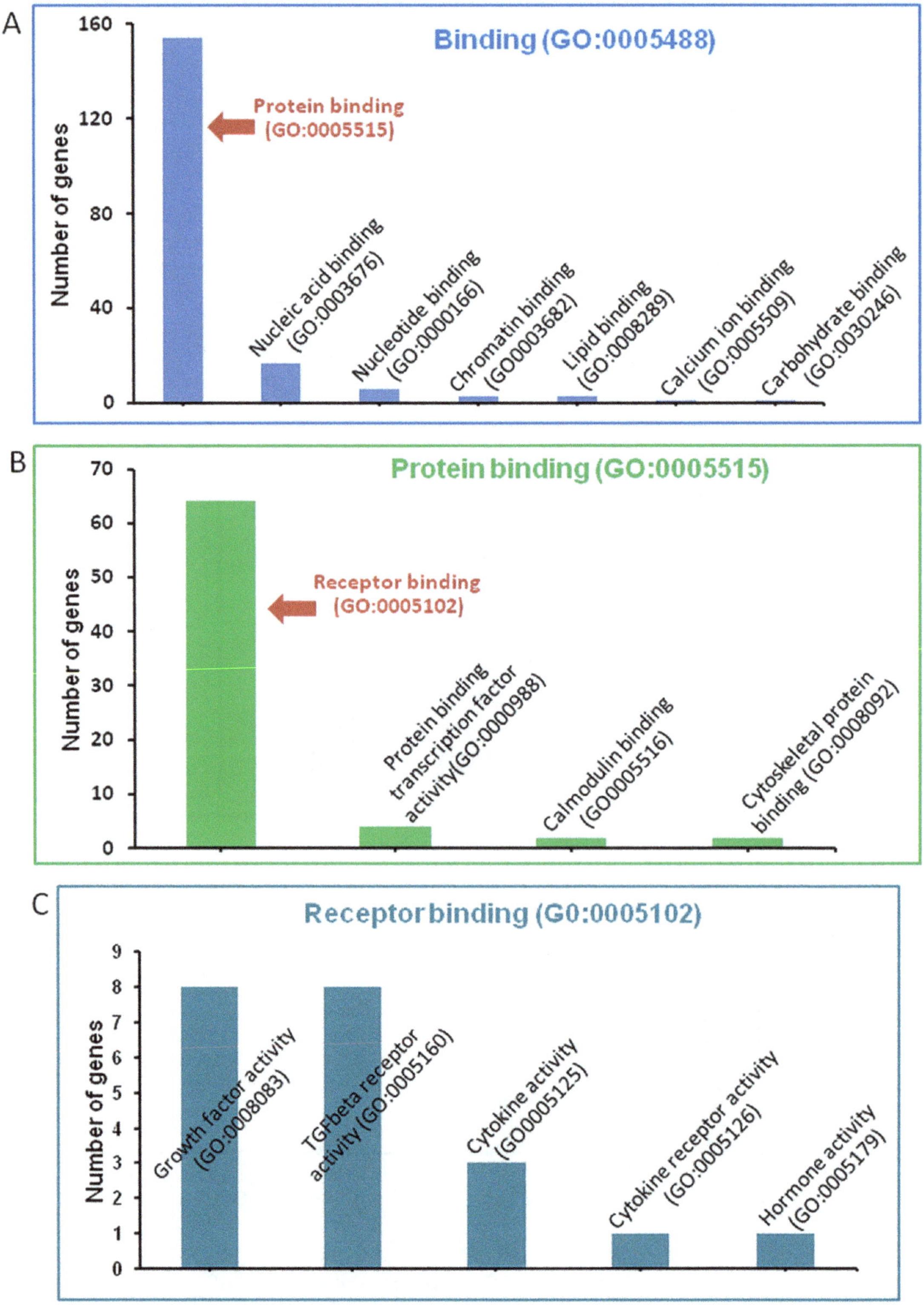

Fig. 8 An example of step-by-step gene ontology analysis in terms of molecular function, performed using Panther Classification System. A-C—Subsequent levels of analysis present gene ontology groups enriched in genes from the hypothetical list. Gene category indicated by red block arrows and described in red serve to drill down to child gene categories

Table 2
A simplified example of gene list classified to "Receptor binding" (GO:0005102, total gene number $n = 64$) retrieved from PANTHER classification system

No	Mapped IDs	Gene name; Gene symbol	Panther protein class (PC)
1.	MSTN	Growth/differentiation factor 8; Mstn	Growth factor (PC00112)
2.	NRG1	Pro-neuregulin-1, membrane-bound isoform; Nrg1	Growth factor (PC00112)
3.	GNA12	Guanine nucleotide-binding protein subunit alpha-12; Gna12	Heterotrimeric G-protein (PC00117)
4.	LRIG2	Leucine-rich repeats and immunoglobulin-like domains 2; Lrig2	Extracellular matrix protein (PC00102); receptor (PC00197)
5.	PROK2	Prokineticin-2; Prok2	Peptide hormone (PC00179)
6.	SEMA6D	Semaphorin 6D; Sema6d	Membrane-bound signaling molecule (PC00152)
7.	ITGB3	Integrin beta; Itgb3	Cell adhesion molecule (PC00069); extracellular matrix glycoprotein (PC00100); receptor (PC00197)
8.	BDNF	Brain-derived neurotrophic factor; Bdnf	Neurotrophic factor (PC00163)
9.	CBLB	E3 ubiquitin-protein ligase CBL-B; Cblb	Ligase (PC00142)
10.	PLCL2	Phosphoinositide phospholipase C; Plcl2	Guanyl-nucleotide exchange factor (PC00113); signaling molecule (PC00207)

Mapped ID, Gene symbol and Protein class name for individual genes are linked to appropriate pages for additional informations (*see* Table 3 for an example)

detailed exploration of gene/protein biological roles (*see* Table 3 for an example).

10. Choose interesting genes for further investigation and for validation microarray analysis using quantitative PCR (see below).

3.9 Functional Gene Annotation Using DAVID Program

1. Let us consider a hypothetical set of genes associated with signal transduction, indicated in Fig. 7b ($n = 303$).
2. Upload the list of genes to the DAVID database and perform functional annotation analysis in terms of molecular function or cellular component (*see* **Note 46**). The classification reveals numerous gene categories and the functional annotation chart created by DAVID provides information as follows: Name of annotation category (BP—Biological process, MF—Molecular

Table 3
A simplified example of summary information retrieved from PANTHER classification page for myostatin (Mstn)

PANTHER family	TGFbeta family (PTHR11848)
PANTHER subfamily	Growth/Differentiation Factor 8 (PTHR11848, SF150)
PANTHER GO Molecular function	Transforming growth factor beta receptor binding
PANTHER GO biological process	MAPK cascade, apoptotic process, cell differentiation, protein phosphorylation, regulation of phopshate metabolic process, response to stimulus, transmembrane receptor protein Ser/Thr kinase signaling pathway
PANTHER GO cellular component	Extracellular space
PANTHER protein class	Growth factor
Pathway categories	TGFbeta kinase signaling pathway>> transforming growth factor beta

function, CC—Cellular component), name of the gene term, number of genes from uploaded gene list that are involved in the corresponding term, percentage (percent of genes involved in the category against total genes), *P*-value (the smaller, the more enriched terms associated with uploaded gene list). The simplified example of data presentation after the DAVID analysis is given in Table 4.

3. Perform functional clustering of the hypothetical gene list using DAVID database (*see* **Note 47**). Biological significance of each cluster is indicated by enrichment score value (the higher value, the more enriched cluster) and *P*-value for particular annotation members within a cluster (*see* **Note 48**). A simplified example of functional annotation clustering report including three-top clusters and the two last clusters from a total list is presented in Table 5.
4. Analyze the hypothetical gene list in terms of gene contribution to signaling pathways. The DAVID pathway mapping enables allocating individual genes (provided on a gene list) on static pathway map generated by BioCarta or KEGG databases. Functional annotation analysis of the hypothetical gene list ($n = 303$) reveals 101 KEGG pathways enriched with uploaded genes (*see* Table 6 for examples). Individual name of each pathway in the report is linked to the static map, where the genes from uploaded gene list that belong to the pathway are blinking as red asterisks (*see* Fig. 9 for an example).

Table 4
Ten-top list of gene ontology terms (annotation category–cellular component) enriched in genes provided on a hypothetical gene list. The analysis performed using a functional annotation chart (DAVID)

No	Term	Gene count	Percentage	*P* value
1.	Cytoplasm	124	41.3	1.0E-7
2.	Nucleus	90	30.0	4.2E-2
3.	Plasma membrane	88	29.3	3.0E-4
4.	Extracellular exosome	58	19.3	6.5E-3
5.	Extracellular space	29	9.7	5.9E-2
6.	Focal adhesion	25	8.3	1.6E-8
7.	Cell surface	25	8.3	3.5E-5
8.	Perinuclear region of cytoplasm	25	8.3	7.5E-5
9.	Golgi apparatus	20	6.7	6.0E-2
10.	Early endosome	13	4.3	1.4E-4

Functional annotation analysis of the uploaded gene list ($n = 303$) reveals 67 terms within the category "Cellular component". The most enriched ten-top list is presented as an example

3.10 Analysis and Visualization of Gene Interactions Using Pathway Studio (See Note 49)

1. Import the hypothetical list of genes ($n = 303$) to pathway studio in order to find gene networks and interactions among genes, proteins, other molecules, cellular processes, and diseases (*see* **Note 50**). The first step of gene network analysis (hardly readable, even enlarged) is presented in Fig. 10a, b. However, there are numerous highly connected genes in this network that can be selected for further investigation. Each node and particular gene is linked to specific information provided by the database.
2. Choose elements or interactions involved in regulation of myogenesis for further exploration. Let us consider myostatin (Mstn), a negative growth factor for muscle cells and a known inhibitor of myogenesis, as an example. Find Mstn in the first-step network presented in Fig. 10a (in the middle of the network, encircled in black).
3. Translocate Mstn symbol outside the network to clearly visualize interactions with other network elements. Such modification reveals 9 genes/proteins interacting with Mstn (Fig. 10b, lower panel). According to this picture Mstn: (1) positively regulates CASP3 (caspase 3), NOS1 (nitric oxide synthase 1), CBLB (Cbl proto-oncogene B) expression, (2) negatively regulates IRS1 (insulin receptor substrate 1) and IGFR1 (insulin-like growth factor 1 receptor) expression, (3) its expression is

Table 5
A short list of selected functional annotation clusters enriched in genes provided on the hypothetical gene list ($n = 303$). The analysis performed using a functional annotation clustering (DAVID)

Annotation cluster 1 enrichment score: 13.31	Count	*P* value
Protein kinase-like domain	51	3.9E-22
Protein kinase, catalytic domain	49	3.6E-22
Kinase	49	7.6E-20
Protein kinase, ATP binding site	40	6.8E-19
Protein phosphorylation	41	1.2E-13
Nucleotide-binding	62	3.2E-14
Serine/threonine protein kinase	28	1.7E-13
Transferase	51	6.2E-9
ATP binding	57	3.4E-10
Annotation cluster 2 enrichment score: 7.35		
Chemokine signaling pathway	27	6.7E-13
Signal transducer activity	20	2.7E-9
Cholinergic synapse	17	5.6E-8
Glutamaergic synapse	17	7.7E-8
Heterotrimeric G-protein complex	8	3.2E-5
Dopaminergic synapse	15	8.2E-6
Annotation cluster 3 enrichment score: 6.98		
Pleckstrin homology domain	26	2.8E-11
Pleckstrin homology-like domain	28	1.9E-8
Positive regulation of GTPase activity	25	8.1E-6
Rho guanyl-nucleotide exchange factor activity	10	1.3E-4
Regulation of rho protein signal transduction	9	4.1E-3
Annotation cluster 59 enrichment score: 0.29		
Zinc-finger, RING-type	6	3.8E-1
Zinc-finger, RING/FYVE/PHD-type	8	4.5E-1
Protein ubiquitination	5	8.3E-1
Ubiquitin mediated proteolysis	3	8.4E-1
Annotation cluster 60 enrichment score: 0.05		
Ankyrin repeat	3	8.7E-1
Ankyrin repeat-containing domain	3	8.8E-1

Functional annotation clustering analysis of uploaded gene list reveals 60 clusters. The three-top and the two last clusters are presented for comparison of analysis indices. Count—the number of genes from the hypothetical list that enrich particular category

Table 6
A short list of selected KEGG pathways enriched in genes provided on hypothetical gene list ($n = 303$). The Functional Annotation analysis in terms of involvement in KEGG pathways performed using DAVID

Category	Term (pathway name)	Count	Percentage	*P* value
KEGG_Pathway	Chemokine signaling pathway	27	9.0	2.5E-14
KEGG_Pathway	Regulation of actin cytoskeleton	28	9.3	5.3E-13
KEGG_Pathway	Insulin signaling pathway	16	5.3	7.1E-7
KEGG_Pathway	Cytokine-cytokine receptor interaction	12	4.0	1.1E-2
KEGG_Pathway	TGF-beta signaling pathway	7	2.3	1.4E-2

Functional Annotation analysis of uploaded gene list reveals 101 KEGG pathways enriched with uploaded genes. Individual name of each pathway in the report is linked to the appropriate static map, where the genes from uploaded gene list are indicated as red asterisks (*see* Fig. 8 for examples)

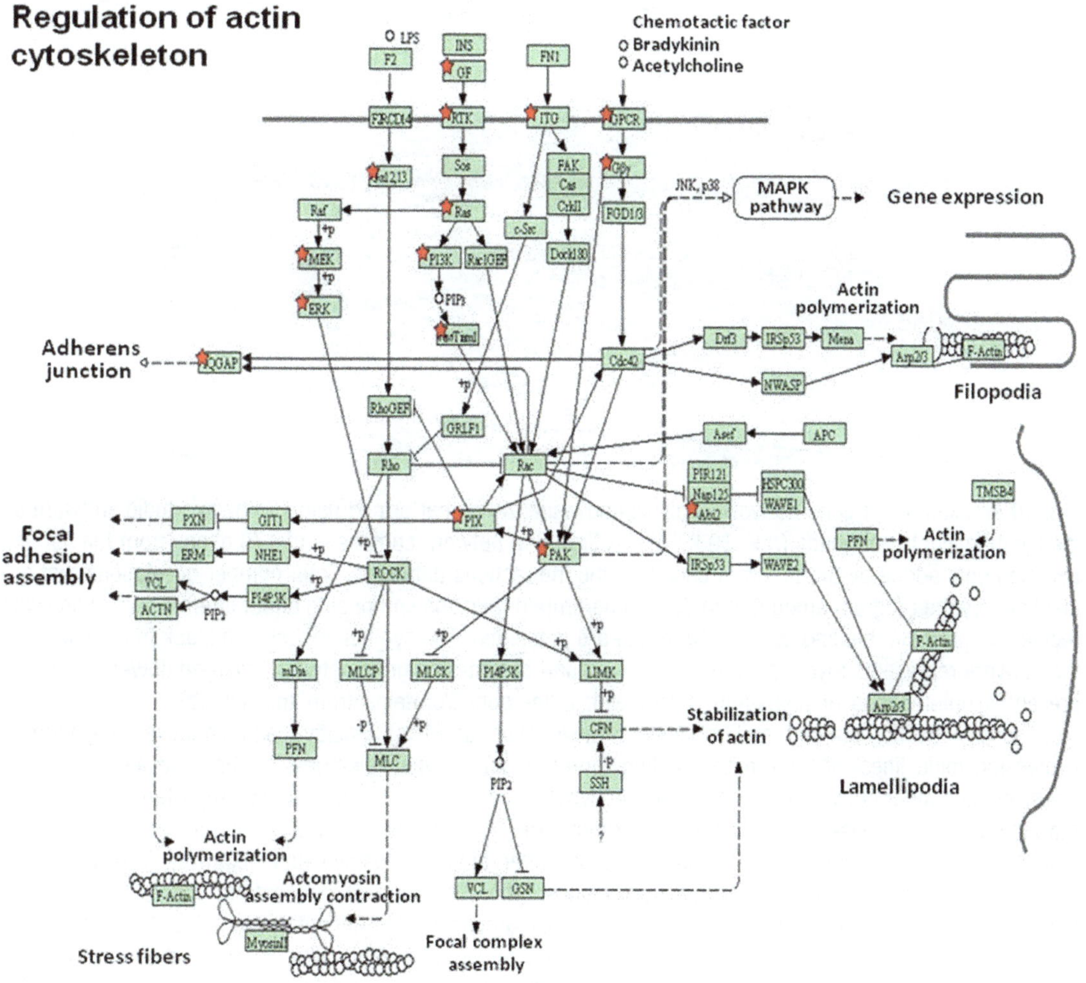

Fig. 9 An example of DAVID Pathway Mapping, allocating individual genes from the hypothetical list on a static pathway map generated by KEGG databases. Presented pathway map is retrieved from DAVID [4]. Individual genes from uploaded list contributed to the pathway are indicated as red asterisks

A

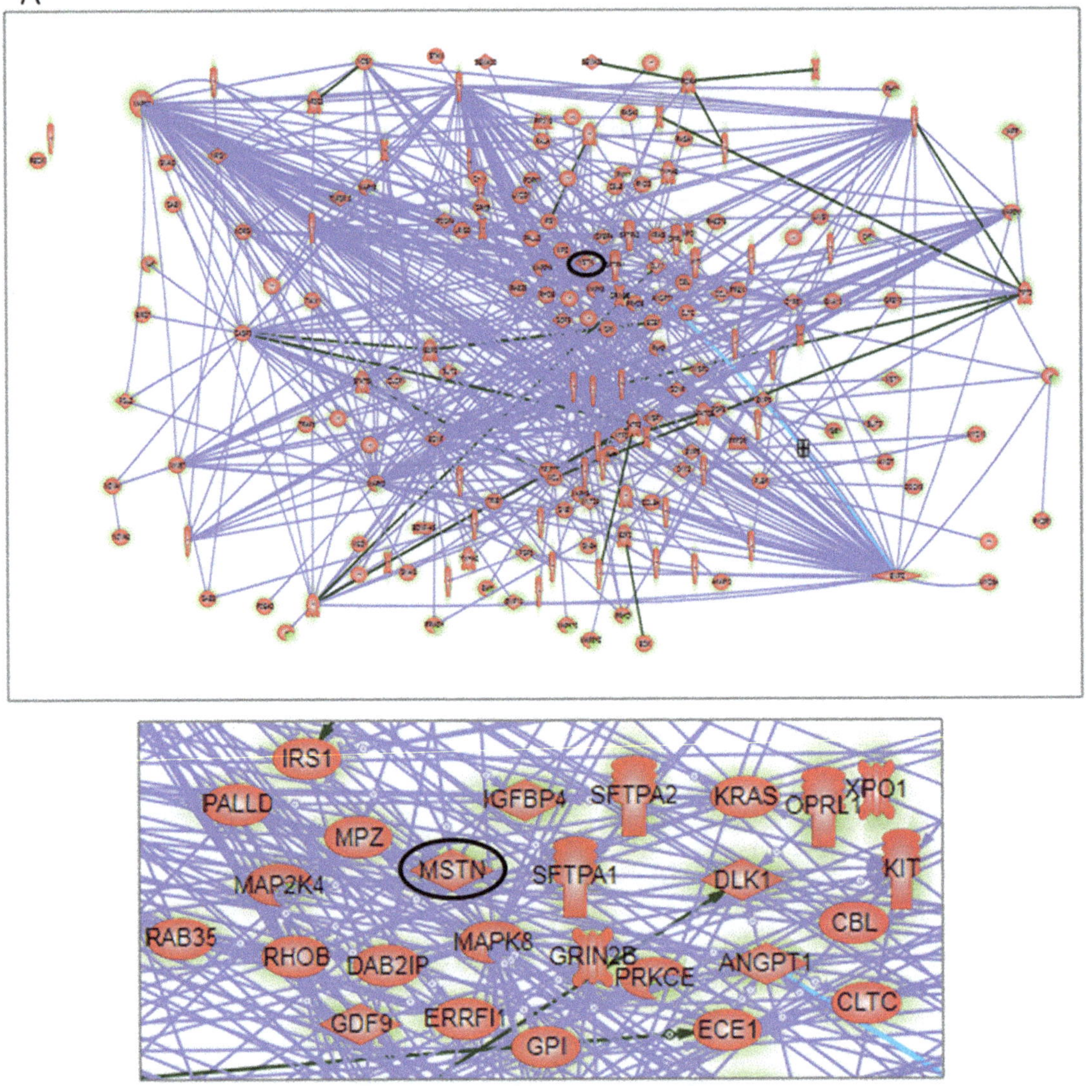

Fig. 10 An example of gene network analysis and visualization performed using Pathway Studio software on the hypothetical list of genes ($n = 303$). (**a**)—A first-step network composed of 179 genes from the list and two elements added by the program based on the interactions database. Total number of relations that are identified by the program amounted to 550. Gene/protein relations comprising direct interactions, expression regulation, promoter binding, and miRNA effect are presented. The symbol encircled in black in the middle of the network represents myostatin (upper panel). A part of first-step gene network enlarged (lower panel) and presenting interactions of myostatin (Mstn) with other genes/proteins from the list. (**b**)—A part of first-step gene network enlarged and modified (translocation of Mstn). Positive and negative regulation of expression (blue lines) and promoter binding (green lines) between elements of the network are visible. Myostatin and other network elements from the hypothetical gene list connected with myostatin according to the interactions database are highlighted in orange. (**c**)—A simplified schema of interactions among some genes from the hypothetical list, identified as promoter binding (positive and negative). This picture of interactions presents 25 elements and 18 relations. (**d**)—Allocation of genes from the hypothetical list (highlighted and encircled in blue) in FGFR (Fibroblast Growth Factor Receptor) signaling, as an example, and localization of proteins encoded by these genes in the cell. Molecular function, different types of interactions between network elements, and their contributions to functional protein classes are indicated

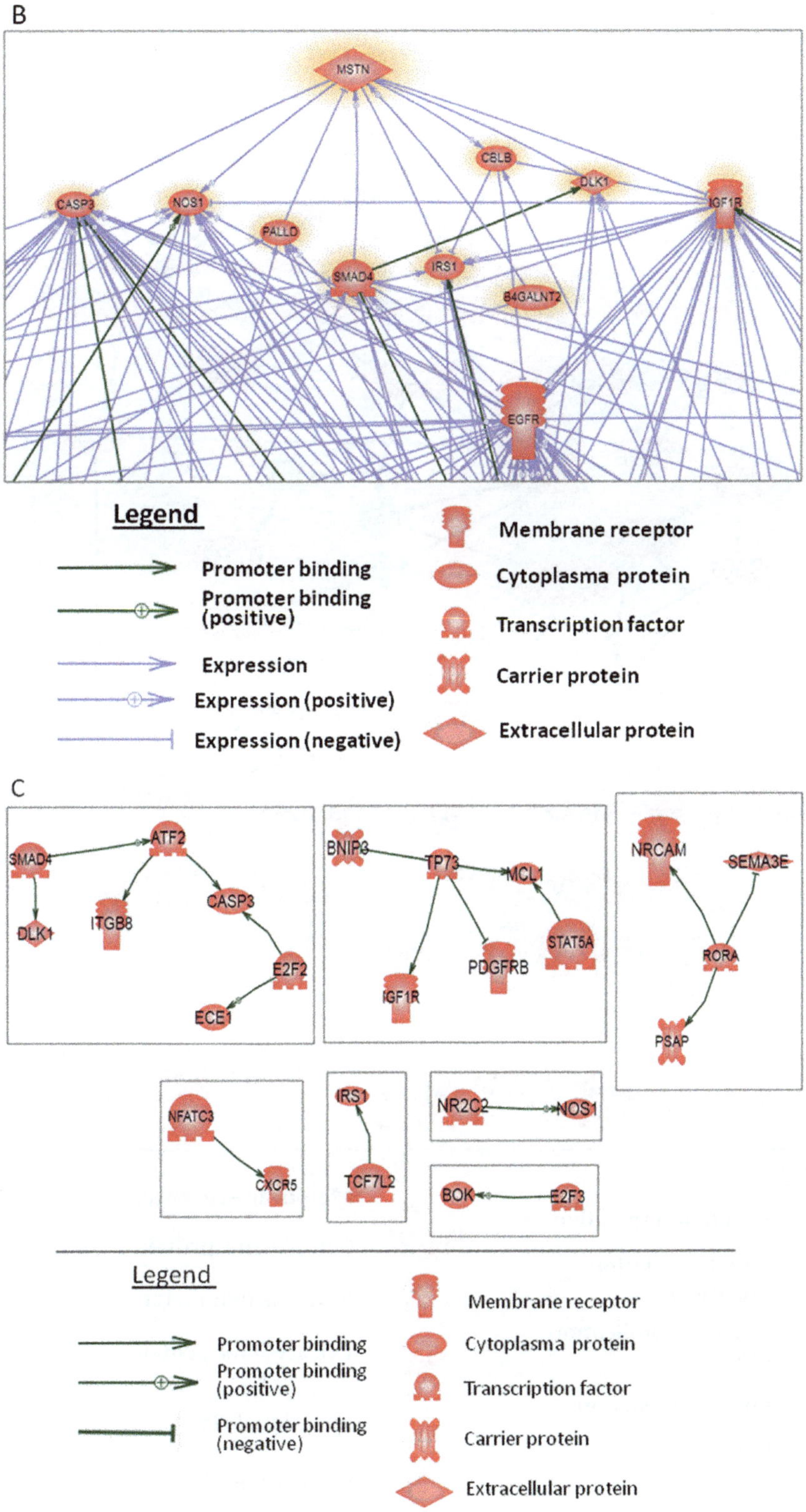

Fig. 10 (continued)

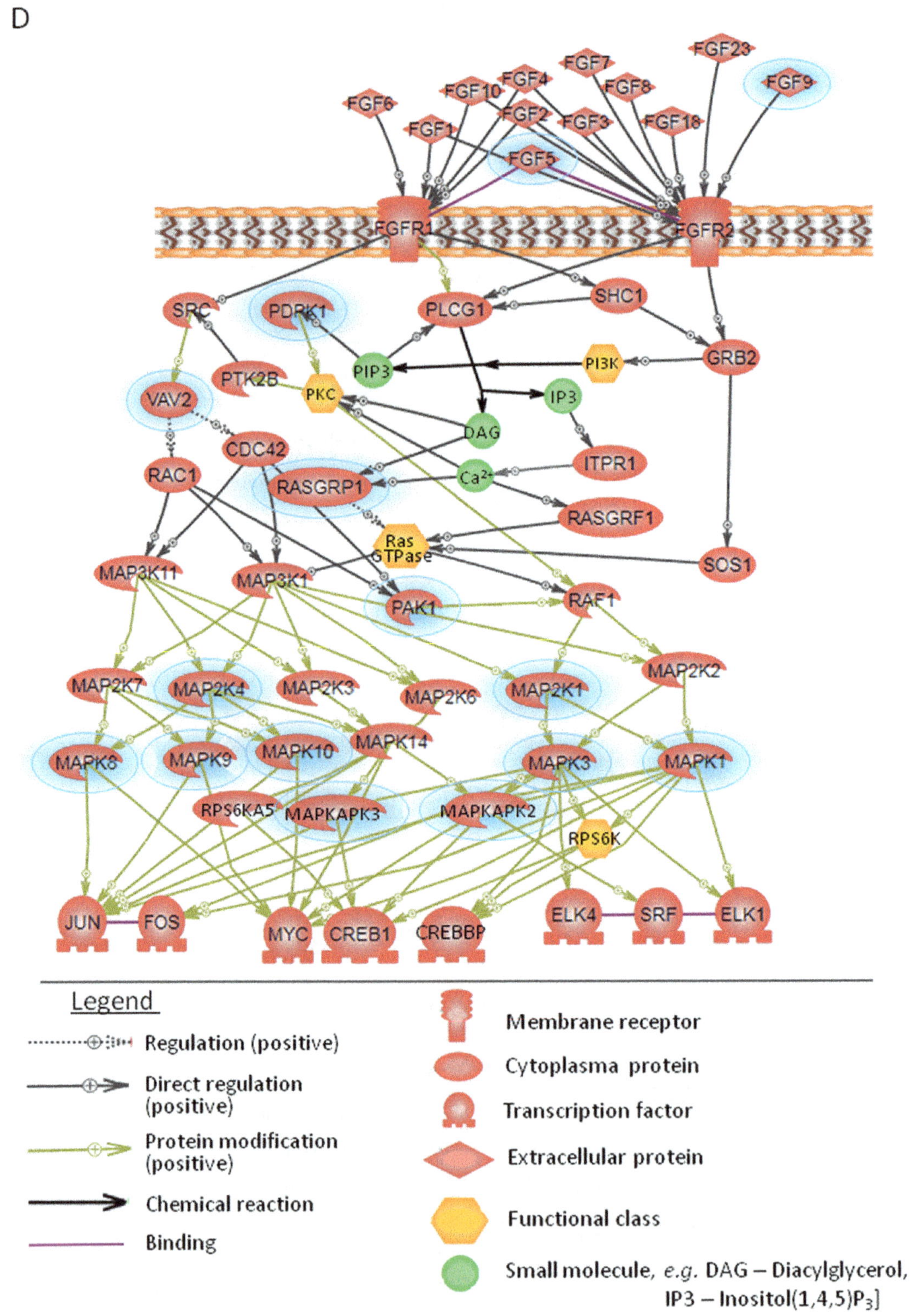

Fig. 10 (continued)

positively regulated by SMAD4 (Sma and Mad family member 4) and B4GALNT2 (beta-1,4-N-acetylgalactosaminyltransferase 2), (4) its expression is negatively regulated by PALLD (palladin). These genes/proteins are highly connected with the rest of the network, for example, with EGFR (epidermal growth factor receptor), SMAD4 regulates DLK1 (delta-like 1 homolog) via promoter binding etc.

4. Simplify the network of interactions by removing elements of minor interest in the context of myogenesis.
5. Consider one type of interaction, for example, promoter binding, and extract it from the first-step network. The simplified picture of gene network based on promoter binding is presented in Fig. 10c.
6. Allocate proteins encoded by genes from the hypothetical list in signaling pathways potentially involved in regulation of myogenesis. Localize these proteins in the cell, in order to establish their particular importance and mechanisms of action (Fig. 10d).
7. Retrieve detailed description, functional classification, and interaction of gene/protein of interest from database. Each element and interaction is linked to appropriate articles in PubMed, providing important information, relevant sentences, and further references.

3.11 Validation of Microarray Data—Further Biological Analyses

In order to verify the hypothesis based on microarray data and to warrant conclusions drawn from this analysis, gene array results should be confirmed by additional biological assays directed to experimentally regulated pathways or processes identified by the gene expression. First, transcriptomic profiling obtained by microarray analysis should be validated by quantitative PCR.

1. Select some genes, which manifest altered expression according to microarray data (*see* **Note 51**).
2. Apart from qPCR analysis of selected genes, consider performing: (1) immunoblotting analysis to verify potential changes at the level of protein products (*see* **Note 52**) or (2) cellular protein phosphorylation (e.g., protein kinase activity) or (3) cellular transport (e.g., glucose uptake) or (4) protein synthesis and/or proteolysis or (5) cell migration ability or (6) myotube size. The analyses listed above are linked with several processes associated with myogenesis and the results can serve for proving potential pro-myogenic or anti-myogenic effects of experimental factor(s).

4 Notes

1. You have to use fresh gloves and clean the area where you will work (the bench/laminar flow hood and all equipment) with 70% ethanol, and wipe the surfaces with RNase removing cleaning agent.
2. Employ reliable biological assay to justify the stage of myogenic cell development chosen for the study. For example, if the microarray data obtained in the experiment may be considered as a transcriptional signature of differentiating myoblasts, proving of the onset and the progress of myogenic differentiation is indispensable. To monitor and support proliferation dynamics assess the increment of myoblast number after subsequent time-points (i.e., 24 and 48 h of culturing), using phase-contrast microscopy. Alternatively, examine intracellular distribution (nuclear vs. cytoplasm) of proteins controlling cell cycle progression (i.e., cyclin A1 or cyclin D1) in myoblast culture, using immunoblotting and/or immunofluorescence [6]. Visualize cells by nuclei or contour staining, count fused cells under a phase-contrast microscope, and calculate fusion index to prove progression of myogenesis. Appearance of cells formed from at least 2 myoblasts is considered onset of fusion [7]. Alternatively, examine cellular content of well-established markers of myogenesis, that is MyoD1, myogenin, myosin heavy chain, using immunofluorescence and/or immunoblotting [8].
3. The size of cell-culture vessel cannot be too big (35–60 mm diameter petri dishes will be right) because usually the protocol recommends using small volumes of lysis solution (350–800 μL—depending on the protocol of used kit), and the amount of lysis solution should cover the total area of the vessel and all cells. Begin the lysis process immediately after removing the cell culture medium.
4. Number of cells from which RNA is isolated depends on the efficiency of kit used. You can find this information in the specific protocol.
5. It is important to completely remove cell culture medium, because residues will inhibit lysis and dilute the lysate, and it may affect/reduce the RNA yield. Chomczyński and Sacchi [9] state that washing cells with PBS is not necessary.
6. Before using RLT buffer check if there is no precipitate in a solution. If necessary, redissolve the buffer by warming at 37 °C, and then place it at room temperature.

7. Frozen samples have to be incubated at room temperature until completely thawed just before extraction, and salts are dissolved, not longer than 5 min.
8. RNA isolation is based on Chomczyński's method [9]. Many commercial kits use more or less modified Chomczyński's method. In our experiments we use the following commercial kits for RNA isolation from cell culture (or animal tissues): Total RNA Mini or Total RNA Mini Plus (A&A Biotechnology) or RNeasy Mini Kit (Qiagen). In most cases RNA extraction from cell culture does not cause trouble, and you get very good quality RNA. Before you start working with any kits check if every reagent is ready to use. Sometimes, the producers ask you to add some solvent (e.g., ethanol, isopropanol) to reagents before using them. There are some important steps for RNA isolation: cell collection and lyses, RNA extraction, RNA precipitation, Wash and RNA solubilization. You can also prepare reagents for RNA extraction by yourself. The very precise instruction wrote by Chomczyński and Sacchi [9] will guide you through it; however, commercial kits substantially reduce the time for reagent setup and RNA extraction.
9. Each commercial kit has specific procedure, it depends on the reagent. In this description we will follow the RNeasy Mini Kit (Qiagen) procedure. All the steps of this procedure done on the bench or in centrifuge should be done at room temperature (20–25 °C).
10. You should have maximum 700 μL of solution, because you added 350 μL of RLT buffer and 350 μL of 70% ethanol to the samples.
11. The column cannot contact the flow-through. Always after centrifugation carefully take out the column with collection tube from centrifuge and then remove the column from the collection tube.
12. Removing genomic DNA contamination is optional. You can find description of DNAse solution preparation in Qiagen protocol. It is important not to vortex the DNase solution.
13. Pipet the DNase I mix precisely onto the membrane of column, be careful not to pipet it on the wall or the O-ring of the column.
14. Calibrate the NanoDrop with the same solution or water used for RNA dissolving.

    The efficiency of downstream reactions (e.g., cDNA synthesis) strongly depends on the purity of the RNA sample used. After measuring RNA concentration by NanoDrop notice two parameters: absorbance ratios at: 260 nm/280 nm and 260 nm/230 nm, which indicate the purity of RNA. For pure RNA the

A260/A280 ratio ranges between 1.8 and 2.0. If there is contamination with proteins, phenol or other contaminants that strongly absorb light at or near 280 nm the A260/A280 ratio will be much lower. The A260/230 ratio should be 2–2.2. Lower values indicate possible contamination with carbohydrates, peptide, phenols, chaotropic salts, or aromatic compounds [10]. When using Qiagen kits the increased absorbance at 230 nm in RNA samples is almost always due to contamination with guanidine thiocyanate present at very high concentrations in the lysis buffer or extraction reagent used in most RNA purification procedures [11]. When the concentration of RNA is high the trace amount of guanidine thiocyanate will have no considerable effect on the A260/A230 ratio. One has to remember that the presence of contaminants cannot inhibit the downstream reactions and procedures performed on RNA. Cicinnat et al. [12] did not report a correlation between A260/A230 ratio and the level of qRT-PCR inhibitors in RNA samples. The Qiagen [11] affirms that the concentrations of guanidine thiocyanate of up to 100 mM in a RNA sample do not compromise the reliability of real-time RT-PCR, even when using PCR reagents that are sensitive to inhibitors. Remember that if you decide on additional RNA purification you have to consider the fact of losing some of the RNA yield.

15. We use Agilent RNA 6000 Nano Kit to check the quality of RNA isolated. This kit is adapted to concentration 5–500 ng/μ L of total RNA, but is recommended to use at least 50 ng/μL. If the concentration of RNA (result-value from NanoDrop) is higher than 500 ng/μL dilute the sample adequately. Agilent 2100 Bioanalyzer software gives the possibility of determining the quality of RNA by visualized form as electropherogram and pseudo-gel image. Figure 11 shows electropherogram with a detailed description of the individual regions (pre-, 5S, fast-, inter-, precursor-, and post-region) and peaks (marker, 18S, 28S). The data from Bionalyzer give also quality assessment metrics such as ribosomal RNA ratios, concentration estimates, and algorithmic interpretation of the sample RNA integrity number (RIN). RIN estimates the integrity of total RNA samples based on entire electrophoretic trace of RNA sample including also degradation RNA products. RIN values range from 10 (intact) to 1 (totally degraded). Figure 12a presents a very good quality samples with RIN equal 10 and Fig. 12b shows a lower quality sample, partially degraded with RIN equal 5.2. When RNA is of good quality (not degraded) then two peaks representing the 18S and 28S ribosomal RNA should be observed on the electropherogram. The peak of 28S should be higher than the peak of 18S (Fig. 12b). The gradual degradation of RNA is reflected by a continuous shift

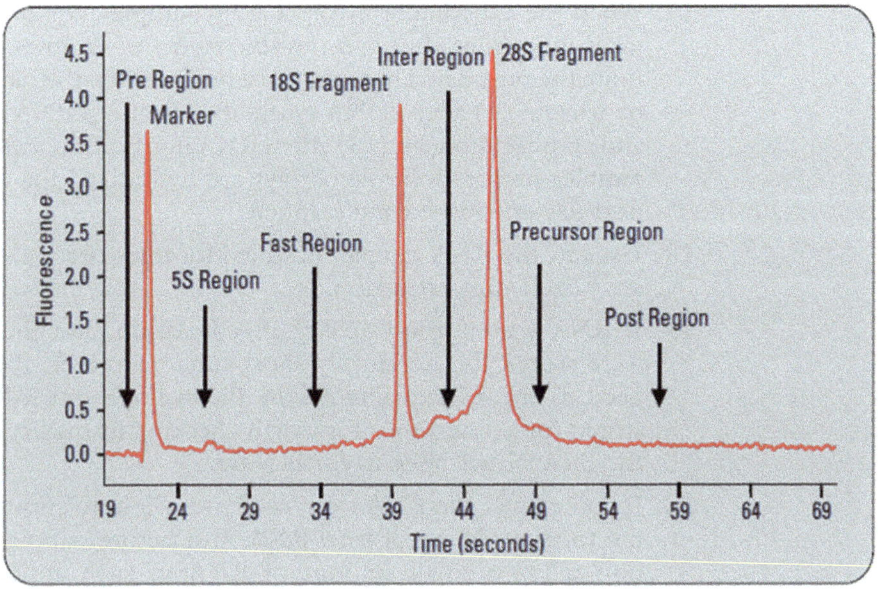

Fig. 11 Electropherogram for RNA analysis from Agilent 2100 Bioanalyzer software, specified the regions, which are indicative of RNA quality (RNA Integrity Number (RIN) Standardization of RNA Quality Control Application–Agilent Material)

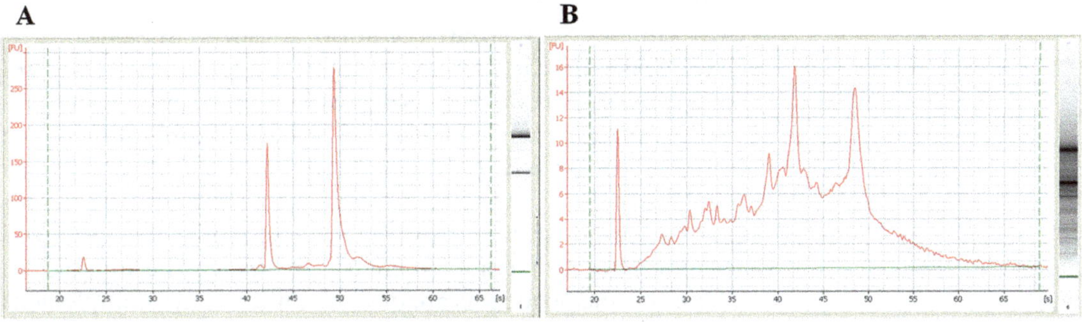

Fig. 12 Examples of electropherogram and pseudo-gel image RNA analysis from Agilent 2100 Bioanalyzer software: (**a**)—of perfect RNA quality RIN 10 and (**b**)—lower quality of RNA RIN 5.2

toward shorter fragment sizes and then there are disordered peaks in the first part of graph, in region between 25 and 35 s (Fig. 12b). For microarray analysis RNA samples have to be good quality. Do not take samples with RIN lower than 7.5.

The RNA extracted from cell culture generally is of very good quality (RIN 9–10). More problematic could be RNA isolated from tissues.

16. Very well-described protocols are available on the website of the producer (Two—Color Microarray-Based Gene Expression Analysis Low Input Quick Amp Labeling Protocol version 6.9.1, August 2015).

17. When the experiment involves many samples, the preparation takes more time, but it is possible to freeze samples and continue the next day. There are three points when the samples can be frozen: (1) after cDNA synthesis, (2) after cRNA synthesis and amplification, and (3) after cRNA purification. After all day samples preparation microarrays are hybridized for 17 h and next day are washed and scanned.
18. Usually, the RNA samples destined for microarray experiment are frozen after extraction.
19. If RNA was of good quality after isolation, and the samples were stored for a relatively short time in freezer, there is no need to check its quality on the Bioanalyzer, but if RNA was stored for a long time, it is worth checking its quality again on the Bioanalyzer prior to further use.
20. If you extract the RNA from cells you will have enough material to take 100 ng of total RNA. Just before sample preparation dilute a small amount of RNA to such concentration which will enable you to transfer it to a microcentrifuge tube in a volume not smaller than 1 μL and not larger than 1.5 μL. The perfect concentration will be 100 ng/μL. The RNA should be constantly kept on ice.
21. A positive control consisting of exogenous RNA of known quantities, different from the investigated RNA, is added to each sample. Such positive control is called "spike-in." The RNA Spike-In Kit consists of two sets of positive control transcript mixtures one for samples labeled with Cy3 and second for samples labeled with Cy5. Each mixture contains ten in vitro synthesized, polyadenylated transcripts derived from the Adenovirus E1A transcriptome that are premixed at various ratios. The positive control transcript mixtures are optimized to anneal to their complementary probes on the microarray. These specific probe sets are designed not to have complementary sequences with the biological samples being analyzed. After hybridization to the complementary probes on the microarray, the log ratios of red to green signal intensities for each Spike-In transcript can be used to monitor the sample amplification and labeling and microarray processing procedures used in the experiment, which are independent from the quality of the starting RNA sample.
22. Store the First Dilution for up to 2 months in a non-defrosting freezer at −70 °C to −80 °C. Do not freeze/thaw more than eight times. After using, discard the rest of the second, third, and fourth dilution.
23. During preparation of Master Mixes calculate the appropriate volumes of reagents suitable for total number of samples used

in your experiment plus one or two more samples than you need.

24. Make sure that the buffer does not contain any precipitates. If the precipitates are visible repeat the procedure. Keep at room temperature until needed.
25. Pay attention when taking the 5× First Strand Buffer. Pipet very slowly and carefully to take an adequate volume of it.
26. Remember to add ethanol before the first use of the RPE buffer.
27. Be careful, after each measurement observed the graph. If the baseline deviates from 0, repeat the measurement and make sure that the baseline forms a flat horizontal line. If the baseline is not flat at 0, then the values of concentration are not proper.
28. This is an important step of the procedure, in which the performance of the transcription reaction and incorporation of cyanines into the newly synthesized cRNA strands are verified. The sufficient yield and labeling of cRNA allows you to continue the procedure. The yield has to be at least 0.825 μg cRNA and $\geq$ 6 pmol Cy3 or Cy5 per μg cRNA. The values are for slides with four or eight microarrays (4-pack, 8-pack).

    Formulas for calculations yield of cRNA and cyanine incorporation:
    (Concentration of cRNA ng/ μL) * 30 μL (elution volume)/ 1000 = μg of cRNA
    (Concentration of Cy3 or Cy5 pmol/μL)/(Concentration of cRNA ng/μL)*1000 = pmol Cy3 or Cy5 per μg cRNA
29. Using thermomixer with thermotop preserves from condensation on tube lid and tube wall.
30. If the samples were incubated in thermoblock spin shortly to drive down the contents from the tubes' walls and lid.
31. Loading microarray slide on gasket slide is an important moment which requires precision and concentration. By one incorrect movement we can lose the microarray. "Active side" (the surface with probes) of slide is the side with labeled barcode. Remember to touch only the barcode end of the microarray slide or its edges! Do not touch the surface of the slide. Put very carefully and slowly microarrays slide on gasket slide. Do not drop the slide on gasket slide. Put "active side" down, parallel to the gasket slide so that the liquid does not infringe the gasket. Make sure that the sandwich-pair is properly aligned.
32. Transfer the slide quickly to avoid premature drying of the slides.
33. Ozone level in environment of laboratory is about 5 ppb (approximately 10 μg/m^3), which can influence adversely the

stability of microarrays and affect the Cy5 signal. To protect microarray from degradation of cyanine dyes and formation of artifact we use ozone-barrier slide cover.

34. It is recommended to have dishes designate and dedicate only to two-color gene expression experiments.
35. Agilent DNA Microarray Scanners are equipped with Scan Control Agilent software and Agilent Feature Extraction software. These programs are dedicated for scanning and results extraction.
36. After excitation of the two fluorophores (Cy3 and Cy5) with a laser beam of defined wavelength, detector in the scanner reads the intensity of fluorescence of hybridized RNA molecules.
37. A grid analysis file is created, containing data related to annotation of the genes or products represented by each feature and general geometry about the microarray (number of rows, columns, subgrids, feature spacings), which are used to find spots before data analysis takes place. FE executes spot quality assessment, data quantification, and normalization. The software rejects outlier pixels, accurately calculates feature intensities and ratios, flags outlier features, and calculates statistical confidences (Feature Extraction 12.0 Quick Start Guide). FE quantifies feature signals and their background, performs background subtraction, Lowess normalization, and calculates feature log ratios (Agilent's Processed Signal value) and error estimates. FE generates two types of files. A text file (.txt) contains raw pixel intensity data expressed either as mean or median values, and the log ratio. This file will be used for further data analysis. The second file, called a QC report—Quality control report includes statistical results which evaluate the reproducibility and reliability of single microarray data. Agilent defines some metrics which are useful to follow the microarray processing and determine whether there were problems at the various laboratory steps (label, hybridization, wash, scan steps) during microarrays processing. The most important metrics and their thresholds are gathered in one table (Table 7). The first line in the table of Evaluation metrics informs if the grid is placed correctly. Agilent has found thresholds that indicate if the data is in the expected range (classified as "Good"—blue color, or out of the expected range—classified as "Evaluate"-red color). These metrics evaluate the quality of microarray, hybridization, presence of wash artifacts, background, non-uniform or population outliers and microarray reproducibility. The first point is to check if the grid is placed correctly. The metric: median % coefficient of variation of Background Subtracted Signal (MedCVBkSubSignal) is used as a measure of intra-array signal reproducibility. A lower

Table 7
Example of table presenting data from FE QC report—evaluation metrics. Red and green boxes mark metrics described in the text

Evaluation Metrics for GE2_QCMT_Sep10 :
Good (11) ; Evaluate (1)

Metric Name	Value	Excellent	Good	Evaluate
IsGoodGrid	1.00		>1	<1
AnyColorPrcntFeatNonUn...	0.00		<1	>1
gNegCtrlAveBGSubSig	-3.61		-20 to 10	<-20 or >10
gNegCtrlSDevBGSubSig	2.11		<15	>15
rNegCtrlAveBGSubSig	-3.34		-20 to 4	<-20 or >4
rNegCtrlSDevBGSubSig	2.32		<6	>6
gNonCntrlMedCVBkSubSig...	12.91		0 to 18	<0 or >18
rNonCntrlMedCVBkSubSig...	13.69		0 to 18	<0 or >18
gE1aMedCVBkSubSignal	12.45		0 to 18	<0 or >18
rE1aMedCVBkSubSignal	10.79		0 to 18	<0 or >18
absE1aObsVsExpCorr	0.91		>0.86	<0.86
absE1aObsVsExpSlope	0.66		>0.85	<0.85

♦ Excellent ♦ Good ♦ Evaluate

%CV indicates better reproducibility of signal intensity and hybridization uniformity of both non-control probes and control E1a probes (Spike-ins), which are termed respectively: Non-Control %CV Background Subtracted Signal (NonCntrlMedCVBkSubSig) and E1a Median %CV Background Subtracted Signal (E1aMedCVBkSubSignal) for both channels green and red (Table 7—metrics in red box). The microarray with the wash artifacts are not expected to show a reproducible signal across the replicate probes within the microarrays, which is observed with this metric as expected.

Synthetic E1A Adenoviral sequences (Spike-ins) serve as internal process controls for linearity, dynamic range, and log ratio fidelity. The graphs in Fig. 13 present the following relationship: the observed average log ratio (axis Y) vs. the expected log ratio (axis X) for each of the spike-in probes. A linear regression analysis is done using these values and the metrics are shown beneath the plot. A slope of 1, y-intercept of 0 and R^2 of 1 is the ideal of such a linear regression. A slope < 1 may indicate compression, such as having under-corrected for background. The regression coefficient (R^2 or R^2) reflects reproducibility. The standard deviation for each data point is shown on the plot by an error bar extending above and below the point. Figure 13a presents correct relationship between the observed average log ratio and the expected log ratio. Observe that the

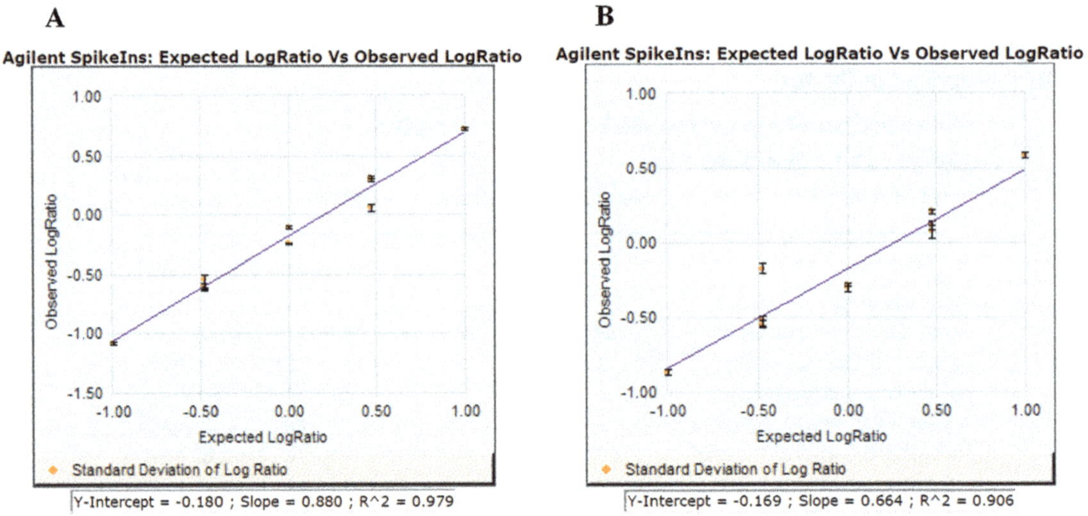

Fig. 13 Example of plots from FE QC report: (a)—example of a plot of linear regression for Spike-ins showing the correct relationship between the expected LogRatio (axis X) vs. observed LogRatio (axis Y); (b)—example of a plot of linear regression for Spike-ins showing an incorrect relationship between the expected LogRatio (axis X) vs. observed LogRatio (axis Y)

values of slope and R^2 are close to 1 (0.880; 0.979 respectively). In Fig. 13b the value of slope is <1 (0.664) and not all points align with the regression line. This suggests that an error could occur during preparation or hybridization step. The same information about the slope of regression line can be found in the last line of Evaluation Metrics table (Table 7—metrics marked in green box). The metric: absE1aObsVsExpSlope (Absolute of slope of fit for Observed vs. Expected E1a LogRatios) is evaluated (0.66).

38. Pay attention to the order of conditions. The direction of changes in gene expression is important (if the expression increases—is upregulated) or decreases—is downregulated). If we analyze condition A vs. condition B the fold change will be calculated as A/B, which means that the gene expression increases or decreases in condition A. If in microarray experiment the control samples are labeled with Cy3 and treated samples are labeled with Cy5 and both samples are loaded on one microarray then after Lowess normalization we have one result for each microarray (Cy5/Cy3). In that case GeneSpring software offers a possibility to use t-test against zero.

39. The aim of microarray experiments is to identify the genes that are differentially expressed between experimental groups. The "null hypothesis" states that a given gene on the array is not differentially expressed between the two conditions under study (e.g., treated and non-treated—control cells). The "alternative hypothesis" is that the expression level of that

gene is different between the two experimental groups. The statistical analysis involves testing tens of thousands of hypotheses simultaneously. Setting *P*-value <0.05 as statistically significant indicates that 5% of genes is false positive. When it concerns thousands of tested genes this statistical assumption is multiplied, e.g., when 20,000 probes is on a microarray then the number of false positive results (falsely positively regulated genes) will equal 1000. In order to avoid two types of errors occurring due to multiple comparisons the correction of statistical analysis needs to be applied. The “Type I” error corresponds to false positive results, which means genes that are regarded as differentially expressed, while in reality they are not. The “Type II” error corresponds to false negative results, thus genes that are not identified as differentially expressed even though they are. The FDR [13] is introduced to evaluate falsification of the results for multiple hypothesis tests. The FDR is one of the methods of correction for multiple comparison, and it shows the percentage of regulated genes that are falsely recognized as differentially expressed genes.

40. Fold-change value describes the level of an increase or a decrease of particular gene expression in experimental group, in comparison with the control group. Positive fold change values indicate higher gene expression under experimental condition(s), whereas negative fold change values indicate lower expression under experimental treatment(s), in comparison with control group.
41. The minimum detectable fold-change value used in Agilent microarray technology is below 1.3 [14]. The fold-change cutoff ranging between 1.6 and 2.0 is often used in analysis of microarray data [15]. This “strategy” allows us to include in further analyses several interesting genes which manifest moderate experimentally-induced changes of expression, but can play important biological functions [16, 17]. It concerns particularly the analysis of potential interactions among genes with experimentally-modified expression (network analysis). It can appear that numerous highly connected genes (key genes in the network) which play critical roles in myogenesis, are altered by experimental condition with absolute fold-change values below arbitrarily chosen cutoff, and thus are excluded from further data analysis and interpretations.
42. The PANTHER (Protein Analysis Through Evolutionary Relationships) Classification System classifies genes/proteins according to: (1) involvement in biological processes, that is the function of a protein in a set of proteins, which accomplish and/or regulate a process at the level of the cell or whole organism; (2) molecular function, which is played by a protein itself or with directly interacting proteins at a biochemical level;

(3) contribution to cellular components structure and/or organization; (4) involvement in signaling pathways that specifies potential interactions among a protein and other numerous proteins; (5) contribution to protein classes of evolutionarily related proteins, which possess the same functions [18]. PANTHER enables dividing general gene categories to child categories, retrieving gene list for each category at any level of classification and provides statistical tools to analyze microarray data [3].

43. Ontology terms presented in Table 1 are overlapping. Notify that sum of all genes in column entitled "Gene count" markedly exceeds the number of genes provided on the hypothetical list ($n = 2463$). Obviously, gene terms such as "Cellular process" and "Metabolic process" possess numerous common members. Moreover, some quite specific gene terms such as "Biological adhesion" can be, at least partly, included in quite general gene term, such as "Biological regulation."
44. The other gene term obtained at this step, that is "Cell-cell signaling" (GO:0007267, $n = 80$), which is interesting in the context of myogenesis, may be selected for further analysis as well.
45. This activity is possible at any previous step of the analysis, however, shorter list of genes obtained at the end of classification is easier to further searching.
46. The DAVID (Database for Annotation, Visualization and Integrated Discovery) web accessible programs provide tools, which, similarly to the PANTHER, allow identification of gene categories/subcategories and finding enriched functionally-related gene groups. For transcriptional profiling in myogenesis particularly interesting and useful for biological interpretation are other tools, like: (1) grouping redundant annotation terms into functionally-related clusters, and (2) creating gene-disease associations, and (3) mapping the genes of interest on known signaling and protein interaction pathways provided by databases of gene/protein interaction models, as BioCarta (Charting Pathways of Life) or KEGG (Kyoto Encyclopedia of Genes and Genomes) [4].
47. This tool is very useful in reducing redundancy in gene classification. Regarding the same hypothetical gene list ($n = 303$) this type of functional analysis groups similar annotation terms into clusters, based on the rule that similar annotations share similar or common gene members.
48. Annotation groups on the top of the cluster list usually display relatively low P-values for their annotation terms.
49. For using Pathway Studio (https://mammal.pathwaystudio.com/) the license is needed.

50. Pathway Studio builds a network of provided genes that comprises hundreds of interactions like: direct or indirect regulation (positive or negative), expression regulation (positive or negative), promoter binding (positive or negative), protein modification (positive or negative), binding and chemical reaction.
51. If you determine transcriptional signature of myogenic cells undergoing differentiation it is advisable to examine the expression of genes involved in this stage of skeletal muscle development. Alternatively, you can choose genes associated with your experimental hypothesis. For example, in order to support the evidence that experimental factor(s) of your interest can regulate myogenic differentiation through selected genes, you should evaluate the expression of the same genes in proliferating myoblasts and compare genes expression in two stages of myogenesis [19].
52. The results obtained by immunoblotting could not correspond to expression data due to post-transcriptional regulatory mechanisms that would be valuable for further study.

References

1. Oberthuer A, Juraeva D, Li L et al (2010) Comparison of performance of one-color and two-color gene expression analyses in predicting clinical endpoints of neuroblastoma patients. Pharmacogenomics J 10:258–266. https://doi.org/10.1038/tpj.2010.53
2. Slonim DK, Yanai I (2009) Getting started in gene expression microarray analysis. PLoS Comput Biol 5:e1000543. https://doi.org/10.1371/journal.pcbi.1000543
3. Mi H, Huang X, Muruganujan A et al (2017) PANTHER version 11: expanded annotation data from gene ontology and Reactome pathways and data analysis tool enhancements. Nucleic Acids Res 45:D183–D189. https://doi.org/10.1093/nar/gkw1138
4. Huang DW, Sherman BT, Lempicki RA (2009) Systematic and integrative nalysis of large gene lists using DAVID bioinformatics resources. Nat Protoc 4:44–57
5. Huber W, Carey VJ, Gentleman R et al (2015) Orchestrating high-throughput genomic analysis with bioconductor. Nat Methods 12 (2):115–121. https://doi.org/10.1038/nmeth.3252
6. Grabiec K, Gajewska M, Milewska M et al (2014) The influence of high glucose and high insulin on mechanisms controlling cell cycle progression and arrest in mouse C2C12 myoblasts: the comparison with IGF-I effect. J Endocrinol Investig 37:233–245
7. Dedieu S, Mazcres G, Cottin P et al (2002) Involvement of myogenic regulator factors during fusion in the cell line C2C12. Int J Dev Biol 46:235–241
8. Chomczynski P, Sacchi N (2006) The single step method of RNA isolation by acid guanidinium thiocyanate-phenol-chloroform extraction: twenty-something years on. Nat Protoc 1:581–585
9. Wieteska-Skrzeczyńska W, Grzelkowska-Kowalczyk K, Tokarska J et al (2011) Growth factor and cytokine interactions in myogenesis. Part I. The effect of TNF-α and IFN-γ on IGF-I-dependent differentiation in mouse C_2C1_2 myogenic cells. Pol J Vet Sci 14:417–424
10. T042-TECHNICAL BULLETIN NanoDrop Spectrophotometers. http://www.nhm.ac.uk/content/dam/nhmwww/our-science/dpts-facilities-staff/Coreresearchlabs/nanodrop.pdf
11. QIAGEN Newsletter March 15, 2010
12. Cicinnati VR, Shen Q, Sotiropoulos GC et al (2008) Validation of putative reference genes for gene expression studies in human hepatocellular carcinoma using real-time quantitative RT-PCR. BMC Cancer 8:350. https://doi.org/10.1186/1471-2407-8-350

13. Benjamini Y, Hochberg Y (1995) Controlling the false discovery rate: a practical and powerful approach to multiple testing. J Roy Statist Soc Ser B (Methodological) 57:289–300
14. Stafford P (2008) Methods in microarray normalization. CRC Press Print ISBN: 978-1-4200-5278-7
15. Morey JS, Ryan JC, Van Dolah FM (2006) Microarray validation: factors influencing correlation between oligonucleotide microarrays and real-time PCR. Biol Proced Online 8:175–193
16. Tarca AL, Romero R, Draghici S (2006) Analysis of microarray experiments of gene expression profiling. Am J Obstet Gynecol 195:373–388
17. Grabiec K, Majewska A, Wicik Z et al (2016) The effect of palmitate supplementation on gene expression profile in proliferating myoblasts. Cell Biol Toxicol 32:185–198. https://doi.org/10.1007/s10565-016-9324-2
18. Mi H, Muruganujan A, Casagrande JT et al (2013) Large-scale gene function analysis with the PANTHER classification system. Nat Protoc 8:1551–1566. https://doi.org/10.1038/nprot.2013.092
19. Grzelkowska-Kowalczyk K, Wicik Z, Majewska A et al (2015) Transcriptional regulation of important cellular processes in skeletal myogenesis through interferon-γ. J Interf Cytokine Res 35:89–99

Chapter 10

Coculture Method to Obtain Endothelial Networks Within Human Tissue-Engineered Skeletal Muscle

Dacha Gholobova, Melanie Gerard, Lisanne Terrie, Linda Desender, Janet Shansky, Herman Vandenburgh, and Lieven Thorrez

Abstract

Skeletal muscle tissue engineering aims at creating functional skeletal muscle in vitro. Human muscle organoids can be used for potential applications in regenerative medicine, but also as an in vitro model for myogenesis or myopathology. However, the thickness of constructs is limited due to passive diffusion of nutrients and oxygen. Introduction of a vascular network in vitro may solve this limitation. Here, we describe tissue engineering of in vitro skeletal muscle consisting of human aligned myofibers with interspersed endothelial networks. To create bio-artificial muscle (BAM), human muscle progenitor cells are cocultured with human umbilical vein endothelial cells (HUVECs) in a fibrin hydrogel. The cell-gel mix is cast into silicone molds with end attachment sites and cultured in endothelial growth medium (EGM-2) for 1 week. The passive forces generated in the contracted hydrogel align the myogenic cells parallel to the long axis of the contracted gel such that they fuse into aligned multinucleated myofibers. This results in the formation of a 2 cm long and ~1.5 mm tick human BAM construct with endothelial networks.

Key words Coculture, Tissue engineering, Skeletal muscle, Myoblasts, Endothelial cells, Bio-artificial

1 Introduction

Skeletal muscle accounts for 40% of the total body mass and is important for body movement, endocrine release, metabolism, and heat regulation. Skeletal muscle tissue has a high regenerative capacity and is able to renew itself upon injury [1]. This property is due to presence of quiescent satellite cells (SC), stem cells lying between the basement membrane of the muscle fiber and sarcolemma. Upon injury activated SCs migrate to the site of injury and start to proliferate; these cells are also known as myoblasts. Myoblasts are able to fuse with the damaged myofibers or with each other resulting in new myofiber formation and injury repair [2]. However, when muscle loss is too voluminous, endogenous regeneration fails, resulting in scar tissue formation and functional loss. The most commonly used treatment involves muscle flap

Sissel Beate Rønning (ed.), *Myogenesis: Methods and Protocols*, Methods in Molecular Biology, vol. 1889,
https://doi.org/10.1007/978-1-4939-8897-6_10,

transplantation: transfer of an autologous vascularized and innervated muscle graft from a donor site to the damaged acceptor site. However, these tissues exist in limited quantities, necessitate a second operation to obtain the tissue, and risk morbidity at the donor site [3]. Therefore, other strategies are being explored such as skeletal muscle tissue engineering approaches. Skeletal muscle tissue engineering aims to reproduce the structure and function of muscle in vitro and thus offers the potential to avoid removal of a muscle flap at a donor site. To achieve functional tissue-engineered skeletal muscle, one must recapitulate its complex structure of aligned myofibers and connective tissue. Current tissue engineering approaches use myoblasts or other cells with myogenic properties to develop myofibers in vitro in a 3D environment. Parallel alignment of muscle fibers is important for muscle function and use of artificial scaffolds may hinder the alignment [4]. Several strategies have been explored to guide myogenic cell alignment [5]. Tissue engineering approaches in a scaffold-free environment can induce alignment as well. This process is driven by cellular contractile force generation due to the presence of two attachment sites. The resulting unidirectional tension induces cell alignment [6–8].

Still, a major shortcoming is the current size of the constructs. The passive diffusion distance of oxygen and nutrients is a few hundred micrometers and cell survival beyond this distance becomes compromised. Implementation of porous scaffolds and prefabricated hollow tubes in the tissue-engineered construct may improve the delivery [9, 10]. Also, creation of an in vitro vascular network in the construct, a process called pre-vascularization, could be desirable to ensure a higher survival and prolonged function throughout the muscle construct, both for in vitro applications as well as in vivo transplantation studies [11, 12]. We have previously studied pre-vascularization by self-organization of endothelial cells into vascular networks in the bio-artificial muscle (BAM) model [13].

We describe a coculture skeletal muscle tissue-engineering approach to obtain human skeletal myofibers with interspersed endothelial networks. From human skeletal muscle biopsies, satellite cells are isolated and expanded to several million myoblasts. These myoblasts are cast within a fibrin hydrogel in a 25-mm-long silicone mold containing two attachment points. To introduce endothelial networks in the BAM, human umbilical vein endothelial cells (HUVECs) are added to the myoblast–hydrogel mix in a 70–30 myoblast–endothelial cell ratio and cultured in endothelial growth medium (EGM-2) for 1 week. During the course of 1 week, the cell-gel mix contracts to form a muscle bundle with multinucleated myofibers aligned in the direction of the attachment points with interspersed endothelial networks. The resulting construct, with dimensions of 2 cm long and 1.5 mm thick, is called a coculture BAM. Besides further development toward regenerative medicine, such a muscle model can also be used to study mechanisms underlying myogenesis, vasculogenesis, and drug effects.

2 Materials

- Sylgard 184 silicone elastomer kit—component A & B.
- Analytical balance.
- Spatula.
- Table top centrifuge.
- 20G needle.
- Distilled water.
- Platform shaker.
- 37 °C, 5% CO_2 incubator.
- Biology safety cabinet.
- 70% Ethanol solution.
- Scotch masking tape 18 mm wide.
- Small dissecting scissors.
- Curved forceps.
- At least two pairs of sterile dissection forceps.
- Sterile cotton gauze.
- Isolation medium—DMEM +1% penicillin/streptomycin (P/S).
- Ice container.
- 60 mm petri dish.
- Scalpel.
- Stainless steel insect pins.
- 0.1% gelatin solution.
- Dispase II.
- Collagenase II.
- Weigh boats.
- Medium filtration unit with 0.22 μm filter.
- Fetal bovine serum.
- 100 μm cell filter.
- Skeletal muscle cell growth medium (SkGM)—SkGM bulletkit.
- Freezing medium—90% FBS, 10% dimethylsulfoxide (DMSO).
- PBS w/o Ca^{2+} Mg^{2+}.
- 50:50 solution of methanol: acetone.
- Blocking buffer—1× PBS, 1% BSA, 0.2% Triton X-100.

- Primary mouse antihuman desmin antibody.
- Alexa fluor labeled secondary anti-mouse antibody.
- 4′,6-diamidino-2-phenylindol (DAPI) solution—DAPI 1:10000 (w/v) in PBS.
- Fluorescence microscope.
- Skeletal muscle fusion medium (SkFM)—DMEM, high glucose with glutamax and pyruvate, 10 ng/mL animal-free recombinant human epidermal growth factor, 50 μg/mL bovine serum albumin (BSA), 10 μg/mL insulin, 50 μg/mL gentamicin.
- 4% formaldehyde solution.
- Primary mouse antihuman tropomyosin antibody.
- Commercially available GFP labeled HUVECs.
- Endothelial growth medium 2 (EGM-2)—endothelial basal medium 2 with EGM 2 singlequot kit supplements & growth factors.
- Aluminum casting mold and screws.
- Dish soap.
- Drying oven.
- Vacuum aspirator.
- Flat screw driver.
- Pincers.
- Ruler.
- Silicone glue.
- 37 °C warm water bath.
- Thrombin.
- Fibrinogen.
- Aprotinin.
- Tranexamic acid.
- Dentistry tooth hook.
- 15 and 50 mL centrifuge tubes.
- Sterile pipets, various sizes.
- 6 and 24-well cell culture plates.
- T25, T75, and T175 cell culture flasks.

3 Methods

3.1 Coating of Six-Well Plate with Sylgard Silicone Elastomer

1. Weigh 30 g of component A (Sylgard 184 silicone elastomer kit) per six-well plate on an analytical balance (*see* **Note 1**).
2. Pipet component B in a 10:1 (A:B) ratio in the tube and stir with spatula until well mixed.

3. Centrifuge at 4500 × *g*, 5 min at room temperature (RT) to remove air bubbles.
4. Slowly pour elastomer to each six-well, while holding a ruler against the six-well to obtain 5 mm silicone coating.
5. Remove air bubbles with 20G needle.
6. Allow to cure for 24 h at room temperature.
7. Rinse each well three times with distilled water.
8. Fill each well with distilled water and place the six-well plate closed on platform shaker at low setting for 1 h.
9. Drain wells, refill with distilled water, and place overnight in 37 °C incubator.
10. Drain molds in biological safety cabinet, refill with 70% ethanol solution, and incubate for at least 1 h. Also sterilize the lids with 70% ethanol.
11. Drain molds and lids well and air dry in biological safety cabinet.
12. Seal the six-well plate with masking tape and store at RT.

3.2 Isolation of Myoblasts from Human Biopsy (See Note 2)

1. Obtain fresh human muscle tissue biopsy (*see* **Note 3**), 200–1000 mg wet weight and transfer muscle strip with sterile dissection forceps in a 50 mL tube with chilled 20 mL DMEM +1% P/S kept on ice in the ice container. Keep the muscle biopsies at 4 °C (*see* **Note 4**).
2. Working under sterile conditions in a biological safety cabinet, transfer muscle tissue with sterile dissection forceps to a 60 mm petri dish in 2 mL DMEM +1% P/S.
3. Using sterile forceps and scalpel, carefully remove excess connective tissue and fat from tissue sample.
4. Weigh tissue in a clean, covered sterile 60 mm petri dish with 1–2 mL sterile DMEM +1% P/S. Tare the balance first using the same 60 mm petri dish.
5. Using sterile scalpel, cut tissue into strips approximately 2 mm × 10 mm.
6. Using sterile dissection forceps and sterile stainless-steel insect pins, pin strips into sterile sylgard coated six-well plate wells (*see* Subheading 3.1) under tension, with one pin at each end of strip, and 3–5 strips in each well.
7. Pipet 6 mL of DMEM +1% P/S into each six-well.
8. Place six-well plate containing tissue strips in 37 °C, 5% CO_2 incubator.
9. After 24 h, remove medium and add fresh DMEM +1% P/S, 6 mL/well.

10. 48–72 h after pinning strips in plate, digest tissue as outlined below.
11. Coat a T25 tissue culture flask with 0.1% gelatin solution (*see* Subheading 3.6).
12. Calculate volume of enzyme solution: For each digestion step, this will be 10 mL for the first 200 mg of tissue, +5 mL for each additional 100 mg. If weight of biopsy is less than 200 mg, use 10 mL of enzyme solution for each digestion. Make up 2× calculated amount (two digestion steps) for each biopsy to be digested.
13. Calculate amount of collagenase needed to give 0.1% final concentration. Calculate amount of dispase to give a final concentration of 4 mg/mL.
14. Working in biological safety cabinet, pipet required amount of DMEM into sterile bottle. Add weighed dispase to DMEM in container, rinsing weigh boat using DMEM in bottle. Add weighed collagenase to DMEM in container, rinsing weigh boat using DMEM in bottle. Mix enzyme solution by inverting gently a few times. Filter through medium filtration unit with 0.22 μm filter. Label bottle and keep on ice.
15. Remove strips from six-well plate using sterile forceps and place in a 15 mL tube with calculated amount of enzyme solution.
16. Digest for 1 h in 37 °C, 5% CO_2 incubator (*see* **Note 5**).
17. Remove the supernatant from 15 mL tube and place in a sterile 50 mL tube. Add 10% (v/v) of chilled FBS to tube. Chill tube on ice.
18. Place fresh dispase/collagenase enzyme solution in 15 mL tube with remaining tissue.
19. Digest for 1 h in 37 °C, 5% CO_2 incubator (*see* **Note 5**).
20. Transfer digest and digested tissue to a 100 μm cell filter and pool filtrate with previous digest (*see* **Note 6**). Add chilled FBS to tube, 10% (v/v) of second digest volume.
21. Rinse filter with another 5 mL DMEM +1% P/S.
22. Pellet cells from digests for 5 min at 200 × *g* in centrifuge.
23. Suspend cells in 37 °C pre-warmed 7 mL SkGM medium and plate in the 0.1% gelatin-coated T25 tissue culture flask for expansion. Refresh medium every 2 days.
24. Expand cells until 70% confluency is reached. Split cells, count the number of cells, freeze cells in freezing medium.

3.3 Human Muscle Biopsy Characterization: Determination of Myoblast Percentage

1. Seed 5000 cells per well (at least two wells) in a 24-well plate.
2. When the cells have reached 50% confluency, remove from incubator.
3. Rinse the cells 3 × 5 min with PBS at RT and fix for 10 min at −20 °C with a 50:50 solution of acetone: methanol. Wash three times for 5 min with PBS.
4. Incubate the cells for 1 h with blocking buffer at RT.
5. Incubate one well overnight (O/N) with primary mouse anti-human desmin antibody (1:100 in blocking buffer) and one well in blocking buffer as negative control.
6. Rinse the cells 3 × 5 min with PBS at RT.
7. Incubate the cells for 30 min with Alexa fluor labeled secondary anti-mouse antibody (1:200 in PBS).
8. Rinse the cells 3 × 5 min with PBS at RT.
9. Incubate the cells for 30 min with DAPI solution at RT.
10. Replace the DAPI solution by PBS.
11. Count the number of desmin positive cells with a fluorescence microscope (*see* Fig. 1a).
12. Count the total number of DAPI stained nuclei in the same field.
13. Repeat previous two steps for a total of at least five microscope fields.
14. Calculate the percentage of myoblasts in each field by dividing the number of desmin positive cells by the total number of nuclei (*see* **Note 7**).

3.4 Human Muscle Biopsy Characterization: Determination of Myoblast Fusion Index

1. Seed 50,000 cells per well (at least two wells) in a 24-well plate.
2. Continue to expand the cells, refresh SkGM medium every 2 days until the wells are 80% confluent.
3. Change the medium to SkFM (1 mL/well) for 4 days, changing medium every second day.
4. Rinse the cells 3 × 5 min with PBS at RT.
5. Fix cells for 10 min at RT with 4% formaldehyde solution.
6. Rinse the cells 3 × 5 min with PBS at RT.
7. Fix cells for 10 min at −20 °C with methanol.
8. Rinse the cells 3 × 5 min with PBS at RT.
9. Incubate the cells for 1 h with blocking buffer at RT.
10. Incubate one well O/N with primary mouse antihuman sarcomeric tropomyosin antibody (1:100 in blocking buffer) and one well in blocking buffer as negative control.
11. Rinse the cells 3 × 5 min with PBS at RT.

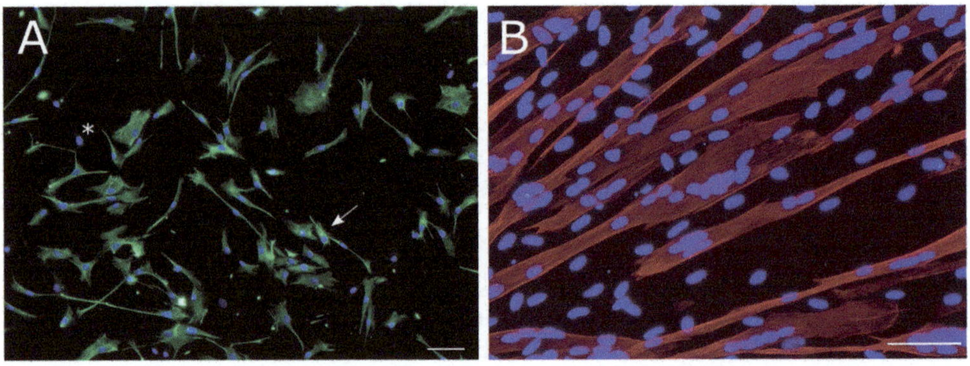

Fig. 1 Human muscle biopsy characterization (**a**) Desmin immunofluorescence staining (green) and nuclear staining (DAPI, blue). The asterisk indicates a nucleus from a desmin negative (non-myogenic) cell, the arrow indicates a nucleus from a desmin-positive myoblast. (**b**) After 4 days in SkFM, positive tropomyosin staining (red) and cell fusion indicates efficient differentiation to muscle fibers. Scale bar represents 50 μm

12. Incubate the cells for 1 h with Alexa fluor labeled secondary anti-mouse antibody (1:200 in PBS).
13. Rinse the cells 3 × 5 min with PBS at RT.
14. Incubate the cells for 30 min with DAPI solution at RT.
15. Count the number of nuclei in tropomyosin positive myofibers with a fluorescence microscope (*see* Fig. 1b).
16. Count the total number of DAPI stained nuclei in the same field.
17. Repeat previous two steps for a total of at least five microscope fields.
18. Calculate the fusion index, the percent of myoblasts which have fused by dividing the number of nuclei in tropomyosin positive cells/field by the total number of myoblasts/field (total number of nuclei/field multiplied by the percent myoblasts in the field as determined by desmin staining) (*see* **Note 7**).

3.5 Human Muscle Biopsy Characterization: Determination of Doubling Time

1. Plates 125,000 muscle cells in a T25 culture flask with 7 mL SkGM.
2. Refresh medium every 2 days.
3. When 70% confluency is reached, split cells, count the number of cells, and plates 125,000 muscle cells in a fresh T25 culture flask with 7 mL SkGM medium.
4. Determine *doubling time = LN (number of cells at passage (X + 1)/number of cells at passage (X) (= 125,000))/LN (2).*
5. Keep repeating the previous four steps until the doubling time exceeds 50 h. This gives the doubling time and the maximum number of doublings the cells may reach (*see* **Note 7**).

3.6 Culture and Expansion of HUVECs

1. Coat a T25 or a T75 tissue culture flask with 0.1% gelatin solution. Therefore add 4 mL of 0.1% gelatin solution in flask and incubate for 1 h in incubator at 37 °C, 5% CO_2. Remove gelatin solution, wash two times with PBS, and add 4 mL cell culture medium to pre-incubate the flask for cell seeding. Keep in the incubator at 37 °C, 5% CO_2.
2. Thaw commercially acquired HUVECs and seed them in a T25 gelatin-coated flask. Change EGM-2 medium every other 2 days until 80% confluency is reached.
3. Subculture the HUVECs on gelatin-coated T75 flasks in a 1:4 ratio until passage 6. Use HUVECs at maximum passage 7 for tissue engineering.

3.7 Production of Six Wells with Silicone Molds and Attachment Sites

1. Soak metal casting mold (two pieces, *see* Fig. 2a) in a 5% dish soap in water.
2. Dry casting mold in 40 °C oven (*see* **Note 8**). Remove mold when it is dry. Assemble mold by using long screws (*see* Fig. 2b) (*see* **Note 9**).
3. Combine 20 g of component A and 2 g of component B (Silicone Elastomer kit) in a 50 mL centrifuge tube and stir with spatula until well mixed (*see* **Note 1**).
4. Centrifuge at 4500 × *g* for 5 min at RT to remove the big air bubbles.
5. Use a vacuum system to remove the small air bubbles.
6. Over-fill the casting mold by pouring slowly silicone elastomer in the casting mold (should be convex meniscus) (*see* Fig. 2c).
7. With 22G needle, remove air bubbles.
8. Dry molds for 15 min in 150 °C oven.
9. Keep the casting molds overnight at RT to let the silicone elastomer cure.
10. Unscrew the screws and separate casting mold pieces by using a flat screw driver to create a lever system. Use a scalpel to remove excess silicone from the casting piece with the silicone molds (*see* **Note 10**). Go underneath the silicone layer with the scalpel and follow the contours of the casting mold (*see* Fig. 2d). Release the silicone molds from the casting mold pieces, resulting in separate silicone tissue molds (*see* Fig. 2e).
11. Remove excess silicone from around tissue molds.
12. To clean silicone molds, rinse each mold three times with distilled water for 5 min.
13. Place 1.4 cm long stainless steel pins at a distance of 2 cm in silicone tissue molds (*see* **Note 11**).

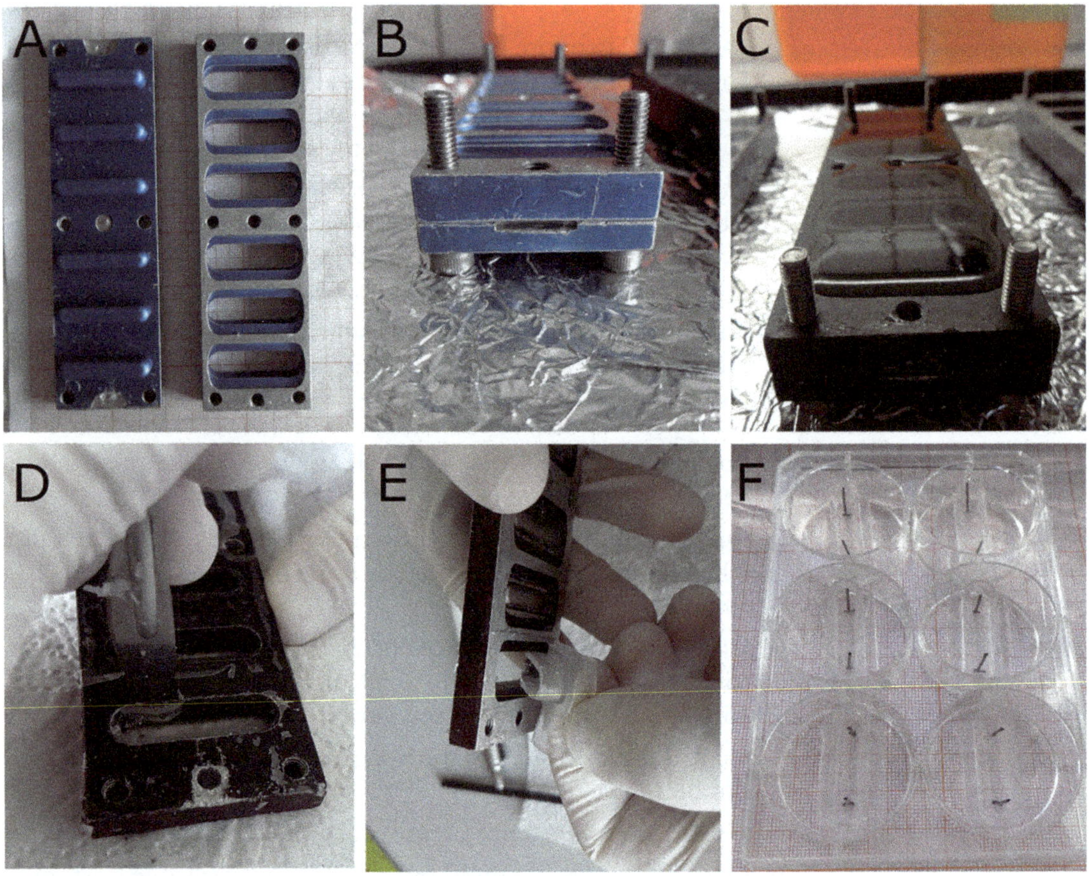

Fig. 2 Production of six wells with silicone molds and attachment sites (**a**) Casting mold consisting of two metal pieces (**b**) Assembled casting mold pieces (**c**) Casting mold overfilled with silicone elastomer (**d**) Detachment of silicone mold from casting mold with a scalpel (**e**) Release of silicone mold from the casting mold (**f**) six-well plate with silicone molds and attachment sites

14. Using a thin film of silicone glue, coat the bottom of the silicone mold and the insertion place in the six-well and place in center of six-well plate (*see* Fig. 2f).
15. Allow 24 h for glue to cure.
16. Rinse each mold three times with distilled water for 5 min.
17. Fill each well with distilled water and place on platform shaker at low setting for 1 h.
18. Drain wells, refill with distilled water, and place overnight in 37 °C incubator.
19. Drain molds well under sterile conditions in biological safety cabinet, refill with 70% ethanol solution, and incubate at least 1 h.
20. Drain molds well and air dry in biological safety cabinet.
21. Seal the six-well plate with masking tape and store at RT.

3.8 Tissue Engineering of Coculture Bioartificial Muscle (BAM)

1. Calculate the required number of cells: Count per BAM 600,000 GFP labeled human umbilical vein derived endothelial cells (HUVECs) and 1.4×10^6 muscle cells (*see* **Note 12**).
2. Final concentration of fibrin in BAM is 1 mg/mL fibrinogen and 2 U/mL thrombin, but ½ of final volume of gel is fibrinogen and ½ of final volume of gel is thrombin, so two times terminal dilution. Final volume of gel per BAM is 1 mL. Calculate the required volume of 2 mg/mL fibrinogen solution and 4 U/mL thrombin solution (*see* **Note 12**).
3. Work at RT and in biology safety cabinet to keep everything sterile.
4. Start with cell pellet of GFP labeled HUVECs and muscle cells.
5. Prepare 2 mg/mL fibrinogen solution:

   Thaw the fibrinogen stock (*see* **Note 13**) in a warm water bath at 37 °C (10 mg/mL) (*see* **Note 14**) and dilute five times in DPBS with Ca^{2+} and Mg^{2+}.
6. Prepare 4 U/mL Thrombin Solution:

   Dilute the thrombin stock (1000 U/mL) 250 times in DPBS with Ca^{2+} and Mg^{2+}.
7. Mix cell pellet with thrombin solution and add 0.5 mL of the cell—thrombin mix to the silicone mold (*see* **Note 15**).
8. Add 0.5 mL 2 mg/mL fibrinogen solution to the silicone mold: pipet three times up and down, and around the attachment sites.
9. Keep the BAMs 10 min on the bench at RT and then transfer them to 37 °C, 5% CO_2 incubator.
10. After 2 h: add 10 mL of EGM-2 medium with aprotinin (final concentration 92.5 μg/mL) and tranexamic acid (final concentration 400 μM) per BAM (*see* **Note 16**) (*see* Fig. 3a).
11. 1 and 2 days post-casting: detach fibrin gel from silicone mold with dentistry tooth hook (*see* **Notes 17 and 18**) (*see* Fig. 3b).
12. Refresh EGM-2 medium every 2 days (*see* **Note 16**).
13. 7 days post-casting a BAM is formed with dimensions of 2 cm long and 1.5 mm thick with aligned multinucleated myofibers and GFP labeled interspersed endothelial networks (*see* Fig. 3c, d).

4 Notes

1. Component A is very viscous and it is not possible to use a pipette. Therefore, pour component A in a (tared) 50 mL centrifuge tube and determine the exact amount by weight; this may be slightly more than 30 g.

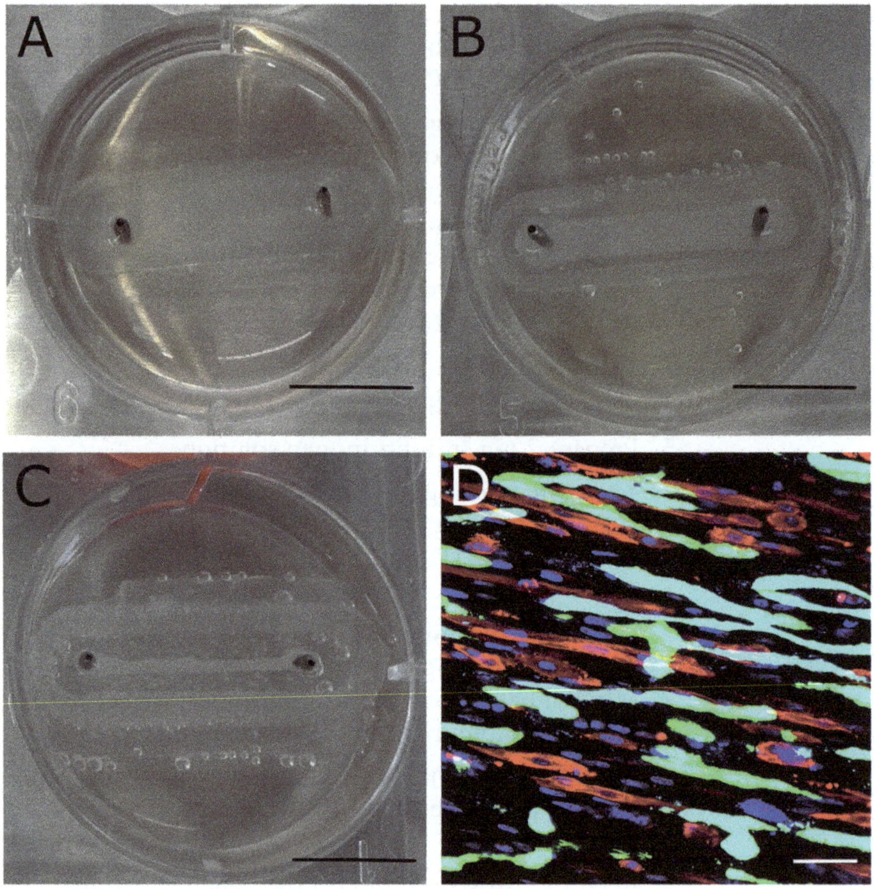

Fig. 3 Tissue engineering of coculture bioartificial muscle over 1 week time-course (**a**) Two hours post-casting, the cell-fibrin solution has gelled and EGM-2 medium is added in the six-well. (**b**) One-day post-casting, detachment of the gelated cell-fibrin solution results in formation of a contracting muscle bundle. (**c**) One-week post casting, cell-driven contraction of the fibrin hydrogel results in the formation of a 2 cm-long and ~1.5 mm-tick human BAM construct. (**d**) Confocal microscopy image of 1 week old 3D coculture BAM shows tropomyosin positive (red) aligned multinucleated (DAPI, blue) myofibers with interspersed GFP-labeled endothelial networks (green). Black scale bars represent 1 cm (**a–c**), white scale bar (**d**) 50 μm

2. This protocol can be omitted if myoblasts are obtained from a commercial source.
3. Perform the muscle biopsy under sterile conditions. Wear sterile gloves, sterile coat, a mouth mask and work with a sterile bed to store your instruments during the biopsy.

   Cut through the muscle fascia very carefully so that you do not damage the underlying muscle fibers.
4. It is best to start the myoblast isolation directly after the muscle biopsy, but it is possible to keep the biopsy up to 48 h in DMEM +1% P/S at 4 °C.

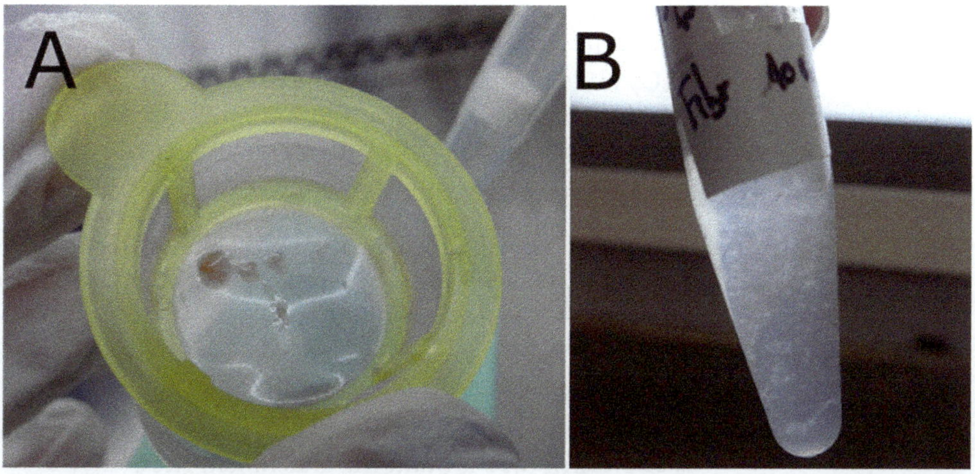

Fig. 4 Troubleshooting: *see* **Notes 6** and **14** (**a**) 100 μm cell filter showing digested human muscle tissue depicted as a thread structure (*see* **Note 6**). (**b**) Thawed fibrinogen solution in warm water bath at 37 °C showing partly clotted fibrinogen visible as white fibrin flakes (*see* **Note 14**)

5. Keep the 15 mL tube on the side so every strip is surrounded by the enzyme solution.
6. If the digestion went well, then the remaining tissue on the 100 μm filter is a thread (*see* Fig. 4a).
7. Advisable requirements for isolated muscle cell population: > 60% desmin positive myoblasts, a fusion index higher than 60%, and a population doubling time lower than 50 h.
8. Transfer the pre-soaked casting molds immediately to the dry oven at 40 °C.
9. The casting mold will feel a little bit greasy.
10. Remove excess silicone from the side of the casting mold that is the upper side of the silicone mold.
11. To obtain stainless steel pins, cut 4 cm long 20G syringe needles in 1.4 cm long pins by using pincers. To keep the steel pins at their place in the silicone molds, prick the pins slightly through the silicone molds.
12. Calculate 10% extra of total amount for cells, thrombin and fibrinogen solution.
13. Fibrinogen stock preparation: Warm 5 mL sterile demi water and 15 mL of DPBS (with Ca^{2+}, Mg^{2+}) in warm water bath at 37 °C. Check label of fibrinogen powder for reconstitution volume. The volume of demineralized water to be added to fibrinogen powder is: reconstitution volume (mL) × (mg fibrinogen)/1000. Working in a biological safety cabinet, add specified reconstitution volume of warm demi H_2O to fibrinogen. Place tube in a warm water bath at 37 °C. Do not disturb until

completely dissolved—this will take 15 min or longer. When dissolved, add 0.316 mL × (mg fibrinogen/100) of 10,000 KIU/mL aprotinin to fibrinogen. Final volume in mL is (mg fibrinogen)/10. Make solution up to the final volume by adding warm DPBS. Mix gently by pipetting up and down—do not vortex! Final concentration of fibrinogen will be 10 mg/mL. Final concentration of aprotinin will be 316 KIU/mL. Aliquot into required volumes for tissue engineering—usually 1 mL—in sterile 1.5 mL eppendorf tubes. Store at -80° C for maximum 1 year.

14. Make sure the thawed fibrinogen solution is clot free. In case of clotting, thaw a new vial (*see* Fig. 4b).
15. When adding the cell-thrombin mix to the mold, make sure it is also distributed around the attachment sites.
16. Don't pipet on the BAM, but pipet into the six-well plate. Make sure the medium in the six-well plate is in contact with the BAM in the silicone mold.
17. Be careful not to damage the BAM. Follow the contours of the silicone mold. Detach also the bottom of the BAM from the mold. If the detachment went well, there will be no little pieces of fibrin coming loose and also 10 min after detachment a bundle will appear (*see* Fig. 3b).
18. Two days post-casting a contracted hydrogel can be observed. In the following days more cell-driven hydrogel contraction will occur.

Acknowledgments

This work was funded by AFM-Telethon trampoline grant 19802 and the Research Foundation-Flanders (FWO) grant 1529817N.

References

1. Frontera WR, Ochala J (2015) Skeletal muscle: a brief review of structure and function. Calcif Tissue Int 96:183–195
2. Abmayr SM, Pavlath GK (2012) Myoblast fusion: lessons from flies and mice. Development 139:641–656
3. Agostini T, Lazzeri D, Spinelli G (2013) Anterolateral thigh flap: systematic literature review of specific donor-site complications and their management. J Cranio-Maxillofac Surg 41:15–21
4. Thorrez L, Shansky J, Wang L et al (2008) Growth, differentiation, transplantation and survival of human skeletal myofibers on biodegradable scaffolds. Biomaterials 29:75–84
5. Cittadella Vigodarzere G, Mantero S (2014) Skeletal muscle tissue engineering: strategies for volumetric constructs. Front Physiol 5:362
6. Thorrez L, Vandenburgh H, Callewaert N et al (2006) Angiogenesis enhances factor IX delivery and persistence from retrievable human bioengineered muscle implants. Mol Ther 14:442–451
7. Vandenburgh HH, Karlisch P, Farr L (1988) Maintenance of highly contractile tissue-cultured avian skeletal myotubes in collagen gel. In Vitro Cell Dev Biol 24:166–174

8. Powell C, Smiley BL, Mills J et al (2002) Mechanical stimulation improves tissue-engineered human skeletal muscle. Am J Physiol Cell Physiol 283:1557–1565
9. Walthers C, Nazemi A, Patel SL, Wu BM, Dunn JC (2014) The effect of scaffold macroporosity on angiogenesis and cell survival in tissue-engineered smooth muscle. Biomaterials 35:5129–5137
10. Takei T, Kitazono Z, Ozuno Y et al (2015) Vascular-like network prepared using hollow hydrogel microfibers. J Biosci Bioeng 121:336–340
11. Kaully T, Kaufman-Francis K, Lesman A et al (2009) Vascularization—the conduit to viable engineered tissues. Tissue Eng Part B Rev 15:159–169
12. Levenberg S, Rouwkema J, Macdonald M et al (2005) Engineering vascularized skeletal muscle tissue. Nat Biotechnol 23:879–884
13. Gholobova D, Decroix L, Van Muylder V et al (2015) Endothelial network formation within human tissue-engineered skeletal muscle. Tissue Eng Part A 21:2548–2558

Chapter 11

Interaction Between Skeletal Muscle Cells and Extracellular Matrix Proteins Using a Serum Free Culture System

Danielle E. Dye, Beverley F. Kinnear, Vishal Chaturvedi, and Deirdre R. Coombe

Abstract

The ability to grow C2C12 myoblasts in a completely defined, serum free medium enables researchers to investigate the role of specific factors in myoblast proliferation, migration, fusion, and differentiation without the confounding effects of serum. The use of defined, animal free in vitro culture systems will improve reproducibility between research groups and may also enhance translation of tissue engineering techniques into clinical applications. Here, we describe the use and characterization of a serum free culture system for C2C12 myoblasts using standard tissue culture medium and readily available, defined growth factors and supplements.

Key words C2C12 myoblasts, Serum free culture, Muscle gene expression

1 Introduction

The C2C12 myoblast cell line is a commonly used cell culture model of myoblast proliferation and differentiation. These cells differentiate rapidly in culture to form contractile myotubes and produce characteristic muscle proteins such as skeletal muscle actin and myosins. These cells were derived from a subclone of a mouse myoblast cell line [1, 2]. C2C12 myoblasts are commonly cultured in DMEM containing 10% fetal bovine serum (FBS), and will begin to differentiate when confluence is reached. Differentiation is enhanced when confluent cultures are changed to medium comprising DMEM supplemented with 2% horse serum. However, animal sera is ill defined and differs from batch to batch, containing variable mixtures and quantities of different growth factors [3]. This limits the validity of in vitro culture systems that use serum containing medium as tools for exploring the roles of specific factors in myoblast proliferation and differentiation. It also means data from different laboratories addressing this question may not be

Sissel Beate Rønning (ed.), *Myogenesis: Methods and Protocols*, Methods in Molecular Biology, vol. 1889, https://doi.org/10.1007/978-1-4939-8897-6_11,

comparable. Moreover, in the context of tissue engineering, defined, xenogeneic free cultures will be more favorably received by regulatory agencies and so are likely to translate more easily into a clinical scenario [4].

Thus, the use of completely defined medium and a reduction in the use of animal products in tissue culture is becoming increasingly important. Here, we describe a method for proliferation and differentiation of C2C12 myoblasts using a defined serum free culture medium and purified extracellular matrix proteins. An important contribution of serum in tissue culture is the provision of extracellular matrix (ECM) proteins like vitronectin and fibronectin to the cultures. The deposition of these proteins onto the tissue culture plastic surface provides a substrate for the engagement of myoblast expressed integrins and triggers myoblast adhesion, enabling proliferation, alignment, and fusion [5].

We and others have shown that in the absence of adhesion, myoblasts die. Therefore, a key component of the culture system described here was the provision of a matrix protein substrate. Interestingly, once myoblast adhesion is initiated, these cells rapidly begin to synthesize their own extracellular matrix [6], which is assembled at the cell surface and deposited onto the substrate. Others have solved the problem of providing a matrix protein substrate by initially using serum in their cultures to stimulate proliferation and then replacing this with serum free medium [7] or using commercial serum substitutes [8, 9] during differentiation. However, these culture systems cannot be classed as completely defined, or as serum free.

A physical effect of adding serum to tissue culture medium is that space is occupied by serum proteins (such as albumin, fibronectin, or vitronectin) within the solution of ions, amino acids, vitamins, sugars, and, in some cases, growth factors that comprise the culture medium. These low molecular weight components of the medium are excluded from the volume which is occupied by the globular proteins and the available space in which they diffuse is reduced compared to that of medium that lacks serum proteins. Thus, the effective concentration of ions, amino acids, vitamins, and growth factors is higher in medium with serum. In a tissue, cells reside in a very crowded environment surrounded by extracellular matrix, interstitial fluid proteins, and other cells. The typical protein concentration of interstitial fluids is 30–70 g/L, that of most serum containing medium is 1–10 g/[10], with serum free medium generally being lower again. Hence, tissue culture medium places cells in a very dilute environment relative to that encountered within a tissue. In an effort to mimic the tissue environment some researchers have added macromolecules like Ficoll to tissue culture medium to produce a crowding effect. The effect of using "macromolecular crowders" in cell culture medium has been to attenuate cell proliferation, increase cell differentiation, and increase the organization of the extracellular matrix secreted by cells in culture [10, 11].

In this chapter we describe the use of serum free (SF) culture and macromolecular crowders to support C2C12 myoblast proliferation and differentiation. Proliferation and differentiation assays are described, followed by methods to characterize differentiation. These include measuring myotube width, immunofluorescent staining of myotubes, and gene expression analyses of C2C12 myoblasts as they progress through differentiation.

Although we are able to support C2C12 proliferation and differentiation with our defined medium and ECM substrate, serum free proliferation is less rapid and the myotubes produced are thinner compared to those observed in serum containing culture. However, the myotubes formed in serum free culture demonstrate muscle-specific gene expression, robust expression of slow myosin protein, and can respond to growth signals. We have also demonstrated that the addition of macromolecular crowding molecules into serum free medium leads to improved myotube formation and alignment.

2 Materials

2.1 Tissue Culture and Cell Counting

1. C2C12 mouse myoblasts (American Tissue Culture Collection, ATCC® CRL-1772™).
2. Dulbecco's Modified Eagles medium (DMEM).
3. Hams F10 medium.
4. Phosphate buffered saline (PBS) pH 7.4.
5. 200 mM L-glutamine.
6. 1 M HEPES.
7. 100 mM Sodium pyruvate.
8. 1 M Monothioglycerol (MTG).
9. 100 μg/mL epidermal growth factor (EGF), prepared in PBS/0.1% BSA (*see* **Note 1**).
10. 100 μg/mL fibroblast growth factor-2 (FGF-2), prepared in PBS/0.1% BSA (*see* **Note 1**).
11. 100 μg/mL insulin like growth factor I (IGF1), prepared in PBS/0.1% BSA (*see* **Note 1**).
12. Dexamethasone, prepared in distilled water.
13. 10% bovine serum albumin prepared in PBS (*see* **Note 1**).
14. Insulin.
15. 1 mg/mL heparin (Celsus Laboratories) prepared in distilled water.

16. Fetal bovine serum (FBS).
17. 1 mg/mL fibronectin from human plasma.
18. 1 mg/mL type I collagen.
19. Nunc tissue culture plastic ware (25 cm^2 flasks; 12, 24, or 96-well plates, as desired).
20. Trypsin: 0.05% Trypsin, 1 mM EDTA.
21. 2.5 mM EDTA/PBS: 2.5 mM EDTA prepared in PBS.
22. Diluted trypsin: 0.05% trypsin-EDTA diluted 1:10 with 2.5 mM EDTA/PBS.
23. Cell counter, e.g., Coulter Z2 Particle Count and Size Analyser or hemocytometer.
24. Light microscope for phase contract inspection.
25. Imaging software.
26. Ficoll 70.
27. Ficoll 400.
28. 3 mg/mL ascorbic acid (Wako; L-Ascorbic Acid Phosphate Magnesium Salt *n*-Hydrate) (*see* **Note 2**).
29. Standard growth medium: 90 mL DMEM, 10% (10 mL) FBS, 2 mM L-glutamine, 10 mM HEPES, 1 mM sodium pyruvate.
30. Serum free (SF) base medium: 25 mL DMEM, 25 mL F10, 2 mM L-glutamine, 10 mM HEPES, 1 mM sodium pyruvate, 3 μL MTG.
31. SF-Proliferation (SF-P) medium: 9.5 mL SF base medium, 50 ng/mL EGF, 25 ng/mL FGFII, 1 μM dexamethasone, 0.12 IU/mL insulin, 1 μg/mL heparin, 0.5% BSA, 30 μg/mL ascorbic acid (*see* **Note 3**).
32. SF-Differentiation (SF-D) medium: 9.5 mL SF base medium, 50 ng/mL EGF, 25 ng/mL IGF-1, 1 μM dexamethasone, 0.5% BSA, 30 μg/mL ascorbic acid (*see* **Note 3**).
33. Control growth medium: 9 mL of SF base medium, 1 mL FBS (*see* **Note 4**).
34. Control differentiation medium: 9.8 mL SF base medium, 0.2 mL horse serum (*see* **Note 4**).
35. Macromolecular crowding (MMC) medium: 1.875 g Ficoll 70, 1.35 g Ficoll 400, resuspended in DMEM:F12 medium at 1:1 ratio to make a 50 mL solution. Dissolve Ficoll with rocking for 30–60 min and sterilize through a 0.22 μm syringe filter. For MMC experiments use MMC medium to prepare SF-base, SF-P, SF-D, and control medium.

2.2 Quantification of Myoblast Formation

1. Phase contrast and/or fluorescent microscope and acquisition software.
2. Analysis software (e.g., ImageJ) [12].

2.3 Quantitative RT-PCR (qRT-PCR)

1. RNA purification method: we commonly use TriReagent, but any phenol or column-based RNA extraction protocol can be used.
2. Isopropanol.
3. 70% ethanol.
4. 50× Tris-Acetate-EDTA electrophoresis buffer diluted to 1× strength for use. 50× stock contains 242 g Tris, 57.1 mL acetic acid, 100 mL 0.5 M EDTA pH 8, made up to 1 L with distilled water.
5. 2% DNA grade agarose: 2 g agarose dissolved in 100 mL 1× TAE buffer.
6. 6× DNA gel loading buffer: 30% (v/v) glycerol, 0.25% (w/v) bromophenol blue, 0.25% (w/v) xylene cyanol FF.
7. Small horizontal gel electrophoresis apparatus.
8. DNA spectrophotometer or similar instrument to assess RNA purity.
9. cDNA synthesis kits or reagents. We use Bioline Tetro cDNA synthesis kit.
10. SYBR-green-based quantitative PCR (qPCR) enzyme and buffer mix. We use Bioline SensiFAST™ SYBR® Lo-ROX Kit. Check the qPCR instrument to be used for recommended qPCR reagents.
11. 96-well plates and sealing film recommended for the qPCR instrument that is to be used.
12. Oligonucleotides for reference genes and markers of differentiation (Table 1). Oligonucleotides can be ordered from many different companies.

2.4 Immunofluorescent Staining

In addition to the materials listed in Subheading 2.1

1. Round coverslips (13 mm).
2. 95% ethanol.
3. 2.5 M NaOH in 95% ethanol.
4. Sterile forceps.
5. Sterile needle with a bent tip.
6. HEPES Buffered Saline (HBS), pH 7.4 containing 1 mM $MgCl_2$ and 1 mM $CaCl_2$ (HBS/Mg/Ca).
7. 4% paraformaldehyde in HBS/Mg/Ca.
8. 0.1% Triton-X-100 in HBS.

Table 1
Oligonucleotides for q-PCR

Gene		Sequence (5′-3′)	NCBI reference	Primer bank ID	T_M (°C)	Length (bp)
Reference genes						
SDHA	F	TGGGGAGTGCCGTGGTGTCA	NM_023281		60	100
	R	TGCCCCGTAGCCCCCAGTAG			60	
ACTAB	F	GTGACGTTGACATCCGTAAAGA	NM_007393	145966868c1	60	245
	R	GCCGGACTCATCGTACTCC			61	
TBP	F	CCTTGTACCCTTCACCAATGAC	NM_013684	172073170c2	61	119
	R	ACAGCCAAGATTCACGGTAGA			61	
GAPDH	F	AGGTCGGTGTGAACGGATTTG	NM_008084		54	139
	R	TGTAGACCATGTAGTTGAGGTC			50	
Markers of early differentiation						
MYOG	F	CGA TCT CCG CTA CAG AGG C	NM_031189	162287254c3	62	137
	R	GTT GGG ACC GAA CTC CAG T			61	
MYF5	F	GCC TTC GGA GCA CAC AAA G	NM_008656	240120094c2	61	187
	R	TGA CCT TCT TCA GGC GTC TAC			61	
Markers of late differentiation						
ACTA1	F	CCCAAAGCTAACCGGGAGAAG	NM_009606	133893192c1	62	89
	R	GACAGCACCGCCTGGATAG			62	
MYH1	F	CGGAGTCAGGTGAATACTCACG	NM_030679	82524273c2	57	153
	R	GAGCATGAGCTAAGGCACTCT			54	
MYH7	F	ACTGTCAACACTAAGAGGGTCA	NM_080728	18859641a1	53	114
	R	TTGGATGATTTGATCTTCCAGGG			53	

SDHA Succinate dehydrogenase complex, subunit A; *ACTAB* β-actin; *TBP* Tata binding protein; *GAPDH* Glyceraldehyde-3 phosphate dehydrogenase; *MYF5* Myogenic factor 5; *MYOG* Myogenin; *ACTA1* Actin, alpha 1, skeletal muscle; *MYH1* Myosin heavy chain-1; *MYH7* Myosin heavy chain-7

9. Goat serum.
10. Primary antibodies, isotype control, and secondary antibodies (Table 2).
11. 1.0 μg/mL DAPI (4′,6-diamidino-2-phenylindole) in HBS.
12. Anti-fade mounting medium.
13. Fluorescent or confocal microscope and imaging acquisition software.

Table 2
Antibodies for immunofluorescent staining

Antibody	Species, isotype	Reacts against	Clone name	Manufacturer, catalogue no.
Primary antibodies[a,b]				
Myosin (slow)	Mouse IgG	Human, rat, cat	NOQ7.4.D	Millipore (MAB1628)
Desmin	Mouse IgG1	Rat, cow, human	DE-U-10	Abcam (ab6322)
Collagen type I	Rabbit IgG,	Mouse, rat, horse, human	Polyclonal	Abcam (ab292)
Collagen type IV	Rabbit IgG	Mouse, rat, cow, human	Polyclonal	Abcam (ab6586)
Fibronectin	Rabbit IgG	Mouse, rat, cow, human	Polyclonal	Abcam (ab2413)
Isotype control antibodies				
Mouse IgG1				Dako (X0931)
Rabbit IgG				Invitrogen (02–6102)
Secondary antibodies[b]				
Goat anti-rabbit Ig	Rabbit Ig	Alexa-fluor 488		MP[c] (A-11008)
Goat anti-rabbit Ig	Rabbit Ig	Alexa-fluor 546		MP (A-11035)
Goat anti-mouse Ig	Mouse Ig	Alexa-fluor 488		MP (A-11001)
Goat anti-mouse Ig	Mouse Ig	Alexa-fluor 546		MP (A-11003)

[a]Antibodies against myosin and desmin are used to identify differentiated myoblasts. We routinely use anti-myosin
[b]Antibodies recognizing collagens type 1 and type IV, and fibronectin, can be used to assess matrix deposition
[c]Secondary antibodies need to recognize the primary antibody. The fluorescent conjugate used needs to be compatible with the fluorescent microscopes available. AF488 and AF546 are two examples

3 Methods

3.1 Plate Preparation

1. Fibronectin coating (2 μg/cm^2) or type I collagen coating (6 μg/cm^2). We prefer fibronectin but type I collagen can also be used. Using sterile conditions, dilute fibronectin or collagen in cold PBS according to Table 3 (*see* **Note 5**). Volumes are given for one well, scale up for the number of wells required.
2. Apply diluted fibronectin or collagen to each well and ensure the bottom of the well is completely covered.
3. Incubate the sterile plate containing fibronectin or collagen coating overnight (O/N) at 4 °C or for 2 h (h) at 37 °C.
4. Following incubation, remove the excess coating solution by pipetting.
5. Wash each well twice with 0.1–1 mL of sterile PBS (depending on well size).

Table 3
Fibronectin/Collagen coating in wells

	12-well plate (1 well, 4 cm^2)	24-well plate (1 well, 2 cm^2)	96-well plate (1 well, 0.3 cm^2)
Fibronectin			
PBS	996 μL	498 μL	99 μL
Fibronectin (1 mg/mL)	4 μL	2 μL	0.3 μL
Total volume	1 mL	0.5 mL	0.1 mL
Type 1 collagen			
PBS	976 μL	488 μL	98 μL
Collagen type I (1 mg/mL)	24 μL	12 μL	1.8 μL
Total volume	1 mL	0.5 mL	0.1 mL

6. Add either 1 mL, 0.5 mL, or 50 μL of SF-P medium to the wells of a 12, 24, or 96-well plate, respectively; and incubate the plate at 37 °C for 1 h prior to the addition of cells.

3.2 C2C12 Myoblast Routine Subculture

1. Maintain C2C12 myoblasts in standard growth medium in a humidified tissue culture incubator at 37 °C, 5% carbon dioxide (CO_2) (*see* **Note 6**).
2. Routine subculture: seed 1×10^4 cells in a 25 cm^2 flask for cells to reach 70% confluence by day 2. If desired, cells can occasionally be seeded at 1×10^3 to reach 70% confluence by day 3.
3. For routine subculture, remove medium and wash cells with PBS before addition to the flask of 0.05% trypsin-EDTA (0.5 mL in 25 cm^2 flask). When the cells are detached, add 4.5 mL warmed medium and transfer to a 10 mL v-bottom tube.
4. Centrifuge cells at $200 \times g$ for 5 min and remove the supernatant. Loosen the cell pellet by tapping the tube and add 1 mL of medium to the cell pellet.
5. Count an aliquot of the cells and seed 1×10^4 cells into a fresh 25 cm^2 flask for a 2 day split and add 5 mL of medium. If a larger number of cells are required, 3×10^4 myoblasts can be seeded into a 75 cm^2 flask with 12 mL medium. Incubate flasks in a humidified tissue culture incubator as described above.

3.3 C2C12 Myoblast Harvest and Seeding for Serum Free Experiments

1. Remove medium and wash cells with PBS before the addition of the diluted trypsin-EDTA (*see* **Note 7**). Use 0.5 mL diluted trypsin for a 25 cm^2 flask or 1.0 mL for 75 cm^2 flask.

2. Incubate cells in the diluted trypsin mixture at 37 °C for 2–3 min. Tap the flask gently and check under the microscope to see if the cells are detached. Return to incubator for 1–2 min if not detached.
3. When the cells are detached, add 4.5 mL warmed standard growth medium (25 cm^2 flask) or 9.5 mL (75 cm^2 flask) and transfer to a 10 mL v-bottom tube (*see* **Note 8**).
4. Centrifuge the cells at 200 × *g* for 5 min and remove the supernatant.
5. Loosen the cell pellet by tapping the tube and add 5 mL of SF-base medium supplemented with 0.5% BSA to the cell pellet.
6. Centrifuge the cells at 200 × *g* for 5 min and remove the supernatant.
7. Repeat **steps 5** and **6** (*see* **Note 9**).
8. Loosen the cell pellet by tapping the tube, resuspend cells in 1 mL of SF-base medium containing 0.5% BSA, and store cells on ice.
9. Count an aliquot of cells and resuspend cells in SF-P medium to a concentration of 1×10^4 cells/mL.
10. Seed C2C12 myoblasts dropwise into 12, 24, or 96-well plates as shown in Table 4. The total number of cells for the wells of a 12, 24, or 96-well plate will be 10,000, 5000, and 500, respectively. These wells have previously been coated with either fibronectin or Type I collagen (Subheading 3.1) (*see* **Note 10**).
11. Incubate plates in a tissue culture incubator at 37 °C and 5% CO_2.
12. Cells seeded in this way can be used for proliferation assays, differentiation assays, and to prepare RNA or cell lysates.

3.4 Proliferation Assays

1. Seed cells into the wells of a 12, 24 or 96-well plate (Table 4) as described in Subheading 3.3
2. SF-P medium should be made fresh and replaced every 2 days (*see* **Notes 11** and **12**).

Table 4
Seeding cells for SF assays

	12-well plate (1 well, 4 cm^2)	24-well plate (1 well, 2 cm^2)	96-well plate (1 well, 0.3 cm^2)
SF-P medium	1 mL	0.5 mL	50 μL
Cells in SF-P at 10000 cells/mL	1 mL	0.5 mL	50 μL

3. For direct cell counts, count cells seeded into 12 or 24-well plates at 24, 48, 72, 96, and 120 h post seeding. We routinely count 4 wells of a 24-well plate (*see* **Notes 13** and **14**).
4. For each time point, carefully remove medium from 3 to 4 wells and rinse with 0.5 mL of PBS. Add 100 μL diluted trypsin per well for the wells of a 24-well plate. Increase the volume if required for the wells of a 12-well plate.
5. Incubate at 37 °C for 2–3 min. Check under the microscope to see if cells are beginning to detach. Return to incubator for 1–2 min if not detached.
6. When cells are detached, add 400 μL of control growth medium (i.e., medium containing FBS, *see* **Note 8**) to each well, and pipette up and down gently before transferring the 500 μL to a 1.5 mL microtube.
7. Return the plate containing the remaining wells of growing cells to the incubator.
8. For direct cell counts, we use a Coulter Z2 Particle Count and Size Analyzer (Beckman Coulter) but a hemocytometer could also be used. For days 1 and 2, use minimal dilutions to count cells. For later days, larger dilutions can be used to bring cells into an optimal counting range.
9. Repeat this procedure each day to construct a growth curve, with time on the X axis and cell number on the Y axis (*see* **Note 15**). Figure 1 shows a growth curve for C2C12 myoblasts over 96 hours, comparing SF-P to serum containing medium, with and without the addition of ascorbic acid to the medium.

3.5 Differentiation Assays

1. We routinely use 24-well plates to visualize differentiation. Volumes given in this protocol are for the wells of a 24-well plate (Table 4; *see* **Note 16**).
2. Following seeding, allow cells to grow for 2 days and then replace medium with freshly made SF-P medium (*see* **Notes 11** and **12**).
3. Allow the cells to reach 90–100% confluence (approximately 4 days post seeding). If cells are not confluent by day 4, replace the medium with fresh SF-P medium on day 4 and leave until day 5.
4. When cells are confluent, remove SF-P medium, wash wells gently with 0.5 mL PBS, and replace with 1 mL SF-D medium (*see* **Note 12**).
5. Return cells to a tissue culture incubator at 37 °C and 5% CO_2.
6. Monitor cells over the following 6–8 days to assess differentiation, replacing SF-D medium every 2 days (*see* **Notes 11** and **12**).

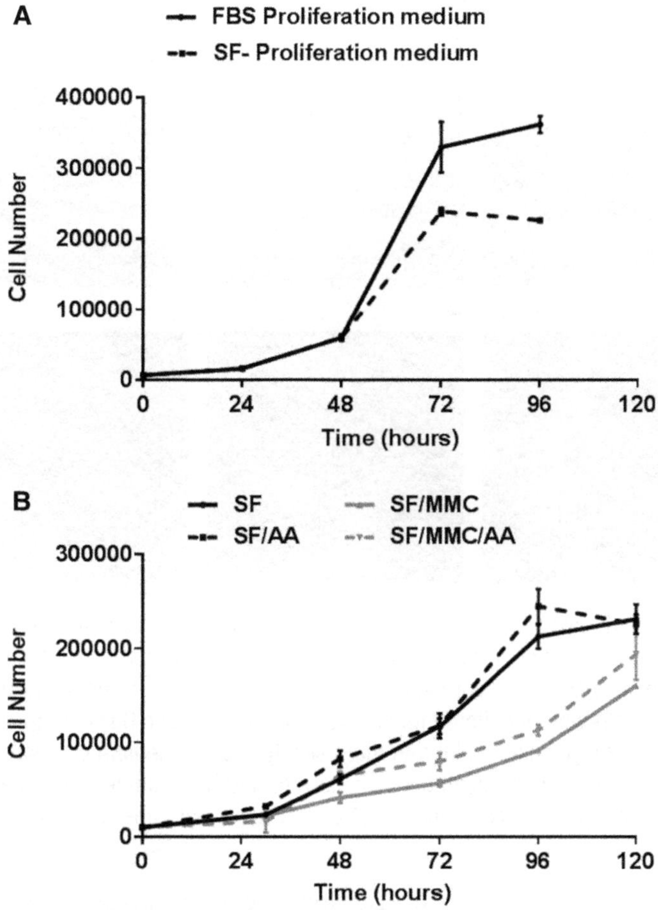

Fig. 1 C2C12 myoblast proliferation. C2C12 myoblast proliferation in (**a**) serum-free vs. serum containing medium; and (**b**) medium containing macromolecular crowding agents and ascorbic acid. Cells in SF-P medium and medium containing serum showed similar proliferative rates for the first 48 h. By 72 h, there was a difference in cell number which persists despite proliferation in both conditions then plateauing (Fig. 1a). This plateauing occurs as the C2C12 myoblasts approach confluence, which stimulates cell cycle exit and differentiation. The addition of ascorbic acid to SF-P medium caused a small (nonsignificant) increase in proliferative rate, whereas macromolecular crowding led to reduced cell proliferation over 4–5 days. This was partly ameliorated by the addition of ascorbic acid to Ficoll containing medium (Fig. 1b)

7. Cell fusion and the formation of multi-nucleated myofibers should be visible by light microscopy 2–3 days post addition of SF-D medium. Cells will appear fully differentiated approximately 6 days post addition of SF-D medium.

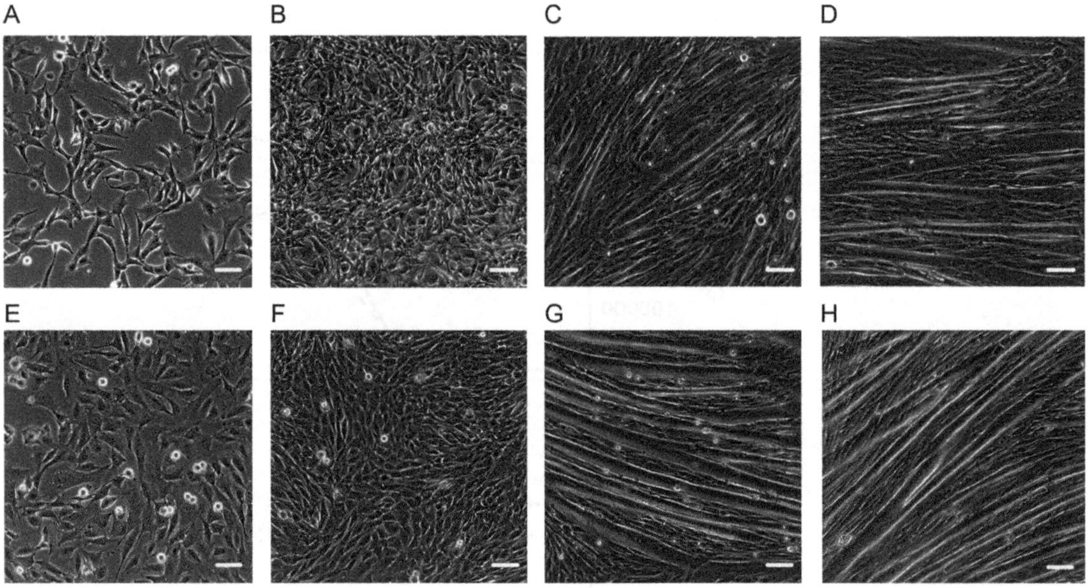

Fig. 2 C2C12 myoblasts in serum free (**a–d**) and serum containing (**e–h**) medium. At day 2 C2C12 myoblasts show similar confluency when grown in the presence or absence of serum (**a**, **e**), but cells in serum reached confluence by Day 3 (**f**), compared to Day 4 for cells in SF-P medium (**b**). After switching into differentiation medium, cells were monitored for an additional 6 days, with Day 4 (**c**, **g**) and Day 6 differentiation (**d**, **h**) shown here. C2C12 myoblasts proliferate more slowly in SF-P medium compared to serum containing medium, and form fewer and thinner myotubes (compare **c**, **d** to **g**, **h**). Images were captured using a Zeiss Axiovert bright field microscope at a magnification of 240×. Scale bars are 100 μm for all images

8. Take phase contrast images at appropriate time points throughout proliferation and differentiation, e.g., Days 2 and 5 of proliferation, and Days 2, 4, and 6 of differentiation. Images should be taken at 100× and 200× magnification and a scale bar (100 μm) included in each image. Fig. 2 is an example of phase contrast images of C2C12 myoblasts during proliferation and differentiation in serum containing and SF medium. Figure 3 shows C2C12 myoblasts cultured in the presence of macromolecular crowding.

3.6 Measuring Myotube Width

1. Import phase contrast images of myotubes (with a scale bar of 100 μm) into ImageJ or a similar analysis package.
2. Draw a line along the scale bar by clicking on the straight line drawing tool, and then click on "analyze" and "measure"—this will indicate the pixel length that corresponds to the 100 μm scale bar.
3. Select 5 nonoverlapping fields of view for each time point—these may be from multiple wells; and select 25 myotubes per field of view.

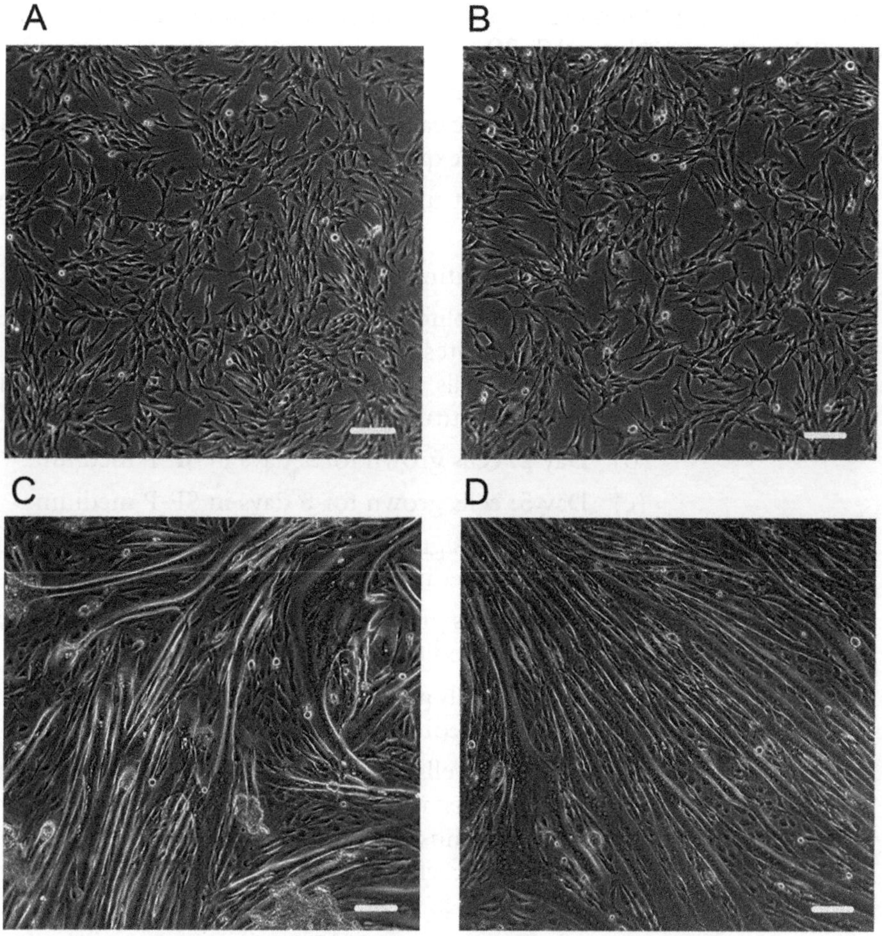

Fig. 3 C2C12 myoblasts in serum free and MMC medium. At Day 3 of proliferation, C2C12 myoblasts were more confluent when grown in SF-P medium (**a**) compared to SF-P-MMC medium (**b**). C2C12 myoblasts in SF-D-MMC medium (**d**) produced better aligned myotubes, compared to those cultured in SF-P/SF-D medium alone (**c**). Images were captured using a Zeiss Axiovert bright field microscope at a magnification of 120×. Scale bars are 200 μm for all images

4. Use the straight line drawing tool to measure the width of each myotube (in pixels). It is good practice to perform this in three places and then calculate a mean pixel width for each myotube; pixels can then be converted to μm using the value obtained from "**step 2**" above.
5. Export the width of each set of myotubes to Microsoft Excel or an appropriate graphing/statistical analysis package. Values can be plotted as means +/− standard errors.

3.7 Assessing Differentiation Using Gene Expression

3.7.1 RNA Extraction

1. Working with RNA and cDNA requires several precautions (*see* **Notes 17–20**).
2. In a 24-well plate, quadruplicate wells per time point can be harvested. These can be treated as independent biological samples within one experiment.
3. We routinely set up as least three different experiments (e.g., 1–2 weeks apart).
4. Time points routinely harvested include:
   (a) Day 0 Proliferation: these are cells used to seed into 24-well plates for proliferation and differentiation assays. Surplus cells harvested and not used for seeding into plates constitute the Day 0 sample.
   (b) Day 2: cells grown for 2 days in SF-P medium.
   (c) Day 5: cells grown for 5 days in SF-P medium.
   (d) Day 7: cells grown for 5 days in SF-P medium plus 2 days in SF-D medium.
   (e) Day 9: cells grown for 5 days in SF-P medium plus 4 days in SF-D medium.
   (f) Day 11: cells grown for 5 days in SF-P medium plus 6 days in SF-D medium.
5. Seed into the wells of a 24-well plate as for proliferation and differentiation assay, but seed each time point into a separate 24-well plate. Thus, there will be five plates containing quadruplicate wells of each sample (*see* **Note 21**).
6. To harvest cells at each required time point, remove medium and wash twice with 0.5 mL PBS.
   (a) To harvest cells for TriReagent extraction add 0.5 mL TriReagent per well, mix by pipetting, transfer sample to a 1.5 microfuge tube, and store on ice (*see* **Note 22**).
   (b) To harvest cells for column-based extraction add 0.2 mL diluted trypsin and return the plate to the incubator for 2–3 min. When detached, add 1 mL of PBS to each well and transfer the cells to a 1.5 mL microfuge tube. Pellet the cells by centrifugation at 3000 × *g* for 5 min. Remove the supernatant.
   (c) Samples in TriReagent or cell pellets can be stored at −80 °C or immediately processed for RNA extraction (*see* **Notes 23** and **24**).
7. For TriReagent extraction, allow cells in 0.5 mL TriReagent to stand at RT for 5 min before adding 0.1 mL chloroform to each sample. Cover the sample, shake vigorously for 15 s, and stand for 5–15 min at RT.

8. Centrifuge the resulting mixture at 12,000 × *g* for 15 min at 2–8 °C to produce three phases: a red organic phase (protein), an interphase (DNA), and a colorless upper aqueous phase (RNA).
9. Transfer the aqueous phase to a fresh tube (*see* **Note 25**), add 0.25 mL of isopropanol and mix.
10. Centrifuge at 12,000 × *g* for 30 min at 4 °C. The RNA will precipitate to form a pellet on the side and bottom of the tube (*see* **Note 26**).
11. Remove the supernatant carefully and wash the RNA pellet by adding a 0.5 mL of 70% ethanol. Flick to mix the sample.
12. Centrifuge at 7500 × *g* for 10 min at 4 °C.
13. Carefully remove the 70% ethanol and air dry samples for 30 min. Check if there is still ethanol remaining around the pellet, as ethanol may interfere with later processes.
14. Resuspend samples in 50–100 μL of RNase free water.
15. Check the RNA concentration and quality by measuring absorbance at 260 nm and the 260/280 and 260/230 ratios using a spectrophotometer (*see* **Note 27**).
16. Check RNA samples via electrophoresis. Prepare a 1% agarose gel in 1× TAE buffer and prepare samples by mixing 5 μL of RNA with 1 μL 6× DNA gel loading dye (*see* **Note 28**). Run the gel at low voltage (e.g., 60 V for a small gel) until the dye front is 2/3 of the way down the gel (*see* **Note 29**).

3.7.2 cDNA Synthesis

1. cDNA synthesis can then proceed according to the manufacturer's instructions. For each sample you wish to compare, you must use the same amount of RNA for cDNA synthesis (*see* **Note 30**).
2. For cDNA synthesis, mix 200 ng–1 μg RNA with 200 units of Maloney Murine Leukemia Virus reverse transcriptase (MMLV-RT), 1 μL of 10 mM dNTPs, 10 units of RNase inhibitor, 1 μL of 50 μM oligo (dT), and/or 1 μL of 50 μM random hexamers. Make up the volume of each reaction to 20 μL with sterile (PCR grade) H_2O (*see* **Note 31**).
3. Incubate the RT reactions at 25 °C for 10 min, 45 °C for 30 min, 85 °C for 5 min, and then chill on ice. The 25 °C step is only necessary if random hexamers are included.
4. Add an additional 20 μL of sterile H_2O to each cDNA sample to bring to a final volume of 40 μL (*see* **Notes 32** and **33**).

3.7.3 Quantitative PCR to Assess Gene Expression During Differentiation (See ***Note 34****)*

Preparation of Master Mixes

1. Prepare qPCR reactions to a total volume of 15 μL, including 1 μL of cDNA.
2. To prepare a 1× PCR master mix, combine the following reagents to give final concentrations of: 1× PCR buffer, 1× SYBR green, 1 unit DNA polymerase, 3 mM $MgCl_2$, 200 μM d NTPs, 400 nM forward primer, 400 nM reverse primer, and make up to 14 μL per sample with sterile water (*see* **Note 35**). Make up additional master mix to allow for pipetting error. For example, if you have eight test samples plus a negative control ($n = 9$), make up a master mix volume sufficient for 10 samples.
3. Prepare separate master mixes containing forward and reverse primers for reference and test genes (Table 1).

Primer Efficiency

1. Mix 15 μL of cDNA produced from the RNA of cells harvested at days 2 and 5 of proliferation, and days 4 and 6 of differentiation to produce a 60 μL pooled cDNA sample.
2. Generate a dilution series of this pooled cDNA sample (e.g., neat, 1/10, 1/50, 1/100, 1/500) to use as a target in qPCR reactions.
3. Prepare master mixes as described above to test all primer sets against the dilution series.
4. Aliquot 1 μL of each sample of the cDNA dilution series into the wells of a 96-well PCR plate (*see* **Notes 36** and **37**).
5. Add 14 μL of master mix to each well (*see* **Notes 36** and **37**) and seal the plate with sealing film.
6. Place the plate in the qPCR instrument of choice and program the PCR cycling conditions, e.g., 95 °C—2 min; followed by 40 cycles 95 °C—5 s; 60 °C—10 s, 72 °C—20 s. Perform melt curve analysis at 95 °C—15 s, 57 °C—2 min, 95 °C—15 s (*see* **Note 38**).
7. Confirm the presence of a single amplicon in each primer set by inspection of the melt curve (*see* **Note 39**) and export the threshold cycle (Ct) values (*see* **Note 40**) for each primer set dilution series for further analyses.
8. Plot the Ct values on the Y axis against the log input of cDNA. e.g., if neat cDNA was generated from 200 ng RNA = log 200 = 2.3; 1/5 cDNA dilution = 40 ng = log 40 = 1.6 etc.). Obtain the slope of the line for each primer set and apply this to the efficiency (E) calculation: $E = -1 + 10^{(-1/\text{slope})}$ to assess the efficiency of each primer set [13]. Efficiency should be 0.9–1.1 (*see* **Note 41**).

Reference Genes

1. Prepare master mixes (described above) containing primers specific to a selection of reference genes (Table 1).
2. Aliquot 1 μL of cDNA of all test samples into the wells of 96-well PCR plate; add master mix and run a qPCR experiment (conditions as described for primer efficiency qPCR experiments).
3. Confirm the presence of a single melt curve profile and export Ct values for all samples for each primer set for analyses.
4. Test reference gene stability using tools such as geNorm, NormFinder, and BestKeeper. There are also programs associated with the statistics package R [14]. These programs will indicate the best combination of reference genes to use for your experiments (*see* **Note 42**).

qPCR Test Gene Analyses

1. Prepare master mixes (described above) containing primers specific to genes expressed during early and late differentiation (Table 1) (*see* **Note 43**).
2. Aliquotl μL of cDNA of all test samples into the wells of 96-well PCR plate; add master mix, seal the plate, and run a qPCR experiment (conditions as described above).
3. Confirm the presence of a single melt curve profile and export Ct values for all samples for each primer set for analyses.

Gene Expression Analyses

1. Calculate the geometric mean of the Ct of the two or three most stable reference genes for each sample. Whether you choose two or three depends on the recommendations of the gene stability program used (e.g. geNorm, Normfinder, Bestkeeper or other) [14].
   (a) E.g., the geometric mean of two Ct values, is the square root of their product, e.g., the square root of $24.1 \times 25.3 = 24.7$; the geometric mean of three Ct values is the cube root of their product, e.g., the cube root of $24.1 \times 25.3 \times 26.0 = 25.1$ [15].
2. Calculate changes in gene expression using the delta delta CT method ($\Delta\Delta C_T$) method of relative quantification, where gene expression is described relative to a control sample and normalized according to reference gene expression [16].
   (a) Step 1: $\Delta C_{T.}$ For each sample, subtract the geometric mean of the Ct value of the reference genes from the Ct value of the test genes. You may have a positive or negative value.
   (b) Step 2: $\Delta\Delta C_{T.}$ Choose a "control sample." In C2C12 proliferation and differentiation experiments the control sample can be proliferating cells at Day 0. Subtract the ΔC_T value of the control day 0 samples from the ΔC_T

value of the later samples to generate the $\Delta\Delta C_T$ value. Again, this may be a positive or negative value.

(c) Step 3: fold change. The ΔΔCT values are in a log 2 base scale because every increase in one Ct number denotes a doubling of gene amplification. To express this as fold change (i.e., compare expression in a Day 4 sample relative to the control Day 0 sample) you need to convert the ΔΔCT value to $2^{\Delta\Delta CT}$ [16].

3. Representative data are shown in Figs. 4 and 5. Expression of *MYOG, MYF5, MYH1, MYH7,* and *ACTA1* were measured in C12C12 myoblasts in serum containing and serum free medium (Fig. 4), with and without the addition of ascorbic acid and macromolecular crowders (Fig. 5). Expression was normalized against the geometric mean of reference genes *SDHA, TBP,* and *HPRT.*

3.8 Immunofluorescence Staining of Slow Myosin

3.8.1 Etching Glass Coverslips (See Note 44)

1. Clean the glass coverslips by placing them in a clean, flat bottomed container and covering with 100% ethanol. Seal the container and gently agitate on a rocker or orbital shaker overnight at RT.
2. Decant the ethanol and replace with 2.5 M NaOH made up in 95% ethanol. Seal the container and gently agitate on a rocker or orbital shaker for 30 min at RT.
3. Remove the NaOH solution and wash the coverslips 5× with ddH_20.
4. Air dry the coverslips on lint free cloth.
5. Sterilize coverslips with UV light for 30 min and store in a sterile container.

3.8.2 Seeding and Staining Cells

1. Place sterile coverslips in the wells of a 24-well plate. Allow four replicate coverslips for each time point to be examined. Three coverslips are required for staining, the additional coverslip is a back-up in case of breakage.
2. Coat coverslips with fibronectin or collagen (Subheading 3.1). Alternatively, uncoated, etched glass coverslips can also be used.
3. Seed C2C12 myoblasts onto the coverslips in SF-P medium. Seed sufficient coverslips for at least three time points, e.g., day 5 proliferation, and days 2, 4, or 6 differentiation.
4. Replace medium every 2 days as previously described.
5. When a time point is reached, gently remove half of the medium from the wells to be harvested.
6. Using sterile forceps, or a sterile bent needle and forceps, remove the coverslip from the well and transfer, cell side up,

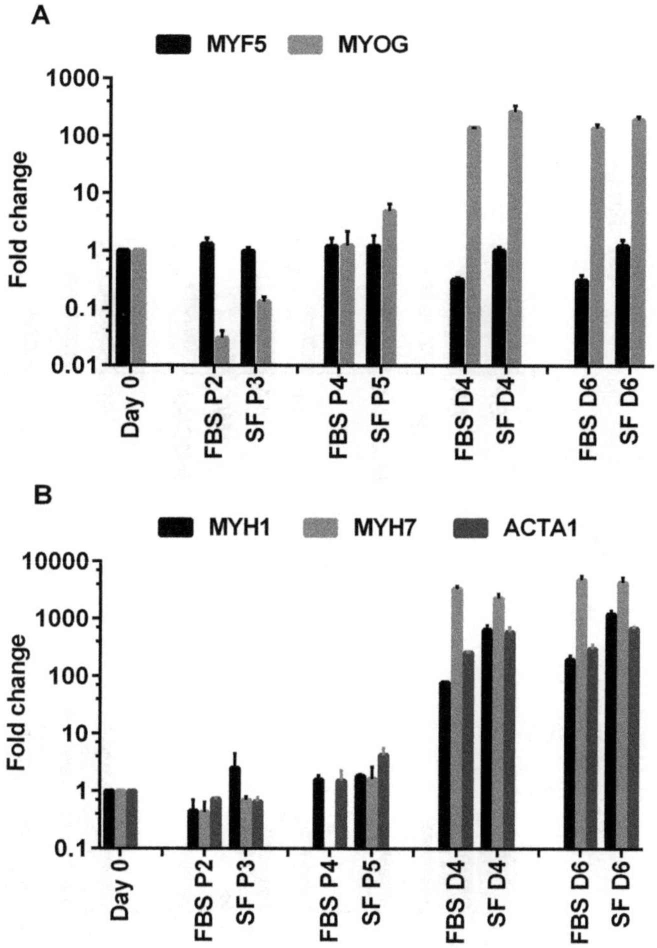

Fig. 4 C2C12 myoblast gene expression in SF-P versus serum containing medium. Gene expression fold changes were obtained by comparing all conditions to cells harvested on Day 0 of the assay, and gene expression trends were similar in cells grown in serum containing or serum free medium. (**a**) *MYF5* shows no significant change in expression in all conditions compared to Day 0, whereas *MYOG* shows a 100 fold increase in cells differentiating in serum and a 200 fold increase in cells in SF-D. (**b**) *MYH7* expression increased 3000–5000 fold in both SF and serum containing medium, whereas *ACTA1* expression increased 300–600 fold; with no significant difference between cells grown in serum vs. SF medium. *MYH1* expression increased 100–200-fold in cells grown in serum, compared to 500–1000-fold increase in SF medium. These gene expression differences may be consistent with morphological differences described in Fig. 2

to a well in a fresh plate containing 0.5 mL warm HBS/Mg/Ca.

7. Replace this solution with 4% paraformaldehyde made up in HBS/Mg/Ca and fix the cells for 10 min at RT.

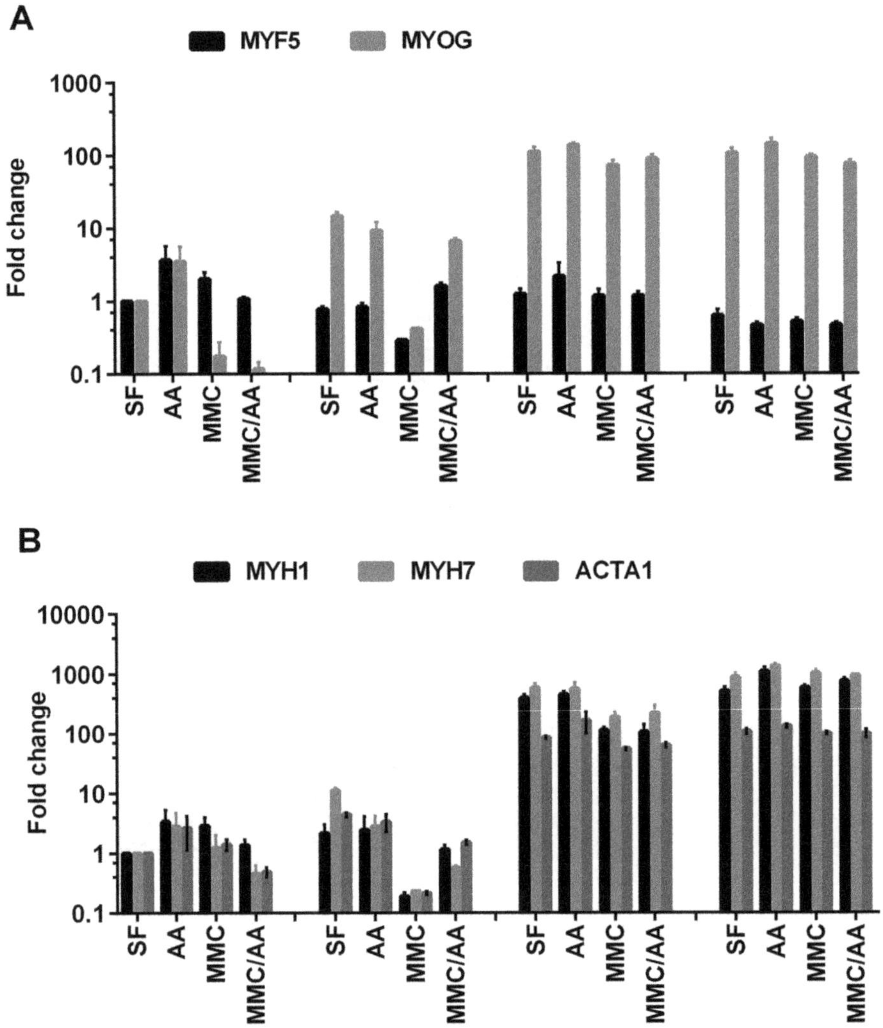

Fig. 5 C2C12 myoblast gene expression. Expression of *MYOG, MYF5, MYH1, MYH7,* and *ACTA1* were measured in C2C12 myoblasts in different culture conditions—serum free [SF], serum free and ascorbic acid [AA]; serum free and macromolecular crowding [MMC]; and serum free, MMC and AA [MMC/AA]. In this assay, gene expression was normalized to proliferating cells in SF medium on Day 3. There were small changes in *MYF5* expression across all conditions while *MYOG* increased 100–200 fold during differentiation. *MYOG* expression also increased in late proliferation; this early activation was reduced in cells grown under MMC. A similar pattern was seen with *MYH7, MYH1,* and *ACTA1*, with MMC associated with a subtle delay in activation of these genes compared to SF medium; before a rapid increase in expression during differentiation. During differentiation, there were no significant differences in expression between the conditions tested. The slight delay in expression of differentiation markers in cells grown in MMC is consistent with the slower proliferation seen in cells grown under these conditions (Fig. 1). However, this delay does not affect overall gene expression later in differentiation, nor the formation of myotubes (Fig. 3). It appears that the slower proliferation rate allows the cells to become more organized and as a result the myotubes are better aligned

8. Remove the paraformaldehyde and wash the cells twice with 0.5 mL HBS, 2 min for each wash, at RT (*see* **Note 45**).
9. Remove HBS and replace with cold 0.1% Triton-X-100 in HBS, incubate for 3–5 min at 4 °C.
10. Remove the 0.1% Triton-X-100 and gently wash 2× with 0.5 mL HBS, 2 min each at RT.
11. Remove the wash solution and add 0.5 mL blocking solution (HBS/10% goat serum/1% BSA) for 1 h at RT or 4 °C overnight.
12. Prepare the minimum volume of antibody required to cover the coverslip surface (e.g., 200 μL). For each time point tested you will require (*see* **Note 46**):
    (a) Primary antibody at optimized concentration. e.g., anti-slow myosin, mouse IgG (clone NOQ7.4.D anti-myosin) is used at 1 μg/mL in blocking solution.
    (b) Appropriate isotype control antibody at a matching concentration. E.g., mouse IgG1 at 1 μg/mL in blocking solution.
    (c) Blocking solution only without primary antibody (control).
13. Incubate primary and control antibodies for 1 h at RT, making sure the slides do not dry out.
14. Remove primary antibody and wash 3× with 0.5 mL HBS, 5 min each at RT.
15. Prepare the secondary antibody for each coverslip—e.g., anti-mouse IgG Alexafluor-488 at 1 μg/mL. Prepare the minimum antibody required to immerse the coverslip. For the three conditions above you will require 600 μL (scale up appropriately).
16. Incubate secondary antibodies for 1 h at RT in the dark, making sure the slides do not dry out.
17. Remove secondary antibody and wash 2× with 0.5 mL HBS, 5 min each at RT.
18. To stain the nuclei, add 200 μL of 1 μg/mL DAPI (4′,6-diamidino-2-phenylindole) in HBS to each coverslip for 5 min at RT in the dark.
19. Remove DAPI and wash 3× with 0.5 mL HBS, 5 min each at RT, in the dark.
20. Remove the coverslips from the wells and mount on to glass slides with an anti-fade mounting medium (e.g., Prolong Gold hard set).
21. Once mounted, stained coverslips may be stored in the dark until it is possible to image the samples. Slides mounted with

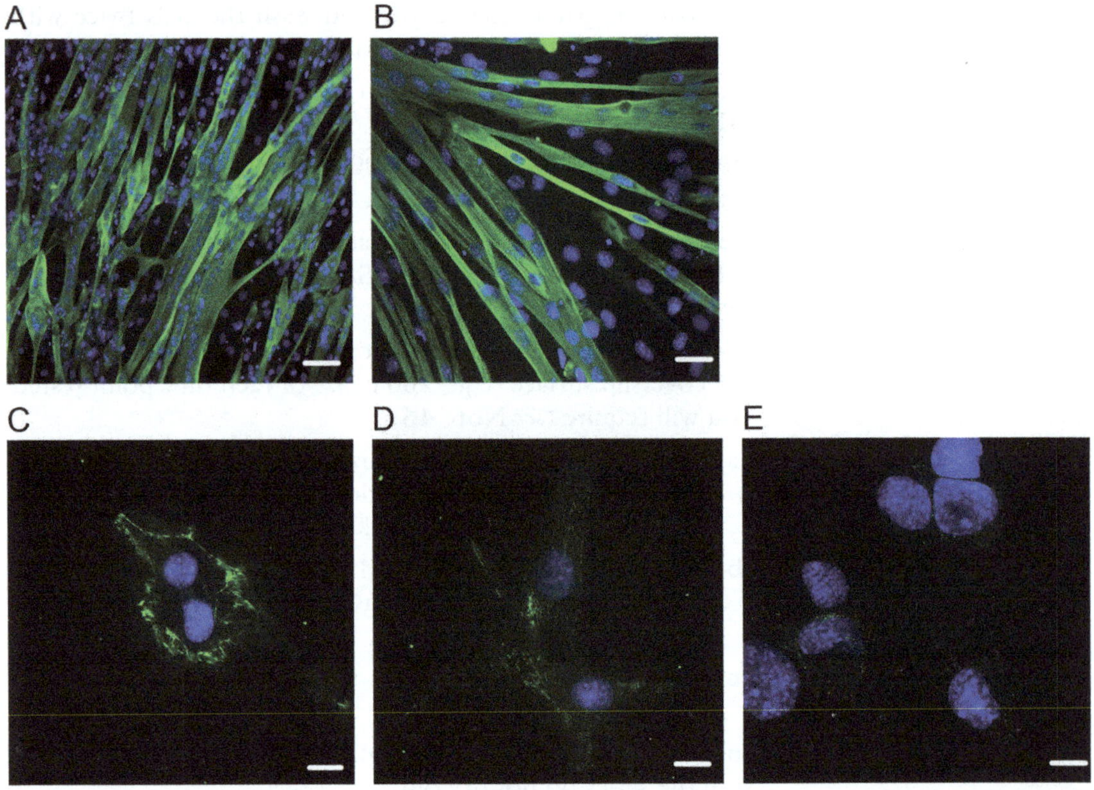

Fig. 6 Immunofluorescent analysis of C2C12 myoblasts. *Myosin expression* (**a**, **b**): C2C12 myoblasts were seeded on to etched glass coverslips in SF-P medium and allowed to reach confluence before being switched into SF-D medium and allowed to differentiate for a further 4 days (**a**) or 6 days (**b**). Cells were stained with an anti-slow myosin antibody followed by anti-mouse IgG Alexafluor488 and co-stained with DAPI. Images were captured using a Zeiss Axioskop fluorescent microscope at a magnification of 240× (**a**) or 480× (**b**) magnification. Scale bars are 100 μm and 50 μm, respectively. Myotubes stain strongly with the anti-slow myosin antibody and at high power, it is possible to see striations (**b**). *Matrix deposition* (**c**–**e**): C2C12 cells were seeded on to etched glass coverslips in SF-P medium and allowed to adhere for 3.5 h before fixing, and staining with antibodies recognizing fibronectin (**c**), perlecan (**d**), and collagen I (**e**); followed by anti-rabbit IgG Alexafluor488 and co-stained with DAPI. Images were captured using a Zeiss Axioskop fluorescent microscope at 1200× magnification, scale bars are 20 μm. C2C12 myoblasts grown in SF medium secrete matrix proteins very quickly after plating and rapidly create their own microenvironment when provided with a substrate that facilitates protein adsorption

aqueous medium should be stored at 4 °C while slides mounted with a hard set medium can be stored at RT.

22. Image slides using a fluorescent or confocal microscope and capture images using an appropriate image acquisition software. Representative results are shown in Fig. 6, where differentiated myotubes stain strongly with the anti-slow myosin antibody.

4 Notes

1. Growth factors and BSA should be prepared and stored in small aliquots to avoid repeated freeze-thaw cycles.
2. Ascorbic acid promotes the secretion and assembly of collagen by cells in culture as well as acting as an anti-oxidant. Ascorbic acid is very unstable and rapidly oxidizes in aqueous solutions which means it is advisable to freshly prepare the stock solution prior to use in tissue culture.
3. Prepare only small volumes of SF-P and SF-D medium as the supplements are stable in medium for only 2–3 days.
4. Control medium is used to compare the proliferation and differentiation of C2C12 myoblasts cultures in standard serum-containing conditions (control) with those of myoblast cultures in serum free conditions.
5. Pipette fibronectin and collagen gently as vigorous pipetting will damage the protein. This is particularly true for fibronectin.
6. Cultures must not be allowed to become confluent as the myoblasts start to differentiate and proliferative potential is lost. Cultures should be maintained at <70% confluence. For example, we seed C2C12 cells at 4×10^4 cells per 25 cm^2 flask for 2 day subculture.
7. The diluted trypsin mixture removes the C2C12 cells from the plastic with minimal disruption, meaning that cells enter the serum free assay in an optimal condition.
8. Standard medium containing FBS is used in this step to neutralize the trypsin.
9. These two washes in SF-base medium/0.5% BSA are designed to remove all traces of FBS from the cells. The 0.5% BSA assists the cells to pellet.
10. The addition of SF-P medium to the plate (Subheading 3.1), followed by the addition of cells in SF-P medium, gives rise to an even distribution of cells over the area of the well.
11. Medium is replaced to replenish growth supplements.
12. When replacing medium, process a small number of wells at one time to minimize the risk of wells drying out.
13. We use direct cell counts to measure cell number. Assays that measure metabolic outputs could also be used, but with caution. This is because as C2C12 myoblasts become confluent, they begin to undergo early differentiation which may affect metabolic readouts. Cell proliferation based on DNA binding dyes could also be used.

14. To measure proliferation in a 96-well plate, consider using metabolic or DNA binding assays (with appropriate validation), as there will not be sufficient cells in a 96-well plate to complete an accurate direct cell count, particularly at the early time points.
15. Using the cell seeding concentrations described here, C2C12 myoblasts tend to proliferate slowly for 1–2 days and then enter an exponential growth phase. Following day 3, proliferation tends to slow as cells become confluent and begin to differentiate.
16. Although C2C12 myoblasts will differentiate in 96-well plates, the surface area of each well does not allow for maximal differentiation. The surface area of the wells of a 24-well plate allows clear differentiation; although larger surface areas (e.g., 6-well plates) will allow the formation of longer myotubes. The use of 24-well plates balances efficiency of myotube formation with the cost of medium and supplements.
17. RNA is prone to degradation by DNases. We recommend using a designated RNA/PCR area for RNA preparation. If this is not possible, use a physical barrier between your samples and the bench.
18. Use racks and tubes that are quarantined for use for RNA and qPCR.
19. Calibrate your pipettes regularly.
20. We recommend using filtered pipette tips to minimize contamination.
21. Separate plates for each time point eliminate the possibility of contamination of remaining wells in a plate and also allow cells to be lysed directly in the plate if required.
22. TriReagent contains phenol and should be treated with care. Use TriReagent in a fume hood and collect phenol containing waste into separate waste containers and dispose of appropriately.
23. Storage of samples prior to RNA extraction enables batch processing of samples and can reduce variability in RNA extracted. However, the number of samples extracted in one procedure varies according to the experience of laboratory personnel; we recommend that between 6 and 12 samples are processed. One extracted, always store purified RNA at −80 °C.
24. We are only describing the TriReagent-based RNA extraction process in this method. We have found this to be economical and if performed with care, it produces high-quality RNA for downstream applications. However, column-based methods can be used and if so, follow the manufacturer's instructions.

25. Take care not to transfer any of the interphase (white material containing DNA) or the pink organic phase to the fresh tube. It is better to leave a small amount of the RNA containing aqueous phase behind than to contaminate the RNA sample with DNA and/or phenol. DNA contamination will lead to errors in nucleic acid quantification.
26. Placing the tubes in the centrifuge with the hinges out means the RNA pellet will collect on the side of the tube in line with the hinge. This allows you to remove supernatant without disturbing the pellet by placing the pipette tip on the opposite side of the tube. RNA pellets can be hard to see.
27. The 260/280 and 260/230 ratios should be between 1.8–2.0 and 2.0–2.2, respectively. A low 260/280 ratio may be due to the presence of protein or phenol contaminants that absorb near 280 nm. A low 260/230 ratio could be due to the presence of EDTA, carbohydrates, or phenol, which all have absorbance near 230 nm. RNA samples with ratios significantly outside these ranges could be re-precipitated with ethanol, although there will be a loss of yield.
28. Traditionally RNA is visualized using denaturing electrophoresis conditions. However, standard agarose gel electrophoresis is sufficient to assess the quality of RNA prior to cDNA synthesis.
29. High-quality RNA will clearly show the 28S and 18S bands with no visible smear below the 18S band. There should be no visible bands in or near the well as this indicates genomic DNA contamination.
30. The amount of RNA used is limited by the RNA yield obtained at the early time points, as there are less cells and smaller yields from these samples. We have had good results using as little as 200 ng RNA per sample.
31. cDNA synthesis can be performed using random hexamers, oligo dT, or a mixture of both of these primers. We routinely use a mixture but it is a matter of personal preference.
32. The addition of 20 μL of RNAse free water to the final cDNA product to make a total volume of 40 μL does not significantly affect downstream amplification and provides additional sample volume to repeat experiments if required.
33. There are many different RNA extraction and cDNA synthesis protocols which will produce good quality cDNA, and you may choose to use a different method to those described here. Whatever method is used, it should remain consistent throughout a series of experiments. It is not appropriate to compare gene expression in samples from different experiments if the RNA was extracted and/or cDNA synthesized differently.

34. We recommend all researchers undertaking qPCR read the MIQE guidelines (minimum information for publication of quantitative real-time PCR experiments) [13].

35. We used a pre-prepared master mix containing buffer, polymerase, dNTPs, $MgCl_2$, and SYBR green, and then added primers and sterile water. However, components can also be sourced and added individually if preferred.

36. We pipette the cDNA first into the bottom of the well and follow this with careful addition of master mix to the side of the well. Change tips between each sample to avoid cross-contamination.

37. We usually have samples arranged in columns and primers sets in rows. E.g., in a 96-well plate, columns 1–12 would contain samples 1–11 plus a negative control (12); and row 1 would contain PCR master mix with primer set 1, row 2 would contain PCR master mix with primer set 2, etc.

38. All primers in Table 1 have an optimal annealing temperature of 60 °C, and the conditions described produce specific products using Bioline SensiFast SYBR Lo Rox PCR mix. Some optimization may be required with different reagents.

39. A single peak in the melt curve indicates amplification of a single product. Multiple peaks or an ill-defined peak indicates the PCR conditions require optimization.

40. Before analyzing qPCR data a threshold level is determined which represents a level above background fluorescence, which also insects the reaction curves at the beginning of the exponential phase. The threshold cycle (Ct) is where the reaction curve for each sample intersects the threshold level and indicates the number of cycles it took to detect signal from each sample. Ct values are inverse to the amount of target nucleic acid in the sample: lower Ct values indicate high amounts of target sequence.

41. If primer efficiencies are not within 0.9–1.1, PCR conditions may need to be altered, or primer sequences adjusted. In our hands, the primers shown in Table 1 ranged from 0.88 to 1.09.

42. The selection of references genes is crucial to accurate qPCR data. Up to 10 potential reference genes may need to be assessed. We have tested *GAPDH, TBP, SDHA, HPRT,* and *ACTAB,* and found the most stable references genes were *SDHA, TBP, and HPRT* in proliferating and differentiating C2C12 cells. Others have reported different genes to be appropriate [17, 18].

43. Test genes we have used to measure C2C12 myoblasts differentiation include markers of early and late differentiation. Markers of early differentiation include myogenic factor

5 (*MYF5*) and myogenin (*MYOG*). Markers of later differentiation include skeletal muscle α-actin (*ACTA1;* expressed in all skeletal muscle fibers) and adult myosin heavy chains 1 and 7 (*MYH1, MYH7*). *MYH1* is expressed in fast type IIX/D muscle fibers and *MYH7* in type 1 slow twitch skeletal muscle and cardiac muscle [19].

44. This process "roughens" the coverslip and provides a topography favorable for the deposition of matrix proteins and cell adhesion. We have found cells can adhere to etched coverslips in the absence of a matrix coating, and then rapidly secrete their own matrix proteins including fibronectin.
45. If necessary, coverslips may be stored in HBS in the fridge for 1–4 days, until coverslips from another time point are ready, or they can be processed immediately for immunofluorescent staining.
46. The isotype control, and no primary antibody control, check for specificity of the primary and secondary antibodies. These controls should be present every time immunofluorescence staining is performed.

References

1. Blau HM, Pavlath GK, Hardeman EC, Chiu CP, Silberstein L, Webster SG, Miller SC, Webster C (1985) Plasticity of the differentiated state. Science 230(4727):758–766
2. Yaffe D, Saxel O (1977) Serial passaging and differentiation of myogenic cells isolated from dystrophic mouse muscle. Nature 270 (5639):725–727
3. Brunner D, Frank J, Appl H, Schoffl H, Pfaller W, Gstraunthaler G (2010) Serum-free cell culture: the serum-free media interactive online database. ALTEX 27(1):53–62
4. McAleer CW, Rumsey JW, Stancescu M, Hickman JJ (2015) Functional myotube formation from adult rat satellite cells in a defined serum-free system. Biotechnol Prog 31(4):997–1003. https://doi.org/10.1002/btpr.2063
5. Vaz R, Martins GG, Thorsteinsdottir S, Rodrigues G (2012) Fibronectin promotes migration, alignment and fusion in an in vitro myoblast cell model. Cell Tissue Res 348 (3):569–578. https://doi.org/10.1007/s00441-012-1364-1
6. Chaturvedi V, Dye DE, Kinnear BF, van Kuppevelt TH, Grounds MD, Coombe DR (2015) Interactions between skeletal muscle myoblasts and their extracellular matrix Revealed by a Serum Free Culture System. PLoS One 10(6): e0127675. https://doi.org/10.1371/journal.pone.0127675
7. Gawlitta D, Boonen KJ, Oomens CW, Baaijens FP, Bouten CV (2008) The influence of serum-free culture conditions on skeletal muscle differentiation in a tissue-engineered model. Tissue Eng Part A 14(1):161–171. https://doi.org/10.1089/ten.a.2007.0095
8. Das M, Rumsey JW, Bhargava N, Gregory C, Riedel L, Kang JF, Hickman JJ (2009) Developing a novel serum-free cell culture model of skeletal muscle differentiation by systematically studying the role of different growth factors in myotube formation. In Vitro Cell Dev Biol Anim 45(7):378–387. https://doi.org/10.1007/s11626-009-9192-7
9. Fujita H, Endo A, Shimizu K, Nagamori E (2010) Evaluation of serum-free differentiation conditions for C2C12 myoblast cells assessed as to active tension generation capability. Biotechnol Bioeng 107(5):894–901. https://doi.org/10.1002/bit.22865
10. Patrikoski M, Lee MHC, Makinen L, Ang XM, Mannerstrom B, Raghunath M, Miettinen S (2017) Effects of macromolecular crowding on human adipose stem cell culture in fetal bovine serum, human serum, and defined xeno-free/serum-free conditions. Stem Cells Int 2017:6909163. https://doi.org/10.1155/2017/6909163
11. Magno V, Friedrichs J, Weber HM, Prewitz MC, Tsurkan MV, Werner C (2017)

Macromolecular crowding for tailoring tissue-derived fibrillated matrices. Acta Biomater 55:109–119. https://doi.org/10.1016/j.actbio.2017.04.018

12. Schneider CA, Rasband WS, Eliceiri KW (2012) NIH Image to ImageJ: 25 years of image analysis. Nat Methods 9(7):671–675
13. Bustin SA, Benes V, Garson JA, Hellemans J, Huggett J, Kubista M, Mueller R, Nolan T, Pfaffl MW, Shipley GL, Vandesompele J, Wittwer CT (2009) The MIQE guidelines: minimum information for publication of quantitative real-time PCR experiments. Clin Chem 55(4):611–622. https://doi.org/10.1373/clinchem.2008.112797
14. De Spiegelaere W, Dern-Wieloch J, Weigel R, Schumacher V, Schorle H, Nettersheim D, Bergmann M, Brehm R, Kliesch S, Vandekerckhove L, Fink C (2015) Reference gene validation for RT-qPCR, a note on different available software packages. PLoS One 10 (3):e0122515. https://doi.org/10.1371/journal.pone.0122515
15. Vandesompele J, De Preter K, Pattyn F, Poppe B, Van Roy N, De Paepe A, Speleman F (2002) Accurate normalization of real-time quantitative RT-PCR data by geometric averaging of multiple internal control genes. Genome Biol 3(7):RESEARCH0034
16. Schmittgen TD, Livak KJ (2008) Analyzing real-time PCR data by the comparative C (T) method. Nat Protoc 3(6):1101–1108
17. Hildyard JC, Wells DJ (2014) Identification and validation of quantitative PCR reference genes suitable for normalizing expression in normal and dystrophic cell culture models of myogenesis. PLoS Curr 6. https://doi.org/10.1371/currents.md.faafdde4bea8df4aa7d06cd5553119a6
18. Masilamani TJ, Loiselle JJ, Sutherland LC (2014) Assessment of reference genes for real-time quantitative PCR gene expression normalization during C2C12 and H9c2 skeletal muscle differentiation. Mol Biotechnol 56 (4):329–339. https://doi.org/10.1007/s12033-013-9712-2
19. Weiss A, McDonough D, Wertman B, Acakpo-Satchivi L, Montgomery K, Kucherlapati R, Leinwand L, Krauter K (1999) Organization of human and mouse skeletal myosin heavy chain gene clusters is highly conserved. Proc Natl Acad Sci U S A 96(6):2958–2963

Chapter 12

LC-MS Analyses of Lipid Species in Skeletal Muscle Cells and Tissue

Marta Moreno-Torres, Jesper F. Havelund, and Nils J. Faergeman

Abstract

Liquid chromatography–mass spectrometry (LC-MS) is a widely used methodology for measuring lipids at a global level. Combined with an optimal extraction method LC-MS enables the detection and characterization of a wide range of lipid species even of low abundance. Here, we describe two extraction- and LC-MS-based quantitative analytical methods for lipid, acyl-CoA, and acyl-carnitine analyses from either mouse C2C12 myotubes or mouse skeletal tissue. We also describe the use of $^{13}C_{16}$-palmitate and its incorporation into acyl-carnitines to show how stable isotope tracers are metabolized within cells and therefore can be implemented for lipidomic flux analysis.

Key words Lipidomics, Acyl-CoA, Acyl-carnitine, Liquid chromatography, Mass spectrometry, Stable isotope tracers, C2C12 myotubes, Skeletal muscle tissue, Labeling, Palmitate

1 Introduction

Lipids are hydrophobic or amphipathic small molecules that besides their structural function in cellular membranes also act as key energy sources and bioactive signaling molecules involved in numerous cellular processes. Alterations in skeletal muscle lipid composition have been associated with insulin-resistance and obesity [1–6], muscular dystrophy [7, 8], and tissue damage [9]. It is therefore critical to optimize extraction, identification, and quantification of molecular lipid species from skeletal muscle cells and tissue. The use of isotope-labeled tracers complements understanding of lipid metabolism during specific physiological conditions. As stable isotope tracers (often ^{13}C- glucose or ^{13}C-palmitate) are incorporated into lipids, the difference in mass can be measured by MS and the isotopic labeling patterns over time can therefore provide information on relative fluxes within cells.

The vast diversity in the physicochemical properties of molecular lipid species makes lipidomic analyses very challenging and depending on the lipids of interest, different protocols of extraction

Sissel Beate Rønning (ed.), *Myogenesis: Methods and Protocols*, Methods in Molecular Biology, vol. 1889,
https://doi.org/10.1007/978-1-4939-8897-6_12,

may be applied. There is not a single protocol that allows the extraction of all lipid classes with optimal recovery for all of them. Therefore, the extraction procedure should either aim to recover as many lipids as possible, or favor the recovery of specific lipid classes. Several lipid extraction methods have been utilized since the late 1800s [10–12]. In this chapter we apply the Folch extraction protocol to mouse skeletal tissue [10], which uses a mixture of chloroform/methanol (2:1, v/v) and is the most commonly used and best lipid extraction method for tissues including skeletal muscle [13, 14].

When considering the measurement of highly polar lipids such as gangliosides, a reliable extraction method is required to prevent their loss during the extraction of other lipid species. Svennerholm and Fredman established a protocol that permits a good recovery of gangliosides together with other lipid species [12]. In this chapter we utilize a modified version of this method in mouse C2C12 myotubes [15].

In addition, given the relevance of acyl-CoAs (ACoA) and acyl-carnitines (ACar) as metabolic intermediates in many cellular processes [16, 17], we here include protocols for their extraction, identification, and quantification in skeletal muscle tissue and C2C12 myotubes. ACoA profiling is very challenging due to the wide variety of polarity of the acyl moiety, the low abundance, and typically poor recoveries during sample extraction. Extraction of acyl-CoA can include solid-phase extraction (SPE) after liquid-liquid extraction (LLE) [18], yet here we present an extraction protocol with methanol/acetonitrile/water (5:3:2) as an organic solvent that allows the extraction of ACoA and ACar of different chain lengths and polarities in a single fraction [19]. An advantage of this extraction method is that this polar solvent also extracts small polar metabolites that can be utilized for further studies of central carbon metabolism.

For sensitive detection and quantification of molecular lipid species we use liquid chromatography-mass spectrometry (LC-MS), which offers improved sensitivity and selectivity over more traditional methods like thin-layer chromatography (TLC) and HPLC coupled to an UV detector (HPLC-UV). Mass spectrometry provides increased sensitivity over NMR [20] and is the most commonly used analytical technique for lipid, ACoA, and ACar analysis (*see* Fig. 1). Separation of lipid species by LC reduces sample complexity prior detection by MS, and thus reduces ion-suppression that consequently increases ionization efficiency and hence improved quantification of low abundant and isomeric lipid species.

Besides describing the above-mentioned extraction and analytical methods, we also present acyl-CoA and acyl-carnitine levels in mouse skeletal muscle tissue, and lipid class composition and incorporation of $^{13}C_{16}$-palmitate derived carbon atoms into acyl-carnitines in cultured C2C12 myotubes (*see* Fig. 2).

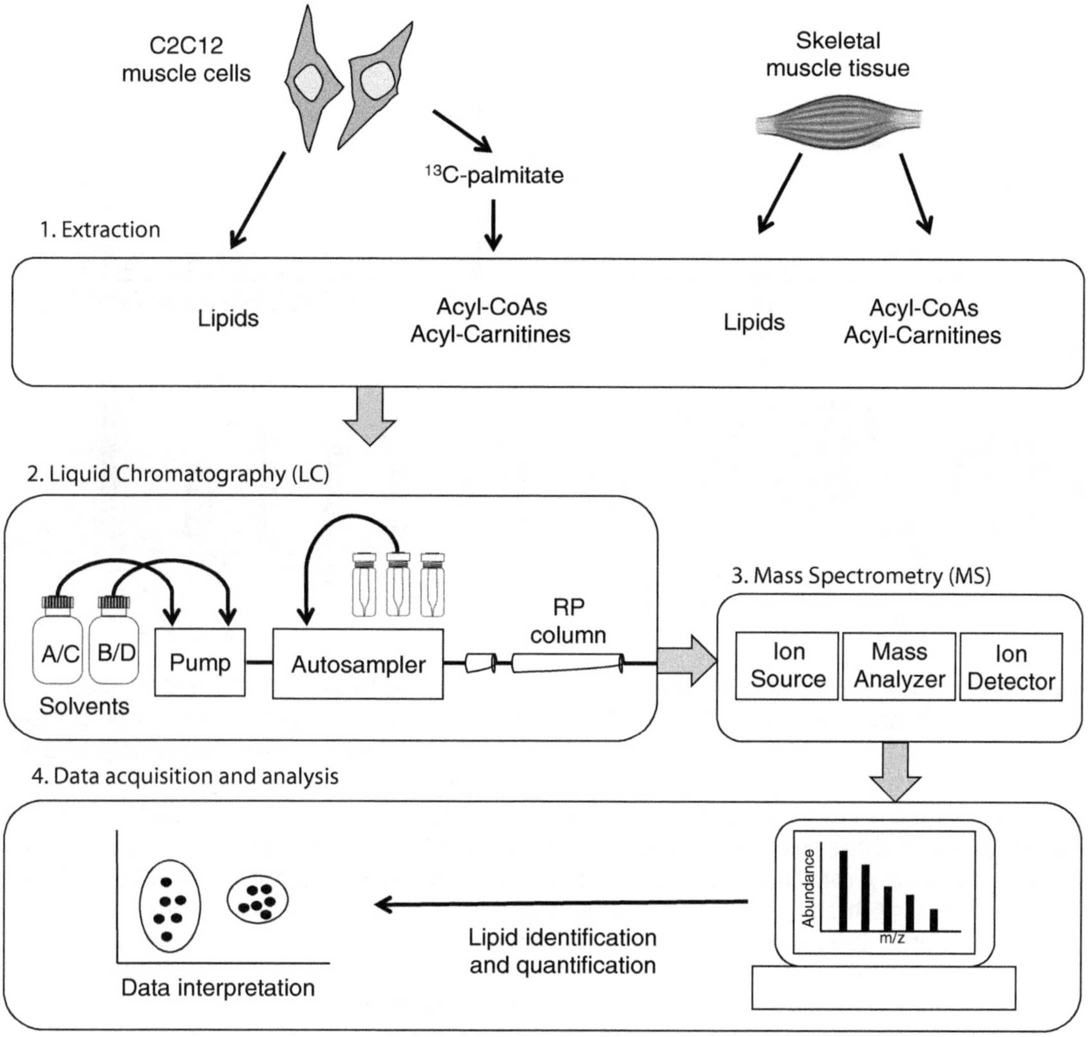

Fig. 1 Flow diagram showing the workflow for lipid, acyl-CoA, and acyl-carnitine analysis from mice C2C12 myotubes and skeletal muscle tissue. To visualize flux of fatty acids, $^{13}C_{16}$-palmitate (100 μM) can be added to cultured C2C12 cells medium for 6 h. Two different extractions methods, one for acyl-CoAs and acyl-carnitines, and another for general lipidomics, are presented for each biological sample. Analytes are separated by reverse phase LC prior to MS analysis. Following raw data acquisition, analytes are identified and quantified for further statistical analysis and data interpretation

2 Materials

2.1 C2C12 Cell Growth and Differentiation

Prepare all solutions using ultrapure and sterile water. C2C12 mouse myoblasts are grown on 10 cm cell culture FALCON petri dishes and maintained at 37 °C in a humidified atmosphere containing 5% CO_2. Carefully follow all waste disposal regulations when discarding cell waste residues.

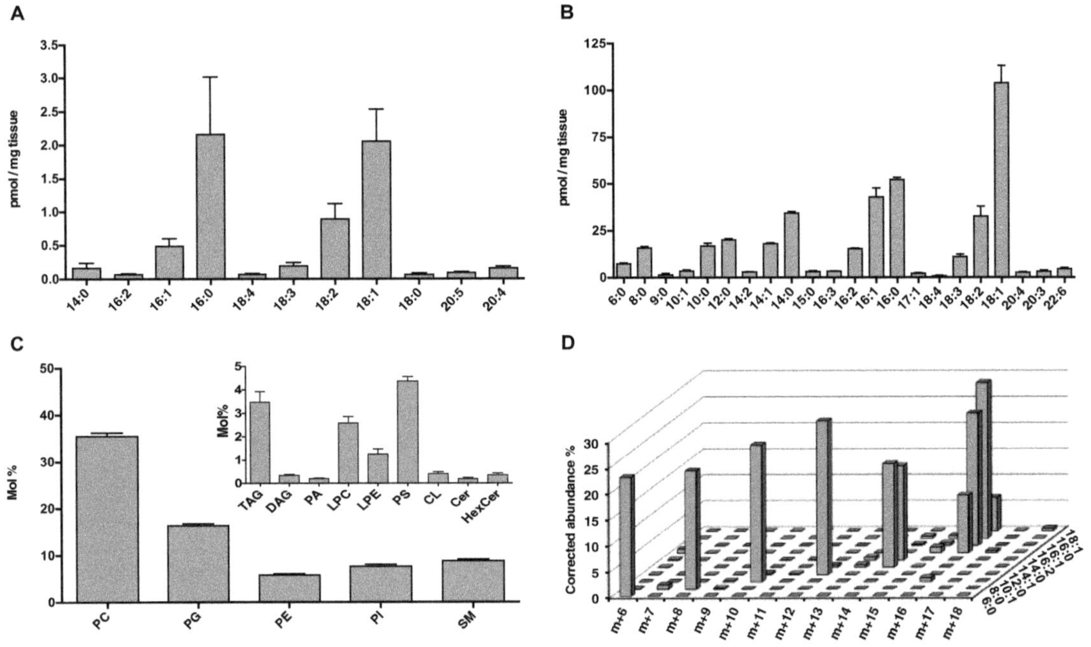

Fig. 2 Acyl-CoAs and acyl-carnitine levels in mouse skeletal tissue and lipid composition and flux analyses in C2C12 myotubes. Acyl-CoAs (**a**) and Acyl-carnitines (**b**) were extracted from approximately 80 mg of tissue and quantified using the internal standards 13:0-CoA and Butyryl (4,4,4,-D3) carnitine, respectively. Data are displayed as pmol/mg tissue. (**c**) Lipid class composition of mouse C2C12 myotubes, extracted from C2C12 cells corresponding to approximately 1 mg of protein. Data is displayed as mol % per all lipid species. The inserted bar graph shows lipid species of less than 5 mol %. (**d**) Incorporation of $^{13}C_{16}$-palmitate-derived carbon atoms into acyl-carnitines in C2C12 cells. C2C12 myotubes were cultured in the presence of 100 μM $^{13}C_{16}$-palmitate for 6 h and acyl-carnitines were subsequently extracted and isotopologue analysis was performed. Corrected abundance percentage (Y axis) of each individual isotope (X axis) is shown for each ACar (Z axis). All figures represent data from three independent biological samples ($n = 3$) and have been normalized by protein content, mg of tissue, and/or by internal standard

1. Dulbecco's modified eagles medium (DMEM): DMEM containing 100 units/mL penicillin, 0.1 mg/mL streptomycin, 44.04 mM $NaHCO_3$, 32.75 μM biotin, 33.58 μM pantothenic acid.
2. C2C12 growth medium: DMEM, 10% Fetal Bovine serum (FBS), 1× GlutaMAX.
3. C2C12 differentiation medium: DMEM, 2% horse serum (HS).
4. 10 cm FALCON petri dishes.
5. Trypsin 1×.
6. Phosphate buffered saline.

2.2 Mice Dissection

1. Hypnorm.
2. Dormicum.
3. 20 mL syringe.
4. PBS.
5. Liquid N_2.
6. Forceps.
7. Scalpel.
8. Scissors.
9. Eppendorf tube.

2.3 Extraction Methods

All solutions in this section contain LC-MS grade reagents. Work always under the fume hood due to the toxicity of volatile solvents. Carefully follow all waste disposal regulations when discarding solvent waste residues.

1. Balance.
2. 2 mL Eppendorf tubes.
3. High performance dispersing instrument (IKA-T10 basic ULTRA TURRAX).
4. Chloroform.
5. Methanol.
6. Chloroform/methanol (2:1).
7. Chloroform/methanol (1:2).
8. Sonication device (Bioruptor).
9. Thermomixer.
10. Centrifuge (that can be set up at 4 °C).
11. Pipette tips.
12. Water.
13. Vortexer.
14. Speed vacuum concentrator.
15. N_2 gas.
16. PBS.
17. Rubber policeman.
18. 1.5 mL Eppendorf tubes.
19. Liquid N_2.
20. Rotating shaker.
21. 0.01 M potassium chloride (KCl).
22. Methanol/water (1:1).
23. Acetonitrile.

24. Methanol/acetonitrile/water (5:3:2).
25. Lipid, acyl-CoA and acyl-carnitine internal standards.
26. $^{13}C_{16}$-palmitate.
27. Bovine serum albumin (BSA, fatty acid free).
28. Ethanol.
29. Water bath.
30. Oven (that can be set up at 55 °C).

2.4 LC-MS

1. HPLC glass vials.
2. 100 μL glass inserts.
3. Lids for glass vials.
4. Pipette tips.
5. Methanol.
6. Isopropanol.
7. Mobile phase A: 5:1:4 isopropanol/methanol/water with 5 mM ammonium acetate (NH_4OAc) and 0.1% acetic acid (CH_3COOH).
8. Mobile phase B: 99:1 isopropanol/water with 5 mM ammonium acetate (NH_4OAc) and 0.1% acetic acid (CH_3COOH).
9. Mobile phase C: 10 mM ammonium acetate (NH_4OAc) (pH 6.8).
10. Mobile phase D: Acetonitrile.
11. Agilent 1290 infinity HPLC system.
12. C18 reverse phase column (Agilent Zorbax eclipse plus C18 RRHD, 2.1 × 50 mm, 1.8 μm).
13. C18 reverse phase column (Agilent Zorbax eclipse plus C18 RRHD, 2.1 × 150 mm, 1.8 μm).
14. Pre-column (Agilent UHPLC guard 3PK Zorbax eclipse plus C18, 2.1 × 5 mm, 1.8 μm).
15. Agilent 6530 quadrupole time of flight (Q-TOF) mass spectrometer.

2.5 Lipids, ACoA, and ACar Identification and Quantification

1. Agilent MassHunter qualitative Analysis software.
2. Agilent MassHunter PCDL manager software.
3. Agilent Profinder software.
4. METLIN database.
5. ALEX^{123} database [21].
6. MetaboAnalyst [22].

3 Methods

Carry out all procedures including homogenizations, centrifugations, and extractions with samples on ice and/or at 4 °C.

3.1 Standardized Method for Lipid Extraction from Mouse Skeletal Muscle Tissue

3.1.1 Tissue Collection

1. Fast the mice for 2–4 h prior to sacrifice.
2. Anesthetize mice by IP injection with 0.01 mL/g of body weight 25% hypnorm and 25% dormicum in sterile water.
3. Subject the mice to transcardial perfusion with approximately 20 mL of sterile filtered PBS.
4. Dissect the mice and collect the tissues of interest (e.g., muscle), freeze in liquid nitrogen, and store tissues at −80 °C.

3.1.2 Tissue Homogenization

1. Take muscle samples from −80 °C freezer and keep them on dry ice.
2. Cut and weigh approximately 80 mg of frozen tissue and transfer it in a 2 mL Eppendorf tube (*see* **Note 1**).
3. Homogenize the tissue using the ultra-Turrax with power strength of 5 for 30–60 s directly in chloroform/methanol (2:1). Repeat this procedure for further homogenization (*see* **Note 2**). Add chloroform/methanol to a final volume 20 times the volume of the tissue sample (e.g., 50 mg of tissue in 1 mL of solvent mixture) (*see* **Note 3**).
4. To complete the homogenization of the tissue, sonicate the samples in a Bioruptor for 15 min (30 s On–30 s Off) (*see* **Note 4**).

3.1.3 Lipid Extraction

1. Shake the extract for 20 min in a Thermoshaker at 1400 rpm (max speed).
2. Centrifuge samples to recover the liquid phase (2500 × *g*, 2 min).
3. Transfer the liquid phase into a new prelabeled 2 mL Eppendorf tube and discard the pellet.
4. Add 0.2 volume of H_2O to each sample (e.g., 0.2 mL of H_2O in 1 mL of chloroform/methanol (2:1)).
5. Vortex for a few seconds.
6. Centrifuge samples to separate the aqueous and organic phase during 5 min at 2500 × *g*.
7. Transfer the aqueous upper phase to a new prelabeled Eppendorf tube. This phase contains small organic polar molecules and can be used for gangliosides analysis.
8. Evaporate the upper phase containing gangliosides under vacuum in a rotary evaporator and the lower organic phase containing the majority of lipids under a stream of nitrogen (*see* **Note 5**).
9. Keep dried pellets at −20 °C until further analysis.

3.2 Standardized Method for Lipid Extraction from Mouse C2C12 Myotubes

3.2.1 Cell Growth and Differentiation

1. Seed the cells in 10 cm Falcon dishes.
2. Routinely split the cells 1:10–1:14 when they get dense, usually every 2 days (*see* **Note 6**).
3. Do not allow the cells to reach 100% confluence or they will begin to differentiate, even in growth medium.
4. Cells are usually grown to 70–80% confluence. Feed cells with new medium every 2–3 days if they are not split.
5. Before starting treatment with differentiation medium, grow the cells until 100% confluence. When grown in differentiation medium, cells differentiate within 3–4 days.

3.2.2 Labeling of C2C12 Cells with $^{13}C_{16}$-Palmitate

For labeling experiments with $^{13}C_{16}$-palmitate, differentiated cells are treated with: DMEM, 1% FBS, 2% bovine serum albumin (BSA) fatty acid free, 100 μM $^{13}C_{16}$-palmitate (or any other desired concentration).

1. Prepare a 75 mM stock of $^{13}C_{16}$-palmitate in ethanol (*see* **Note 7**).
2. Dilute the 75 mM $^{13}C_{16}$-palmitate stock into 2% BSA/D-MEM/1% FBS to a final concentration of 3 mM (*see* **Note 8**).
3. Incubate the 3 mM stock for 10 min at 55 °C and sonicate in a water bath for 15 min (the medium will appear very turbid).
4. Dilute the 3 mM stock in 2% BSA/DMEM/1% FBS to the final concentration of 100 μM. The medium will appear slightly turbid when the 3 mM stock is added but will clear again after 30–60 min incubation at 37 °C.
5. Add 100 μM of $^{13}C_{16}$-palmitate in the cell culture medium of C2C12 differentiated myotubes, replacing all natural isotopes of palmitate. Collect cells after 6 h incubation. On a typical cell culture experiment with stable isotope tracers, the cells are cultured for different time points until labeling reaches steady-state levels. This allows kinetic examination of the rate of isotope labeling.

3.2.3 Collection of Cells

1. Remove medium from cells by aspiration (*see* **Note 9**).
2. Wash cells with chilled (0–4 °C) PBS (*see* **Note 10**).
3. Wash cells with water (*see* **Note 11**).
4. Add 150 μL of water and scrape off the cells with a rubber policeman.
5. Transfer cells into 1.5 mL Eppendorf tube.
6. Freeze them in liquid nitrogen and store them at −80 °C.

3.2.4 Homogenization of Cells

1. Thaw cells on ice.
2. Sonicate the samples in the Bioruptor during 10 min (30 s On–30 s Off).

3.2.5 Lipid Extraction

The extraction solvent used is chloroform/methanol (1:2) and should be stored at −20 °C before use. All solvents must be MS-grade. Known quantities of commercially available lipid standards are added directly to the solvent (*see* **Note 3**).

1. Transfer 100 μL of cell homogenate to a new prelabeled Eppendorf tube (*see* **Note 12**).
2. Freeze the remaining homogenate at −20 °C for protein determination (*see* **Note 13**).
3. Add 0.81 mL of solvent to each sample.
4. Place the Eppendorf tubes on a rotating shaker for 30 min.
5. Centrifuge at 2500 × *g* for 10 min.
6. Transfer the supernatant into a new prelabeled 2 mL Eppendorf tube.
7. For re-extraction, add 0.1 mL of water and 0.4 mL of chloroform/methanol (1:2) to the pellet.
8. Place tubes again on a rotating shaker for 30 min.
9. Centrifuge at 2500 × *g* for 10 min.
10. Transfer this supernatant to the tube where the first supernatant was collected. Discard the pellet.
11. Add 0.26 mL of H_2O to each sample.
12. Invert the tube gently three to four times.
13. Centrifuge at 2500 × *g* for 10 min.
14. Transfer the upper phase into a new 2 mL Eppendorf tube (note how many μL you transfer).
15. Add 0.1 mL of 0.01 M KCl to the lower phase.
16. Invert the tube gently three to four times.
17. Centrifuge at 2500 × *g* for 10 min.
18. Transfer the upper phase into the Eppendorf tube containing the previous upper phase (note how many μL you transfer) and fill up to 2 mL with methanol/water (1:1).
19. Evaporate the upper phase containing gangliosides under vacuum in a rotary evaporator and the lower organic phase containing the majority of lipids under a nitrogen stream (*see* **Note 5**).
20. Keep dried pellets at −20 °C until further analysis.

3.3 Extraction of Acyl-CoA Esters and Acyl-Carnitines from Mouse Skeletal Muscle

The extraction solvent used is methanol/acetonitrile/water (5:3:2) and should be stored at −20 °C prior to use. All solvents must be MS-grade. Known quantities of commercially available lipid standards are added directly to the solvent (*see* **Note 3**).

3.3.1 Tissue Collection

Follow the procedure described in Subheading 3.1.1.

3.3.2 Tissue Homogenization

1. Take muscle samples from the −80 °C and keep them on dry ice.
2. Cut and weigh approximately 80 mg of frozen tissue and put it in a 2 mL Eppendorf tube for homogenization (*see* **Note 1**).
3. Add 1 mL of solvent and homogenize using the Ultra-Turrax with power strength of 5 for 30–60 s. This step can be repeated for better homogenization (*see* **Note 2**).
4. To complete the homogenization of the tissue, sonicate the samples in the Bioruptor during 15 min (30 s On–30 s Off) (*see* **Note 4**).

3.3.3 Extraction of Acyl-CoA Esters and Acyl-Carnitines

1. Shake the homogenate for 20 min in a Thermoshaker at 1400 rpm (max speed).
2. Centrifuge the samples to recover the liquid phase (16,000 × *g*, 10 min).
3. Transfer the supernatant into a new 1.5 mL Eppendorf tube and discard the pellet.
4. Evaporate the supernatant under vacuum in a rotary evaporator.
5. Store pellets at −20 °C until further analysis.

3.4 Extraction of Acyl-CoA Esters and Acyl-Carnitines from C2C12 Cells

The extraction solvent used is methanol/acetonitrile/water (5:3:2) and should be −20 °C when used. All solvents must be MS-grade. Known quantities of standard lipids are added directly to the solvent (*see* **Note 3**).

3.4.1 Collection of Cells

1. Remove medium from cells by aspiration (*see* **Note 9**).
2. Wash cells with chilled (0–4 °C) PBS (*see* **Note 10**).
3. Wash cells with chilled (0–4 °C) water (*see* **Note 11**).
4. Add 1 mL of solvent to the plate, scrape cells off with a rubber policeman, and transfer cell suspension to a prelabeled 1.5 mL Eppendorf tube.

3.4.2 Homogenization of Cells

1. Sonicate samples in the Bioruptor during 10 min (30 s On–30 s Off).

3.4.3 Extraction of Acyl-CoA Esters and Acyl-Carnitines

1. Shake the cell suspension for 20 min in a Thermomixer at 1400 rpm (max speed).
2. Centrifuge samples to recover the liquid phase (16,000 × *g*, 10 min).
3. Transfer the supernatant into a new prelabeled 1.5 mL Eppendorf tube and discard the pellet.

4. Evaporate the supernatant under vacuum in a rotary evaporator.
5. Keep dried pellets at −20 °C until further analysis.

3.5 LC-MS Conditions for Lipid Analysis

1. Prepare mobile phase A: 5:1:4 isopropanol/methanol/water with 5 mM ammonium acetate (NH_4OAc) and 0.1% acetic acid (CH_3COOH).
2. Prepare mobile phase B: 99:1 isopropanol/water with 5 mM ammonium acetate (NH_4OAc) and 0.1% acetic acid (CH_3COOH).
3. Solubilize the pellets (from Subheadings 3.1.3, **step 9** and 3.2.5, **step 20**) in 100 μL of mobile phase A and transfer to HPLC glass vials (*see* **Note 14**). Inject 5–10 μL of sample into the system (injecting volumes will vary depending on the concentration of the lipids of interest in the sample and the sensitivity of the system).
4. Randomize the loading sequence of the samples prior to the LC-MS analysis. Perform separation on chromatographic system (as Agilent 1290 Infinity HPLC system) using a C18 reverse phase column (Agilent Zorbax eclipse plus C18 RRHD, 2.1 × 50 mm, 1.8 μm) and a guard-column (Agilent UHPLC guard 3PK Zorbax eclipse plus C18, 2.1 × 5 mm, 1.8 μm) kept at 50 °C (*see* **Note 15**).
5. Separate analytes using a 0.35 mL/min flow rate and with the following gradient (time/min, % mobile phase B): (0,0), (3,0), (5,20), (25,30), (35,95), (36,95), (38,0). Add 2 min post-run with the initial conditions for equilibration of the system between samples. Samples are kept at 8 °C in the autosampler during the analysis.
6. The LC flow is coupled to an Agilent 6530 quadrupole time of flight (Q-TOF) mass spectrometer scanning from a mass range of 70–1700 m/z. Run a master mix (pool of all samples) sample in all-ion fragmentation mode with collision energy of 10, 20, and 40 V (or traditional MS/MS) in order to produce fragments for the identification of lipids (*see* **Note 14**). Analyze samples with the MS operated both in positive and negative ion mode, respectively (*see* **Note 16**).

3.6 LC-MS Conditions for ACoA and ACar Analysis

1. Prepare mobile phase C: 10 mM ammonium acetate (NH_4OAc) (pH 6.8).
2. Prepare mobile phase D: Acetonitrile.
3. Solubilize the pellets (from Subheadings 3.3.3, **step 5** and 3.4.3, **step 5**) in 100 μL of methanol and transfer to HPLC glass vials (*see* **Note 14**). Inject 10 μL of sample into the system (injecting volumes will vary depending on the concentration of

the ACoAs and ACar present in the sample and the sensitivity of the system).

4. Randomize the loading sequence of the samples prior to the LC-MS analysis. Perform separation on chromatographic system (as Agilent 1290 infinity HPLC system) using a C18 reverse phase column (Agilent Zorbax Eclipse Plus C18 RRHD, 2.1 × 150 mm, 1.8 μm) and a guard-column (Agilent UHPLC guard 3PK Zorbax eclipse plus C18, 2.1 × 5 mm, 1.8 μm) kept at 32 °C.
5. Separate analytes using a flow rate of 0.2 mL/min and the following solvent gradient (time/min, % mobile phase D): (0,20), (15,100), (22.5,100), (22.51,20), (30,20). Add 2 min post-run with the initial conditions for equilibration of the system between samples. Samples are kept at 8 °C in the autosampler during the analysis.
6. The LC flow is coupled to an Agilent 6530 quadrupole time of flight (Q-TOF) mass spectrometer scanning from 50 to 1500 m/z. Run a master mix (pool of all samples) sample in all-ion fragmentation mode with collision energy of 10, 20, and 40 V (or traditional MS/MS) in order to produce fragments for identification of ACoAs and ACar (*see* **Note 14**). Analyze samples with the MS operated in positive ion mode.

3.7 Targeted Identification and Quantification of Analytes

1. Build a database of the different analytes with the molecular formula, exact mass, and retention time (RT) with Agilent MassHunter PCDL manager (*see* **Note 17**).
2. Identify each lipid compound based on the accurate mass and RT with Agilent MassHunter qualitative analysis. Confirm each identity by comparing the MS fragments obtained from the mastermix with the spectrum fragments from the Metlin MS/MS and MS2 ALEX123 databases. Further validation can be done by generating formulas from spectrum peaks with the Agilent MassHunter qualitative analysis software.
3. Extract ion chromatogram peak areas for all compounds at the appropriate retention times with Agilent Profinder software using a mass tolerance of 20 parts per million (ppm) and a RT tolerance of 0.1 min.
4. Normalize raw data by mg of tissue or protein content of cells and/or by internal standards (*see* **Note 18**).
5. Statistical analysis can be performed using MetaboAnalyst. See Xia and Wishart [22] for details. A minimum of three samples is required for statistical analysis.

4 Notes

1. Keep the tissue on dry ice as long as possible. Cool down Eppendorf tubes on dry ice prior to use to avoid thawing of tissue while weighing.
2. Muscle tissue is difficult to homogenize. Be careful when using the Ultra-Turrax as tissue attaches to the disperser. To avoid spills, always homogenize in less than 500 μL and, when required, add the remaining volume after homogenization.
3. Spike desired amounts of internal standards in the solvent mixture for further analyte normalization and/or quantification. This avoids variability due to pipetting errors if internal standards are added individually to each sample as early as possible in the extraction process. The internal standards used in this chapter are: 17:0 PC (850360P), 17:0 SM (d18:1/17:0) (860585P), 17:0 Cer (d18:1/17:0) (860517P), C12 Glucosyl(β) Ceramide (d18:1/12:0) (860543P), 13:0 Coenzyme A (870713P), 19:0 Coenzyme A (870738P) from Avanti polar lipids and Butyryl (4,4,4,-D3)-L-carnitine (71-1734) and Hexadecanoyl (16,16,16-D3)-L-carnitine (71-1746) from Larodan.
4. To keep the samples cold during the homogenization, homogenize 7 min, cool down the water of the Bioruptor with ice, and homogenize again for the remaining 8 min (when there is no automatic cooling down in the system).
5. Although ganglioside levels have not been evaluated in this chapter, they can be analyzed with the same LC-MS conditions presented for general lipidomics in negative ionization mode.
6. Split C2C12 cells: Remove the medium, wash cells with 1× PBS, add 500 μL trypsin 1×, and incubate 3 min at 37 °C. After incubation, add growth medium to inactivate trypsin and perform the corresponding dilution with growth medium.
7. Palmitate is best dissolved when vortexing and warming the ethanol, e.g., in a warm water bath.
8. Due to the low solubility of palmitate in aqueous solutions, BSA is used as a carrier and stabilizing agent.
9. When grown to 100% confluence, C2C12 cells detach very easily from the plate. Be extremely cautious during aspiration and washes to prevent detachment and ensure cells are not washed away.
10. It has been previously shown that washes with chilled isotonic saline stop the conversion of ATP to ADP and AMP, demonstrating metabolic arrest [23]. This argues that there are no significant metabolic changes during the process of sample collection.

11. Remember to use LC-MS grade water. It is very important to completely remove the PBS, since the excess amount of salts in the HPLC system may precipitate and interfere with the chromatography.
12. Always include at least three blank samples with 100 μL of water as control.
13. Protein concentration can be determined using a BCA protein assay kit.
14. Prepare a mastermix sample containing equal amounts of each sample. This mastermix will be analyzed in all-ion fragmentation mode with collision energy of 0, 10, 20, and 40 V in order to produce fragments for analyte identification.
15. It is very important to condition the column before use to stabilize the system. Conditioning is achieved by repeated injections of blanks or standards.
16. For the present LC-MS methodology, ceramide (Cer), hexosylceramide (HexCer), phosphatidylcholine (PC), lysophosphatidylcholine (LPC), sphingomyelin (SM), diacylglycerol (DAG), triacylglycerol (TAG) are better detected in positive ion mode, whereas cardiolipin (CL), phosphatidylethanolamine (PE), lysophosphatidylethanolamine (LPE), phosphatidylinositol (PI), phosphatidylglycerol (PG), phosphatidylserine (PS), phosphatidic acid (PA) are better detected in negative ion mode.
17. For targeted analysis run standards of all lipids with the mentioned LC-MS conditions. For each compound, annotate the retention time and together with the accurate mass of the singly or doubly charged ions build your own database.
18. If absolute concentrations of lipids present in the sample are required, quantify the concentration of analytes using internal standards. The absolute amounts of endogenous analytes are obtained by normalizing their extracted ion chromatogram peak areas by the ones of the corresponding internal standards, and multiplying by the amount of internal standard spiked into the sample. This will correct for variable parameters such as ion suppression, losses during extraction, or decreased MS sensitivity throughout the analysis. A stable isotope version of the analyte of interest is ideal since it has identical chemical properties as the endogenous analyte but has a different mass. If absolute quantification is not required, the relative comparison of lipid peak areas is sufficient to show changes in lipid abundances between samples.

Acknowledgments

This work was supported by The Danish Council for Independent Research, Natural Sciences. Marta Moreno-Torres was supported by a postdoctoral grant from The Danish Diabetes Academy.

References

1. Coen PM, Goodpaster BH (2012) Role of intramyocelluar lipids in human health. Trends Endocrinol Metab 23(8):391–398. https://doi.org/10.1016/j.tem.2012.05.009
2. Gault CR, Obeid LM, Hannun YA (2010) An overview of sphingolipid metabolism: from synthesis to breakdown. Adv Exp Med Biol 688:1–23
3. Turner N, Kowalski GM, Leslie SJ, Risis S, Yang C, Lee-Young RS, Babb JR, Meikle PJ, Lancaster GI, Henstridge DC, White PJ, Kraegen EW, Marette A, Cooney GJ, Febbraio MA, Bruce CR (2013) Distinct patterns of tissue-specific lipid accumulation during the induction of insulin resistance in mice by high-fat feeding. Diabetologia 56(7):1638–1648. https://doi.org/10.1007/s00125-013-2913-1
4. Bergman BC, Hunerdosse DM, Kerege A, Playdon MC, Perreault L (2012) Localisation and composition of skeletal muscle diacylglycerol predicts insulin resistance in humans. Diabetologia 55(4):1140–1150. https://doi.org/10.1007/s00125-011-2419-7
5. Coen PM, Dube JJ, Amati F, Stefanovic-Racic-M, Ferrell RE, Toledo FG, Goodpaster BH (2010) Insulin resistance is associated with higher intramyocellular triglycerides in type I but not type II myocytes concomitant with higher ceramide content. Diabetes 59 (1):80–88. https://doi.org/10.2337/db09-0988
6. Bozic J, Markotic A, Cikes-Culic V, Novak A, Borovac JA, Vucemilovic H, Trgo G, Ticinovic Kurir T (2018) Ganglioside GM3 content in skeletal muscles is increased in type 2 but decreased in type 1 diabetes rat models: implications of glycosphingolipid metabolism in pathophysiology of diabetes. J Diabetes 10 (2):130–139. https://doi.org/10.1111/1753-0407.12569
7. Albarracin I, Lassaga FE, Caputto R (1974) Changes of gangliosides and other lipids in skeletal muscle from rabbits with experimental dystrophy. J Lipid Res 15(1):89–93
8. Higatsberger MR, Auff E (1984) Gangliosides in rabbit and human skeletal muscle with denervation atrophy. J Neurol 231(2):79–82
9. Stanley CA, Hale DE (1994) Genetic disorders of mitochondrial fatty acid oxidation. Curr Opin Pediatr 6(4):476–481
10. Folch J, Lees M, Sloane Stanley GH (1957) A simple method for the isolation and purification of total lipides from animal tissues. J Biol Chem 226(1):497–509
11. Bligh EG, Dyer WJ (1959) A rapid method of total lipid extraction and purification. Can J Biochem Physiol 37(8):911–917. https://doi.org/10.1139/o59-099
12. Svennerholm L, Fredman P (1980) A procedure for the quantitative isolation of brain gangliosides. Biochim Biophys Acta 617 (1):97–109
13. Schiller J, Arnold K (2002) Application of high resolution 31P NMR spectroscopy to the characterization of the phospholipid composition of tissues and body fluids - a methodological review. Med Sci Monit 8(11):MT205–MT222
14. Iverson SJ, Lang SL, Cooper MH (2001) Comparison of the Bligh and dyer and Folch methods for total lipid determination in a broad range of marine tissue. Lipids 36 (11):1283–1287
15. Fong B, Norris C, Lowe E, McJarrow P (2009) Liquid chromatography-high-resolution mass spectrometry for quantitative analysis of gangliosides. Lipids 44(9):867–874. https://doi.org/10.1007/s11745-009-3327-1
16. Neess D, Bek S, Engelsby H, Gallego SF, Faergeman NJ (2015) Long-chain acyl-CoA esters in metabolism and signaling: role of acyl-CoA binding proteins. Prog Lipid Res 59:1–25. https://doi.org/10.1016/j.plipres.2015.04.001
17. McCoin CS, Knotts TA, Adams SH (2015) Acylcarnitines--old actors auditioning for new roles in metabolic physiology. Nat Rev Endocrinol 11(10):617–625. https://doi.org/10.1038/nrendo.2015.129
18. Haynes CA (2011) Analysis of mammalian fatty acyl-coenzyme A species by mass spectrometry and tandem mass spectrometry. Biochim Biophys Acta 1811(11):663–668. https://doi.org/10.1016/j.bbalip.2011.05.010

19. Mackay GM, Zheng L, van den Broek NJ, Gottlieb E (2015) Analysis of cell metabolism using LC-MS and isotope tracers. Methods Enzymol 561:171–196. https://doi.org/10.1016/bs.mie.2015.05.016
20. Dettmer K, Aronov PA, Hammock BD (2007) Mass spectrometry-based metabolomics. Mass Spectrom Rev 26(1):51–78. https://doi.org/10.1002/mas.20108
21. Pauling JK, Hermansson M, Hartler J, Christiansen K, Gallego SF, Peng B, Ahrends R, Ejsing CS (2017) Proposal for a common nomenclature for fragment ions in mass spectra of lipids. PLoS One 12(11): e0188394. https://doi.org/10.1371/journal.pone.0188394
22. Xia J, Wishart DS (2016) Using MetaboAnalyst 3.0 for comprehensive metabolomics data analysis. Curr Protoc Bioinformatics 55:14 10 11-14 10 91. https://doi.org/10.1002/cpbi.11
23. Dietmair S, Timmins NE, Gray PP, Nielsen LK, Kromer JO (2010) Towards quantitative metabolomics of mammalian cells: development of a metabolite extraction protocol. Anal Biochem 404(2):155–164. https://doi.org/10.1016/j.ab.2010.04.031

Chapter 13

A siRNA Mediated Screen During C2C12 Myogenesis

Alexis Parenté, Luce Pèrié, Laetitia Magnol, Khaled Bouhouche, and Véronique Blanquet

Abstract

Myogenesis is a multistep process taking place during pre- and postnatal stages for muscle formation, growth, and regeneration. It is a highly regulated process involving many molecular factors which act during myoblast proliferation and differentiation. To provide new insights into the molecular mechanisms and interactions behind the regulation of these different steps, RNA interference is an efficient methodology to implement. We developed a high-throughput siRNA screen in C2C12 murine myoblast cells for identification of genes relevant to signaling pathways controlling muscle growth. The proposed protocol is based on (1) the analyses of a maximum number of cells/myotubes to detect and quantify both clear and subtle phenotypes during proliferation/fusion cells and (2) the use of two cellular fluorescent markers, DAPI and myosin, decorating nuclei and myotubes respectively. Four phenotypic criteria were quantitatively assessed: cellular density, myotubes quantity, fusion index, and size and morphology of myotubes.

Key words siRNA, Functional screen, Myogenesis, C2C12, Quantitative-imaging analysis, Proliferation, Differentiation, Immunostaining

1 Introduction

Skeletal muscle is a highly complex, heterogeneous, and plastic tissue, which plays an important role in body movement, posture, and breathing but also in energy and protein balance. The muscle-generating process is defined as « myogenesis » which is a multistep process involving myoblast determination, cycle activation, proliferation, alignment, fusion in multinucleated myotubes, and the formation of mature myofibers [1]. This process takes place during embryonic/fetal and postnatal stages for muscle formation, growth, and regeneration. The intrinsic molecular signals regulating these different waves have been well described in the literature. Myogenesis is controlled by a core network of transcription factors, including Pax3, Pax7, and a set of muscle regulatory factors (MRFs) consisting of Myf5, MyoD, MRF4, and myogenin [2–5].

Sissel Beate Rønning (ed.), *Myogenesis: Methods and Protocols*, Methods in Molecular Biology, vol. 1889,
https://doi.org/10.1007/978-1-4939-8897-6_13,

The prenatal myogenesis can be divided into two phases: an early embryonic phase (E10.5-E12.5 in mouse) and a later fetal phase (E14.5-17.5 in mouse) [6, 7]. The first phase results in the production of the primary myofibers, which derive from dermomyotomal progenitors expressing in specific manner Pax3 [8]. During the second phase, a subset of the $Pax3^+$ myogenic progenitors begins to express Pax7 and downregulates Pax3. These $Pax7^+$ myogenic precursors fuse between themselves or with the primary fibers to give the secondary fibers [9]. Skeletal muscle is therefore composed of a heterogeneous population of muscle fibers, conferring remarkable myoplasticity for adaptive change [10, 11]. In postnatal stage, a subset of $Pax7^+$ progenitors forms the pool of adult muscle stem cells—the satellite cells—essential for muscle regeneration [12]. Furthermore, to control the postnatal muscle mass, a dynamic balance between protein synthesis and degradation is orchestrated by extracellular myokines such as IGF-1 or myostatin and its inhibitors [13–15].

The current state of the research does not explain all the possible interactions between the elements affecting muscular function and the adaptive processes of skeletal muscle classically described. Thus, techniques to identify new partners involved in myogenesis are of great interest. Cell culture systems are indispensable tools for studying regulation of many fundamental biological processes or for clinical in-vitro studies. The field of functional genomics has been re-energized by the discovery of the RNA interference (RNAi) pathway [16].

RNAi is an endogenous cellular process conserved in most eukaryotes [17]. It is discovered for the first time in *Caenorhabditis elegans* in which the introduction of double strand RNA (dsRNA) induces the silencing of endogenous gene [18]. RNAi is accomplished through transcriptional or post-transcriptional gene silencing [19] by chromatin remodeling, RNA destruction, or mRNA translation repression [20]. One form of the post-transcriptional gene silencing is induced by introduction of small interfering RNA (siRNA), which recognizes target sequences by base-pairing interactions inducing target mRNA degradation [21]. In this pathway, silencing is triggered by the introduction of dsRNA into cells, which is processed by the RNase III family enzyme Dicer to produce 21- to 23-nucleotide-long siRNA [22]. One of the two siRNA strands is then loaded into the RNA-induced-silencing complex (RISC), which uses this strand to recognize complementary mRNA targets by base-paring interaction [23]. The Argonaute protein within RISC will catalyze the cleavage of mRNA target leading to its destruction in the processing bodies [24–26].

siRNA-mediated gene-silencing is currently used in a high-throughput manner that can systematically knock down every single gene within a genome in both cell culture and living organisms [27, 28]. Large-scale siRNA screens in mammalian primary cultures [29] or cell lines [30] such as C2C12 myoblasts [31, 32] give opportunity to identify genes that are directly or indirectly responsible for a phenotype or involved in a mechanism of interest. Murine C2C12 myoblasts are used for many years to decipher the molecular mechanisms regulating the different steps of in-vivo myogenesis: myoblasts proliferation, fusion, and multi-nucleated myotubes formation [33]. Here we describe a newly designed protocol based on C2C12 myoblasts, RNAi technology, and quantitative image analysis to develop a siRNA-mediated functional screen, aiming identification of novel actors involved in myogenesis [34]. Improvements were based on the analysis of a maximum number of cells and myotubes to detect and quantify both clear and subtle phenotypes during proliferation/fusion of C2C12 cells. In our protocol, we choose to silence the candidate genes by transient transfection of siRNA duplexes rather than stable transfection with small hairpin RNA (shRNA) in order to prevent technical difficulties related to time selection and position effect. For each candidate gene, two distinct siRNA were used and cells were transfected twice during the whole differentiation stage to ensure the continued efficiency of siRNA silencing. Phenotypic analyses were assessed on four criteria: cellular density, myotube quantity, fusion index, and size and morphology of myotubes.

2 Materials

2.1 siRNA Stock Solution

1. 10 μM siRNA, Qiagen (*see* **Notes 1** and **2**).
2. Negative control siRNA: 3'alexafluor647, Qiagen, AATTCTCCGAACGTGTCACGT (*see* **Note 3**).
3. AllStars Mm Cell Death Control siRNA, Qiagen (*see* **Note 4**).
4. Dulbecco's Modified Eagle Medium (DMEM).
5. Final siRNA concentration: 5 nM of siRNA in DMEM.

2.2 C2C12 Seeding and Culture

1. C2C12 cell line.
2. Standard cell culture growth medium (GM): DMEM, 10% fetal bovine serum and antibiotics (50 units.mL^{-1} penicillin and 50 μg.mL^{-1} streptomycin).
3. Sterile Phosphate buffered saline solution (PBS).
4. Trypsin solution: 0.25% in PBS-0.5 mM EDTA.
5. 10 cm dish for growing.
6. 12-well plates for transfection (*see* **Note 5**).
7. 1.8 cm^2-coverslips placed at the bottom of the 12-well plates.
8. Brightfield microscope.
9. Cell culture incubator (37 °C, 5% CO_2).
10. Laminar flow hood.

2.3 C2C12 Differentiation and Culture

1. C2C12 cells in culture, 80% confluent.
2. Differentiation culture medium (DM): DMEM, 2% horse serum and antibiotics (50 units.mL^{-1} penicillin and 50 μg.mL^{-1} streptomycin).
3. Sterile phosphate buffered saline solution.
4. 12-well plates.
5. 1.8 cm^2 coverslips placed at the bottom of the 12-well plates.
6. Brightfield microscope.
7. Cell culture incubator (37 °C, 5% CO_2).
8. Laminar flow hood.

2.4 C2C12 Transfection

1. C2C12 cells seeded at 2.10^4 cells/cm^2 in 12-well plate for real time quantitative PCR (RT-qPCR) and immunostaining experiment.
2. 5 nM siRNA and siRNA control.
3. DMEM.
4. HiPerfect® transfection reagent, Qiagen.
5. Sterile phosphate buffered saline solution.
6. Standard culture growth medium (GM): DMEM, 10% fetal bovine serum, and antibiotics (50 units.mL^{-1} penicillin and 50 μg.mL^{-1} streptomycin).
7. Differentiation culture medium (DM): DMEM, 2% Horse serum and antibiotics (50 units.mL^{-1} penicillin and 50 μg.mL^{-1} streptomycin).
8. 12-well plates.
9. 1.8 cm^2 coverslips placed at the bottom of the 12-well plates.
10. Brightfield microscope.
11. Cell culture incubator (37 °C, 5% CO_2).
12. Laminar flow hood.

2.5 Immunostaining

1. Sterile phosphate buffered saline solution.
2. Paraformaldehyde 4% in PBS.
3. Permeabilization solution: 20 mM HEPES, 300 mM sucrose, 50 mM NaCl, 3 Mm $MgCl_2$, 0.5% Triton X-100, pH 7.4.
4. Blocking buffer: PBS, 10% goat serum, 1% bovine serum albumin (BSA), 0.1% Triton X-100.
5. Washing solution 1: PBS, 0.2% BSA.
6. Primary antibody: monoclonal anti-myosin skeletal fast, clone My-32, mouse antibody (anti-MyHC, Sigma Aldrich) 1:400 diluted in PBS-1% BSA solution.
7. Washing solution 2: PBS, 0.2% BSA, 0.1% Tween 20.

8. Secondary antibody: Alexa Fluor® 488 Goat anti-mouse IgG (Thermofisher) diluted 1:1000 in PBS-1% BSA solution.
9. Stock solution 1 mg.mL^{-1} DAPI diluted at 1:1000 in PBS.
10. Mounting medium: 9.6% Mowiol 4–88, 12% glycerol, 0.1 M Tris–HCl, pH 8.5.
11. Transparent quick drying nail polish.
12. Automated Leica DMI6000B inverted epifluorescence microscope monitored by MetaMorph software (Molecular Devices, Sunnyvale, USA).

2.6 Nuclei Counting and Fusion Index

1. Image J v1.49 software.
2. Superposed large images with DAPI and Alexa Fluor® 488 staining.
3. Fusion index: nucleus number in myotubes/total nucleus number X100.

2.7 Myotube Measurements

1. Image J v1.49 software.
2. Large images with Alexa Fluor® 488 staining.
3. Myotube Length: distance between the two distal extremities of the myotube.
4. Myotube Width: distance between the two proximal extremities of the myotube.

2.8 RNA Extraction and cDNA Synthesis

1. C2C12 cells in 12-well plates.
2. Sterile phosphate buffered saline solution.
3. Trypsin solution: 0.25% in PBS-0.5 mM EDTA.
4. RNeasy midi kit.
5. 2 μg of extracted RNA.
6. High capacity cDNA reverse transcription mix (10 μL per tube): 2 μL 10× RT Buffer, 0.8 μL 25× dNTPs, 2 μL 10× random primers, 1 μL 50 unit.μL^{-1} multiscribe reverse transcriptase, 4.2 μL nuclease free water.
7. Nanodrop spectrophotometer.

2.9 Real Time Quantitative PCR

1. 50 ng of cDNA/condition.
2. Taqman Gene expression master mix (20 μL per tube): 10 μL 2× Taqman mix, 1 μL 20× Taqman probe.
3. Taqman probes specific to each tested gene.
4. Taqman reference gene probes: GAPDH and β-2 M.
5. Quantstudio® Design and Analysis Software version 1.1 and SDS 2.3.
6. ABI PRISM© 7900 System.
7. 96-well plates.

3 Methods

3.1 C2C12 Cells Seeding and Differentiation

1. Pre-warm GM medium, DM medium, trypsin solution, and PBS at room temperature (RT).
2. Use 70% confluence C2C12 cells for subculturing. Wash cells twice with 5 mL PBS and add trypsin solution (3 mL per 10 cm dish) for detachment. Incubate for 3–5 min at 37 °C.
3. Check for complete detachment under the microscope.
4. Recover cells in 3 mL GM and centrifuge them at 800 × *g* for 5 min.
5. Discard the supernatant, resuspend cell pellet in 1 mL of GM, and count cell number using a Malassez slide.
6. For the RT-qPCR experiment, seed cells at 2.10^4 cells/cm^2 in a 12-well plate with 1 mL GM and incubate for 24 h at 37 °C in a 5% CO_2 in air atmosphere.
7. For the Immunostaining experiment, seed cells at 2.10^4 cells/cm^2 on a 1.8 cm^2 coverslip placed at the bottom of the well of a 12-well plate in 1 mL of GM and incubate for 24 h at 37 °C in a 5% CO_2 in air atmosphere.
8. Next day, shift cells (70% confluence) to DM medium and cultivate them for 48 h for RT-qPCR analysis or 144 h for immunostaining experiment (*see* **Note 6**).
9. Change DM every 48 h (*see* Fig. 1a).

3.2 siRNA Transfection of C2C12 Cells

Transfections conditions proposed here have been adapted for 12-well plate cells and C2C12 cells. The amount of siRNA and HiPerfect® reagent required for successful transfection varies depending on the cell type and passage number. It is better to start any new transfection by testing the recommended concentrations of siRNA and HiPerfect® Reagent to determine an optimum amount.

1. Transfect cells twice at 0 h and 48 h of differentiation (*see* **Note 7** and Fig. 1a).
2. Each candidate gene will be silenced separately with two siRNA. For each gene, negative control siRNA and untransfected cells are used as controls (*see* Fig. 1b). Inactivated cells and controls must be performed in three technical replicates.
3. Pre-warm complete DM and DMEM medium at RT. Keep HiPerfect® transfection reagent at 4 °C until needed.
4. For each well plate, remove cell culture supernatant and replace it by the new pre-warmed one in a laminar flow hood.
5. Dilute 37.5 ng of siRNA in 100 μL DMEM (final concentration 5 nM) containing 3 μL of HiPerfect transfection reagent.

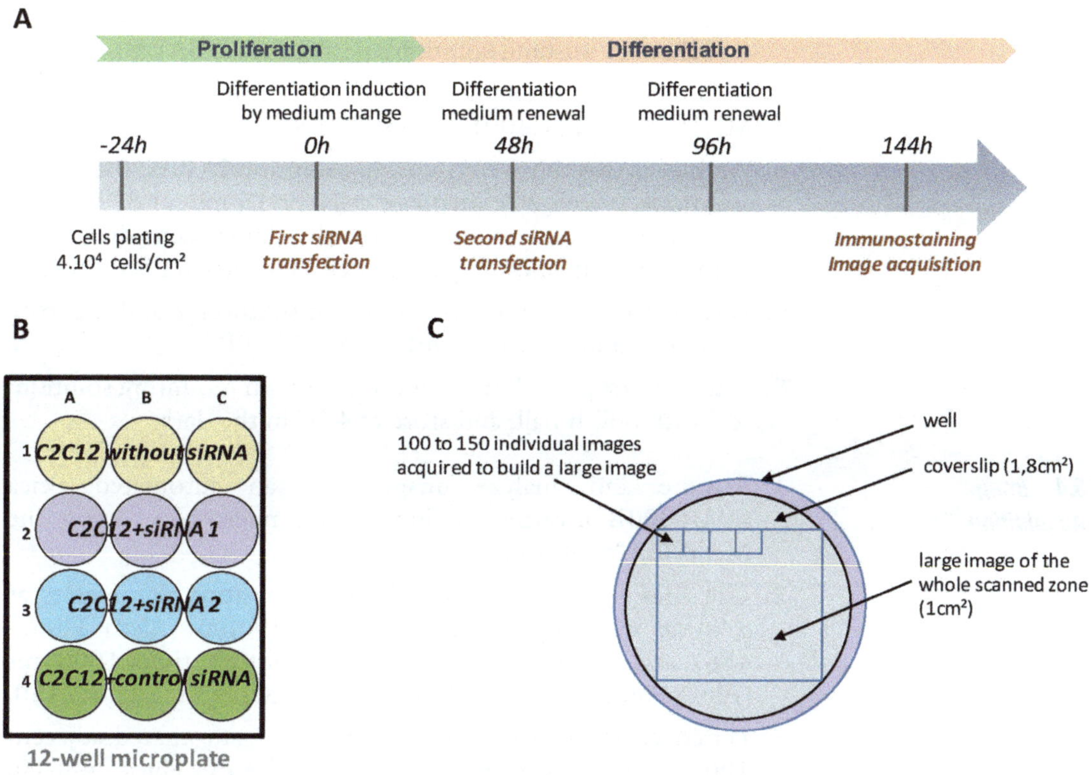

Fig. 1 siRNA-mediated screen experimental approach and reconstruction of the 1 cm^2 large image. (**A**) C2C12 cells are seeded at 2.10^4 cells/cm^2 in 12-well plate. After 24 h, the proliferation medium is replaced by the differentiation medium ($T = 0$ h). To ensure the continued efficiency of siRNA silencing during the whole differentiation stage, we transfected cells twice with the same siRNA, at 0 h and 48 h of differentiation. (**B**) To eliminate false positives, commercial negative control siRNA and two distinct siRNA are used for each candidate gene. (**C**) C2C12 cells are seeded on a 1.8 cm^2 coverslip placed at the bottom of the well. Per coverslip, automated 100–150 fluorescent images were taken using 100× magnification objective with the aim of reconstructing a 1 cm^2 large image of the whole scanned zone

6. Incubate the mix for 10 min at RT. Add the 100 µL transfection complex in a drop-wise manner onto cells in each well. Swirl plates gently to ensure uniform distribution and incubate at 37 °C in a 5% CO_2 in air atmosphere.
7. Repeat this transfection at 48 h of differentiation (*see* **Note 8**).

3.3 Immunostaining

To phenotypically screen the candidate genes, myosin can be immunostained and used as a differentiation marker of myotubes.

1. Wash cells three times in PBS 1×.
2. Fix cells during 15 min in 4% PFA solution.
3. Wash again three times with PBS 1× and permeabilize with HEPES/Triton Buffer.

4. Block cells 1 h at RT with blocking buffer and wash them quickly with washing solution 1 (PBS, 0.2% BSA).
5. Incubate cells overnight at 4 °C with primary antibody anti-MyHC diluted (1:400) in PBS, 1% BSA.
6. Wash cells two times with washing solution 2 (PBS, 0.2% BSA, and 0.1% Tween 20), incubate cells for 15 min at 37 °C with secondary antibody conjugated with a fluorescent dye (Alexa Fluor® 488 anti-mouse IgG) diluted (1:1000) in PBS, 1% BSA.
7. Wash cells three times with washing solution 2 and incubate cells for 5 min with 1 $mg.mL^{-1}$ DAPI in PBS.
8. Place coverslips back to front on slides with mounting solution, seal with polish nail, and store at 4 °C in the dark.

3.4 Image Acquisition

1. Acquire and analyze image with an automated Leica DMI6000B inverted epifluorescence microscope using the MetaMorph software.
2. Use Alexa Fluor® 488 and DAPI acquisition parameters as follows: wavelength spectrum 499–519 nm (Alexa Fluor® 488) and 345–455 nm (DAPI), exposure time 1000 ms (Alexa Fluor® 488) and 50 ms (DAPI) and light intensity 150.
3. Per coverslip, scan a 1 cm^2 zone by taking automated-adjacent 100–150 fluorescent images (DAPI and Alexa Fluor® 488) in 100× magnification.
4. Reconstruct a large image from the 150 images using Metamorph software (*see* **Note 9** and Fig. 1c).
5. Save individual images and reconstruct images in .tif format.
6. Quantify fluorescent area of each marker (DAPI and Alexa Fluor® 488) on the corresponding large image, this quantification consists of the percentage of fluorescent area compared to the whole scanned zone.
7. Eliminate the background fluorescence with statistic correction. The percentage of DAPI and Alexa Fluor® 488 fluorescent areas correspond to the percentage of the fluorescent surface occupied by nuclei or myotubes compared to the total coverslip scanned zone (*see* **Note 10** and Fig. 2).

3.5 Image Analyses

Myosin reconstructed images will be used to evaluate the size and morphology of myotubes by measuring the length and the width of many myotubes. Myosin and DAPI reconstructed images will be superposed in order to calculate the fusion index. Percentages of fluorescent area (DAPI and myosin), myotube size and morphology and fusion index will be defined as phenotypic criteria to classify candidate genes.

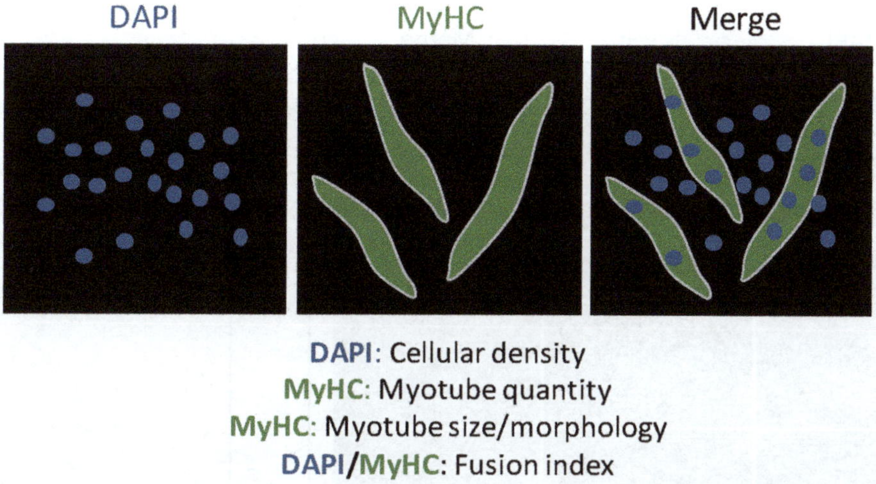

Fig. 2 Schematic representation of the criteria used for the analysis. Phenotypic study was based on two cellular fluorescent markers, DAPI and myosin, decorating nuclei and myotubes respectively. Fluorescence quantification allowed us to define and analyze four phenotypic criteria: cellular density, myotube quantity, fusion index, myotube size, and morphology. The first one occurs during the proliferation stage while the others take place during the differentiation stage

1. Cellular density and myotube quantity: eliminate the background fluorescence with statistic correction. The percentage of Alexa Fluor® 488 fluorescent area represents the surface coverslip occupied by myosin, which reflects the differentiated myotubes. The percentage of DAPI fluorescent area represents the surface coverslip occupied by nuclei (*see* Fig. 2).
2. Myotube size and morphology: measure the length and the width of 30 myotubes from 15 randomly chosen fields of the Alexa Fluor® 488 reconstructed images with the ImageJ Software.
3. Index fusion analysis: measure the total number of nuclei and the number of nuclei inside myotubes on a section of the overlaid reconstructed images using ImageJ Software. Fusion index of C2C12 cells is calculated as the ratio of the number of nuclei in myotubes to the total number of nuclei (*see* Fig. 2). Realize this analysis through the following steps:
   (a) On DAPI image: automatic adjustment of brightness and contrast/Threshold setting with Li method (Minimum Cross Entropy thresholding method)/Realization of binary mask/Glued nuclei separation by watershed/Nuclei enumeration of all image area.
   (b) On Alexa Fluor® 488 image: automatic adjustment of brightness /contrast and Threshold setting as DAPI image/Median filter construction (pixel size =3) /Realization of a new binary mask on Alexa Fluor® 488 and DAPI images /Combination of the two masks/Nuclei counting in myotubes.

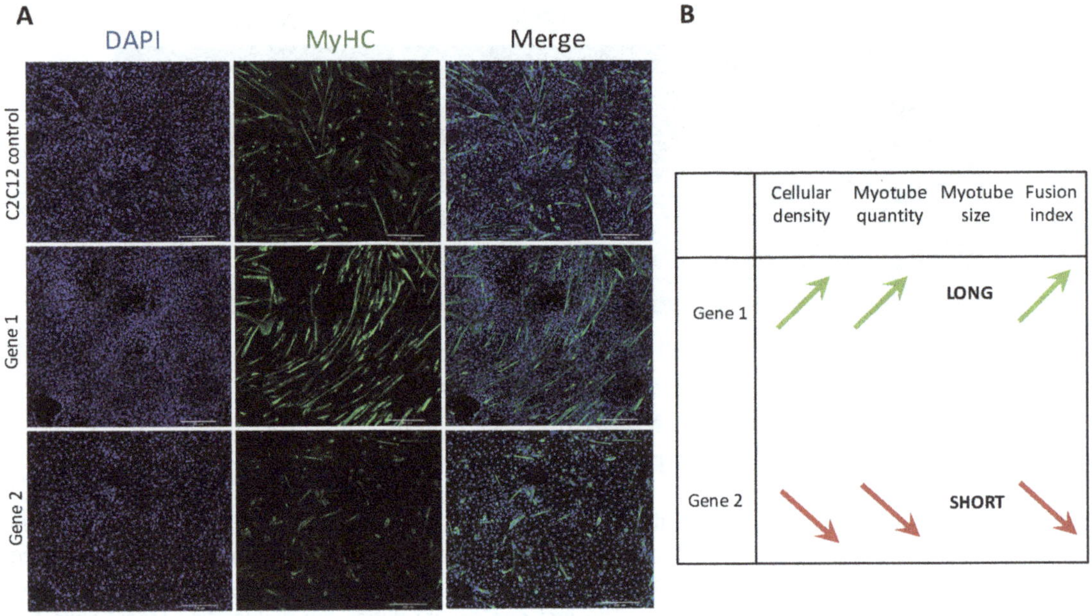

Fig. 3 Example of siRNA-silencing analysis of two candidate genes. (**A**) Fluorescence microscopy images after DAPI staining of nuclei (left) and immunostaining with an anti-myosin antibody (middle) of control cells (siRNA control) and Gene 1 or Gene 2 knockdown cells; merged fluorescence images (right). (**B**) After fluorescence quantification, the phenotypic criteria described in Fig. 2 were assessed for Gene 1 or Gene 2 knockdown cells

3.6 Classification of Observed Phenotypes

For each gene, quantitative data obtained from silencing with the two siRNA will be systematically compared to those obtained from associated controls. Genes for which quantitative data are similar between the two siRNA and significantly different from controls will be retained.

1. Determine the phenotype of inactivated C2C12 cells for each candidate gene (*see* Fig. 3).
2. Classify the inactivated genes according to the phenotypes observed following the extinction by the two siRNA.
3. Eliminate all genes with no phenotype after silencing with the two siRNA (*see* **Note 11**).
4. Separate genes in two groups: a group with genes showing the same phenotype with the two independent siRNA and a group with genes displaying different phenotypes between the two siRNA.
5. The first group will be retained and subdivided into two sets of genes according to the nature of the observed phenotype: increase or decrease of the analyzed phenotypic criteria.

3.7 Cell Lysis for RNA Extraction

siRNA knockdown of selected genes must be confirmed by RT-qPCR.

1. Pre-warm GM medium, trypsin solution, and PBS at RT.
2. Remove cell culture supernatant of 12-well plate at 48 h of differentiation. Wash twice with 1 mL of PBS and add 500 μL of trypsin solution for detachment. Incubate for 3–5 min at 37 °C.
3. Check for complete detachment under the microscope.
4. Resuspend cell pellet in 1 mL GM and centrifuge cells at 800 × *g* for 5 min.
5. Discard the supernatant and resuspend cell pellet with 350 μL RLT lysis buffer from the RNeasy Midi Kit. Mix well and centrifuge for 3 min at 13,000 × *g*, RT.
6. Transfer the lysate in a new eppendorf tube and add 1 volume of ethanol 70%. Mix well by pipetting and transfer up to 700 μL of sample to an RNeasy spin column placed in 2 mL collection tube. Centrifuge for 15 s at 13,000 × *g*, RT. Discard the flow-through.
7. Add 350 μL buffer RW1, and then centrifuge for 15 s at 13,000 × *g*, RT. Discard the flow-through.
8. To avoid DNA contamination, add 10 μL of DNAse solution stock of the kit to 70 μL RDD. Add DNAse incubation mix (80 μL) directly to Rneasy spin column and place on benchtop for 15 min. Add 350 μL buffer RW1, then centrifuge 15 s at 13,000 × *g*, RT. Discard the flow-through.
9. Add 500 μL buffer RPE to RNeasy spin column and centrifuge for 15 s at 13,000 × *g*, RT. Discard the flow-through. Add again 500 μL buffer RPE to RNeasy spin column and centrifuge for 2 min at 13,000 × *g*, RT. Discard the flow-through.
10. Place the RNeasy spin column into a new collection tube and centrifuge for 1 min at 13,000 × *g*, RT to dry the membrane.
11. Place the RNeasy spin column into an Eppendorf tube. Add 30 μL RNase-free water and centrifuge for 1 min at 13,000 × *g*, RT to elute the RNA.
12. Dose RNA by nanodrop spectrophotometer.

3.8 Reverse Transcription

1. Using the High Capacity cDNA reverse transcriptase kit, add 2 μg of RNA in 10 μL of nuclease free water.
2. Prepare 10 μL of reverse transcription mix.
3. Add RNA samples to the reverse transcription mix for a final volume of 20 μL.
4. Place the tubes in a thermocycler and cycle through the following steps: 25 °C for 10 min, 37 °C for 120 min, 85 °C for 5 min and hold at 4 °C.

3.9 RT-qPCR

Relative amounts of transcripts will be determined using specific Taqman probes for each tested gene and the two reference genes.

1. Perform RT-qPCR in triplicate using 50 ng of cDNA and Taqman Gene expression master mix.
2. In each 96-well plate, add 50 ng of cDNA to 20 μL Taqman Gene expression master mix.
3. Perform RT-qPCR reaction according to the following program 40 cycles of 95 °C for 15 s and 60 °C for 1 min.
4. Calculate the relative mRNA expression values by the ΔΔCt method with normalization of each sample to the average change in cycle threshold (Ct) value of the controls using the Quantstudio® Design and Analysis Software version 1.1 and SDS 2.3.
5. For all analyses, perform three independent experiments.

4 Notes

1. All siRNA used in our study are mouse-specific siRNA designed and produced by Qiagen. We decided to silence the candidate genes by transient transfection of siRNA duplexes rather than by generating stables expressing short hairpin RNAs (shRNAs) to prevent technical difficulties related to time selection and position effect associated with the establishment of hundreds of shRNA stable clones.
2. To eliminate false positives, two distinct siRNA targeting either the coding region or the 3'UTR will be used for each candidate gene.
3. Commercial negative Control siRNA so-called "non-targeting" siRNAs is used: you must be sure that such a non-targeting siRNA transfection does not perturb your system.
4. In order to test the efficiency of transfection, C2C12 cells are transfected with an AllStars Mm Cell Death Control siRNA (10 μM Qiagen®) targeting the mRNA of essential genes for C2C12 cell survival.
5. C2C12 cells are seeded in 12-well plates for molecular studies (RT-qPCR experiment) and 12-well plates with 1.8 cm^2 coverslips placed at the bottom of the well for phenotypic studies (immunostaining experiment).
6. RT-qPCR is performed on inactivated cells grown only 144 h in differentiation medium; however, for immunostaining experiments, inactivated cells are cultured in differentiation medium for 144 h.

7. To ensure the continued efficiency of siRNA silencing during the whole differentiation stage, we transfected cells twice with the same siRNA, at 0 h and 48 h of differentiation.
8. The second transfection is only performed for the inactivated cells used for phenotypic analyses.
9. We used an automated-scan microscope capable of very fast image acquisition and the reconstruction of 1 cm^2 large images. This protocol allowed us to perform fluorescence quantification and to track easily four types of phenotypes mediated by the silencing of candidate genes.
10. Phenotypic analysis in our protocol was based on two cellular fluorescent markers (DAPI and myosin) decorating nuclei and myotubes respectively. Quantification of the decorated surface allowed us to define and analyze four phenotypic criteria (cellular density, myotube quantity, fusion index, myotube size, and morphology). The first one occurs during the proliferation stage while the others take place during the differentiation stage.
11. Many silenced genes may show no phenotype with the two siRNA or phenotypes only with one siRNA. Three hypotheses could explain this absence of phenotype: (1) the incomplete target gene suppression due to non- or partially functional siRNA, (2) the functional redundancy that can occur between candidate genes and their paralogues belonging to the same multigenic family, (3) no contribution of these genes in C2C12 myogenesis.

References

1. Buckingham M, Rigby PWJ (2014) Gene regulatory networks and transcriptional mechanisms that control myogenesis. Dev Cell 28:225–238. https://doi.org/10.1016/j.devcel.2013.12.020
2. Berkes CA, Tapscott SJ (2005) MyoD and the transcriptional control of myogenesis. Semin Cell Dev Biol 16:585–595. https://doi.org/10.1016/j.semcdb.2005.07.006
3. Pownall ME, Gustafsson MK, Emerson CP (2002) Myogenic regulatory factors and the specification of muscle progenitors in vertebrate embryos. Annu Rev Cell Dev Biol 18:747–783. https://doi.org/10.1146/annurev.cellbio.18.012502.105758
4. Rudnicki MA, Schnegelsberg PN, Stead RH et al (1993) MyoD or Myf-5 is required for the formation of skeletal muscle. Cell 75:1351–1359
5. Tapscott SJ (2005) The circuitry of a master switch: Myod and the regulation of skeletal muscle gene transcription. Development 132:2685–2695. https://doi.org/10.1242/dev.01874
6. Biressi S, Tagliafico E, Lamorte G et al (2007) Intrinsic phenotypic diversity of embryonic and fetal myoblasts is revealed by genome-wide gene expression analysis on purified cells. Dev Biol 304:633–651. https://doi.org/10.1016/j.ydbio.2007.01.016
7. Stockdale FE (1992) Myogenic cell lineages. Dev Biol 154:284–298
8. Horst D, Ustanina S, Sergi C et al (2006) Comparative expression analysis of Pax3 and Pax7 during mouse myogenesis. Int J Dev Biol 50:47–54. https://doi.org/10.1387/ijdb.052111dh

9. Fougerousse F, Edom-Vovard F, Merkulova T et al (2001) The muscle-specific enolase is an early marker of human myogenesis. J Muscle Res Cell Motil 22:535–544
10. Berchtold MW, Brinkmeier H, Müntener M (2000) Calcium ion in skeletal muscle: its crucial role for muscle function, plasticity, and disease. Physiol Rev 80:1215–1265. https://doi.org/10.1152/physrev.2000.80.3.1215
11. Chargé SBP, Rudnicki MA (2004) Cellular and molecular regulation of muscle regeneration. Physiol Rev 84:209–238. https://doi.org/10.1152/physrev.00019.2003
12. Relaix F, Rocancourt D, Mansouri A, Buckingham M (2005) A Pax3/Pax7-dependent population of skeletal muscle progenitor cells. Nature 435:948–953. https://doi.org/10.1038/nature03594
13. Thomas M, Langley B, Berry C et al (2000) Myostatin, a negative regulator of muscle growth, functions by inhibiting myoblast proliferation. J Biol Chem 275:40235–40243. https://doi.org/10.1074/jbc.M004356200
14. Pèrié L, Parenté A, Brun C et al (2016) Enhancement of C2C12 myoblast proliferation and differentiation by GASP-2, a myostatin inhibitor. Biochem Biophys Rep 6:39–46. https://doi.org/10.1016/j.bbrep.2016.03.001
15. Périè L, Parenté A, Baraige F et al (2017) Alterations in adiposity and glucose homeostasis in adult Gasp-1 overexpressing mice. CPB 44:1896–1911. https://doi.org/10.1159/000485878
16. Agrawal N, Dasaradhi PVN, Mohmmed A et al (2003) RNA interference: biology, mechanism, and applications. Microbiol Mol Biol Rev 67:657–685
17. Cerutti H, Casas-Mollano JA (2006) On the origin and functions of RNA-mediated silencing: from protists to man. Curr Genet 50:81–99. https://doi.org/10.1007/s00294-006-0078-x
18. Fire A, Xu S, Montgomery MK et al (1998) Potent and specific genetic interference by double-stranded RNA in *Caenorhabditis elegans*. Nature 391:806–811. https://doi.org/10.1038/35888
19. Heo I, Kim VN (2009) Regulating the regulators: posttranslational modifications of RNA silencing factors. Cell 139:28–31. https://doi.org/10.1016/j.cell.2009.09.013
20. Castel SE, Martienssen RA (2013) RNA interference in the nucleus: roles for small RNAs in transcription, epigenetics and beyond. Nat Rev Genet 14:100–112. https://doi.org/10.1038/nrg3355
21. Elbashir SM, Lendeckel W, Tuschl T (2001) RNA interference is mediated by 21- and 22-nucleotide RNAs. Genes Dev 15:188–200
22. Bernstein E, Caudy AA, Hammond SM, Hannon GJ (2001) Role for a bidentate ribonuclease in the initiation step of RNA interference. Nature 409:363–366. https://doi.org/10.1038/35053110
23. Hammond SM, Bernstein E, Beach D, Hannon GJ (2000) An RNA-directed nuclease mediates post-transcriptional gene silencing in Drosophila cells. Nature 404:293–296. https://doi.org/10.1038/35005107
24. Meister G, Landthaler M, Patkaniowska A et al (2004) Human Argonaute2 mediates RNA cleavage targeted by miRNAs and siRNAs. Mol Cell 15:185–197. https://doi.org/10.1016/j.molcel.2004.07.007
25. Hutvagner G, Simard MJ (2008) Argonaute proteins: key players in RNA silencing. Nat Rev Mol Cell Biol 9:22–32. https://doi.org/10.1038/nrm2321
26. Kim VN, Han J, Siomi MC (2009) Biogenesis of small RNAs in animals. Nat Rev Mol Cell Biol 10:126–139. https://doi.org/10.1038/nrm2632
27. Mohr SE, Smith JA, Shamu CE et al (2014) RNAi screening comes of age: improved techniques and complementary approaches. Nat Rev Mol Cell Biol 15:591–600. https://doi.org/10.1038/nrm3860
28. Crotty MM, Henderson J, Ward PR et al (2015) Analysis of social networks supporting the self-management of type 2 diabetes for people with mental illness. BMC Health Serv Res 15:257. https://doi.org/10.1186/s12913-015-0897-x
29. Khanjyan MV, Yang J, Kayali R et al (2013) A high-content, high-throughput siRNA screen identifies cyclin D2 as a potent regulator of muscle progenitor cell fusion and a target to enhance muscle regeneration. Hum Mol Genet 22:3283–3295. https://doi.org/10.1093/hmg/ddt184
30. Echeverri CJ, Beachy PA, Baum B et al (2006) Minimizing the risk of reporting false positives in large-scale RNAi screens. Nat Methods 3:777–779. https://doi.org/10.1038/nmeth1006-777
31. Rajan S, Chu Pham Dang H, Djambazian H et al (2012) Analysis of early C2C12 myogenesis identifies stably and differentially expressed transcriptional regulators whose knock-down

inhibits myoblast differentiation. Physiol Genomics 44:183–197. https://doi.org/10.1152/physiolgenomics.00093.2011

32. Ge Y, Waldemer RJ, Nalluri R et al (2013) RNAi screen reveals potentially novel roles of cytokines in myoblast differentiation. PLoS One 8. https://doi.org/10.1371/journal.pone.0068068

33. Burattini S, Ferri P, Battistelli M et al (2004) C2C12 murine myoblasts as a model of skeletal muscle development: morpho-functional characterization. Eur J Histochem 48:223–233

34. Alwan R, Bruel A-L, Da Silva A et al (2017) An siRNA-based screen in C2C12 myoblasts identifies novel genes involved in myogenic differentiation. Exp Cell Res 359:145–153. https://doi.org/10.1016/j.yexcr.2017.07.037

Chapter 14

Fluorescence-Activated Cell Sorting of Larval Zebrafish Muscle Stem/Progenitor Cells Following Skeletal Muscle Injury

Dhanushika Ratnayake and Peter D. Currie

Abstract

This chapter describes a protocol for the isolation of larval zebrafish muscle stem/progenitor cells by fluorescence-activated cell sorting (FACS). This method has been successfully applied to isolate *pax3a* expressing cells 3 days following needle stab skeletal muscle injury. The cell sorting strategy described here can easily be adapted to any cell type at embryonic or larval stages. RNA extracted from the sorted cells can be used for subsequent downstream applications such as quantitative PCR (qPCR), microarrays, or next generation sequencing.

Key words Zebrafish larvae, Skeletal muscle, Muscle stem cells, Satellite cells, Muscle progenitors, Needle stab injury, Muscle injury, Regeneration, FACS

1 Introduction

The remarkable regenerative ability of skeletal muscle is credited to the tissue resident stem cell termed satellite cell [1]. Following injury, satellite cells are activated and migrate into the wound site, where they proliferate and differentiate to form myotubes that fuse and mature into myofibers. This process is firmly regulated by a hierarchy of transcription factors encoded by master regulatory genes [2–4]. Advanced methods allowing the isolation of satellite cells at distinct stages of myogenic commitment present a robust approach to further dissect these molecular pathways.

The larval zebrafish (*Danio rerio*) is a tractable model to investigate the molecular basis of myogenesis. Zebrafish possess a muscle stem/progenitor cell population analogous to mammalian satellite cells that carry out comparable functions [5–7]. Furthermore, there is a host of transgenic zebrafish lines available that specifically label these populations, providing a powerful tool to not only monitor

Sissel Beate Rønning (ed.), *Myogenesis: Methods and Protocols*, Methods in Molecular Biology, vol. 1889,
https://doi.org/10.1007/978-1-4939-8897-6_14,

and track these cells over time, but also, isolate them based on fluorescence at distinct stages of myogenic commitment.

Here, we describe a protocol to isolate stem/progenitor cells from transgenic zebrafish larva using fluorescence-activated cell sorting (FACS) following skeletal muscle injury. With this protocol, we succeed in obtaining a pure population of injury responding *pax3a*$^{+}$ cells from TgBAC(*pax3a*:GFP)^{i150} [5] transgenic larva. The sorted cell population was shown to express known stem/progenitor markers such as *pax7*, *cmet*, *myf5*, *myod1*, and *myog*.

In this chapter we provide a detailed description on inducing a needle stick muscle injury as well as dissociating the injured tissue by sequential collagenase and trypsin treatment to generate a single cell suspension to use in FACS. This protocol gives the user the ability to generate a purified pool of stem and progenitor cells that can then be used for a host of downstream applications.

2 Materials

2.1 Larval Zebrafish Lines

1. Embryos from the transgenic zebrafish reporter line TgBAC (*pax3a*:GFP)^{i150} [5] 4 days post fertilisation (dpf) (*see* Fig. 1b and **Note 1**).

2.2 Needle Stab Skeletal Muscle Injury

1. 30-gauge needles.
2. Glass Pasteur pipettes.
3. 90 mm plastic petri dishes.
4. Stereomicroscope.
5. Incubator set to 28.5 °C.
6. 100 mM 1-phenyl-2-thiourea (PTU) stock solution (use 1:1000 dilution). Stir 1.52 g in 100 mL ethanol overnight to dissolve. Store solution at 4 °C (*see* **Note 2**).
7. 0.4% 3-amino benzoic acidethylester (tricaine) stock solution. Dissolve 400 mg tricaine with 97.9 mL ddH_2O. Adjust to pH 7.2–7.4 with Tris–HCL (pH 9). Store solution at 4 °C.
8. 1× Ringer's solution. 116 mM NaCl, 2.9 mM KCl, 1.8 mM $CaCl_2$ and 5.0 mM HEPES in ddH_2O. Adjust to pH 7.2–7.4 with 5 M NaOH. Store solution at room temperature.

2.3 Cell Dissociation and FACS Sorting

1. Sterile disposable 70-micron cell strainers adaptable on 50 mL Falcon tubes.
2. Sterile disposable 40-micron cell strainers adaptable on 50 mL Falcon tubes.
3. 5 mL Cell strainer tubes with 35-micron mesh.
4. Tabletop microcentrifuge with refrigeration.
5. 50 mL Falcon tubes.

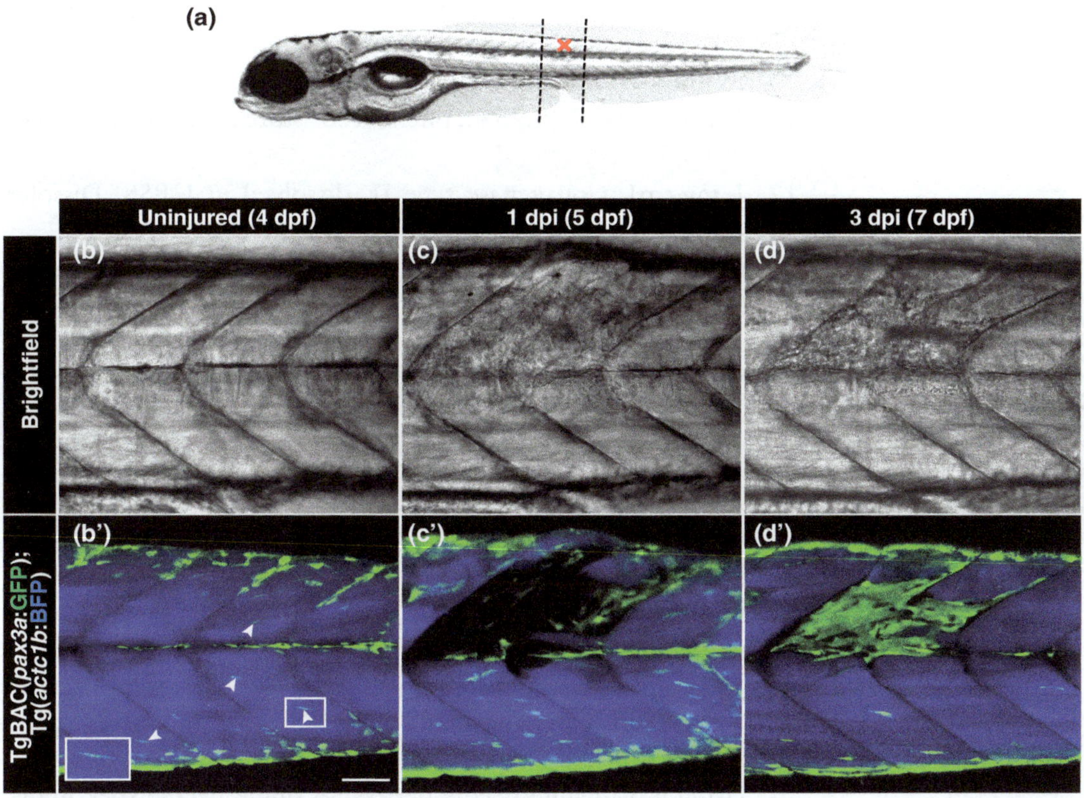

Fig. 1 Muscle stem/progenitors expressing *pax3a* respond to needle stab skeletal muscle injury. (**a**) Schematic showing location of needle stab injury at the dorsal somitic muscle above the anal pore (*cross*). The dotted lines indicate the position of the excision made to isolate the injury region for cell sorting. (**b–d'**) Uninjured skeletal muscle (Tg(*actc1b*:BFP), Blue) contains *pax3a*-positive cells (TgBAC(*pax3a*:GFP)^{i150}, green) (**b**, *arrowheads*), a subset of which are activated following injury. Maximum intensity projection through a representative fluorescent Z-stack of a laterally mounted 4 dpf larvae before injury (**b**), 1 dpi (5 dpf) (**c**) and 3 dpi (7 dpf) (**d**) An enrichment of *pax3a* expressing cells and GFP perduring muscle fibers are seen in the injury site at 3 dpi compared to the adjacent somites. Scale bar, 50 μm

6. Microcentrifuge tubes, 2.0 mL.
7. Scalpel blades, #22.
8. Plastic Pasteur pipettes.
9. DNA LoBind tubes, 1.5 mL.
10. Thermomixer with a 2.0 mL reaction vessel thermoblock.
11. Automated cell counter and disposable cell counting chamber slides.
12. Ice.
13. Pipettes and filtered pipette tips (P1000, P200, and P10).
14. 20 μg/mL 4′,6-diamidino-2-phenylindole dihydrochloride (DAPI) stock solution (use 1:200 dilution). Store solution in the dark at 4 °C (*see* **Note 3**).

15. Cell dissociation solution, 1× Hank's balanced Salt solution (HBSS) (Ca^{2+}, Mg^{2+} free), supplemented with 5% fetal bovine serum (FBS). Store solution at 4 °C.
16. Trypsin-EDTA (0.25%), phenol red. Store solution at 4 °C (*see* **Note 4**).
17. 1 mg/mL Collagenase type II, dissolved in HBSS. Dissolve enzyme immediately before use (*see* **Note 5**).
18. Cell sorting solution, 1× Dulbecco's phosphate buffered saline (DPBS), supplemented with 2% FBS. Store solution at 4 °C.

3 Methods

3.1 Needle Stab Skeletal Muscle Injury

1. Take 4 dpf larvae out of the incubator and anesthetize by adding 0.4% tricaine drop wise using a plastic Pasteur pipette until larvae stop moving but still have beating hearts (*see* **Note 6**).
2. Using a glass Pasture pipette transfer larvae in a drop of liquid onto a dry petri dish under a stereomicroscope (*see* **Note** 7).
3. Manoeuvrer larvae into a lateral lying position using a p200 pipette and remove the remaining liquid surrounding the larvae (*see* **Note 8**).
4. Using a 30-gauge needle perform a single needle stick injury into the dorsal musculature above the anus at a 75° angle (*see* Fig. 1a and **Note 9**).
5. Immediately transfer injured larvae into a petri dish containing Ringer's solution and PTU and place back in the incubator to recover.

3.2 Cell Dissociation and FACS Sorting

1. Take appropriately staged injured larvae out of the incubator and anesthetize (*see* Subheading 3.1, **step 1**).
2. Using a glass Pasteur pipette transfer-anesthetized larvae in a drop of Ringer's solution onto a dry petri dish.
3. Using a scalpel blade make two incisions on the larval trunk excising the head and tail regions and isolating the needle stab injury region (*see* Fig. 1a and **Notes 10** and **11**).
4. Transfer the trunk tissue with the injury region into a 2 mL microcentrifuge tube with 500 μL of cold cell dissociation solution using a glass Pasteur pipette. Keep tissue on ice until tissue collection is completed (*see* **Note 12**).
5. Once tissue collection is completed, remove cell dissociation solution and resuspend in 1 mL of collagenase type II.
6. Incubate in a thermomixer for 15 min at 32 °C with a mixing speed of 750 rpm (*see* **Note 13**).

7. Following incubation gently pipette up and down (10–20 times) using P1000 pipette to encourage cells to break apart (*see* **Note 14**).
8. Resume incubation in the thermomixer for a further 15 min at 32 °C with a mixing speed of 750 rpm (*see* **Note 15**).
9. Spin at 1000 × *g* for 5 min, 4 °C.
10. Discard the supernatant and resuspend pelleted tissue in 1 mL trypsin.
11. Incubate in a thermomixer for 10 min at 30 °C with a mixing speed of 750 rpm.
12. Following incubation pipette up and down (10–20 times) using P1000 pipette to encourage cells to break apart.
13. Resume incubation in the thermomixer for a further 10 min at 30 °C with a mixing speed of 750 rpm (*see* **Note 16**).
14. Following incubation add 1 mL of cell dissociation solution to the tube and spin at 1000 × *g* for 5 min, 4 °C (*see* **Note 17**).
15. Discard the supernatant and resuspend pelleted cells in 1 mL of cell dissociation solution.
16. Filter cells through 70-micron filter into a 50 mL Falcon tube placed on ice.
17. Wash the microcentrifuge tube with 1 mL of fresh cell dissociation solution and filter through the same strainer.
18. Transfer filtered cell suspension into a new 2 mL microcentrifuge tube.
19. Spin at 800 × *g* for 5 min, 4 °C.
20. Discard the supernatant and resuspend pelleted cells in 1 mL of cell dissociation solution.
21. Filter cells through 40-micron filter into a 50 mL Falcon tube placed on ice.
22. Wash the microcentrifuge tube with 1 mL of fresh cell dissociation solution and filter through the same strainer.
23. Transfer filtered cell suspension into a new 2 mL microcentrifuge tube.
24. Spin at 800 × *g* for 5 min, 4 °C.
25. Discard the supernatant and resuspend pelleted cells in 500 μL cell sorting solution.
26. Quantify the number and integrity of cells using an automated cell counter as per the manufacturer's instructions.
27. Resuspend cells in 1–2 mL of cell sorting solution to obtain a maximum concentration of 10^6 cells/mL (*see* **Note 18**).
28. When ready for FACS sorting filter cells into a 5 mL cell strainer tube with 35-micron mesh (*see* **Note 19**).

29. Add viability dye DAPI to the cell suspension in a 1:200 dilution and immediately proceed to cell sorting (*see* **Note 20**).
30. FACS sort samples at 4 °C using 100 μm nozzle at 20 psi and collect different cell fractions into 500 μL cell sorting solution in 1.5 mL DNA LoBind tubes (*see* Fig. 2a–h' and **Notes 21–25**).
31. FACS sorted cells can be used for a variety of downstream applications including qPCR and RNA sequencing (*see* **Note 26**).

4 Notes

1. There are a number of published transgenic zebrafish lines that mark muscle stem and progenitor cell populations. These include *pax7a* [5], *pax7b* [8], *pax3a* [5], *myf5* [9], and *cmet* [7]. For this work we have used the *pax3a* marker to identify stem/progenitor cells following needle stab skeletal muscle injury. In addition, transgenic lines driving fluorophore expression under the control of muscle specific promoters may be of use. These include pan-muscle markers such as *actc1b* [10] and *503unc* [11], or muscle subset-specific markers such as slow muscle (*smyhc*) [12], fast muscle (*fmyhc*) [13], muscle pioneers (*prox1a*) [14], and differentiated muscle (*myod*) [5].
2. PTU is added to inhibit melanisation of zebrafish larvae, which may ease screening of transgenic lines. If PTU is being added it should be carried out between 50% epiboly—24 hpf and maintained for the required duration. However, pigmentation does not hinder FACS and PTU can be omitted from this protocol. Caution should be taken when using PTU as some studies have shown side effects that may interfere with gene expression and function [15–17].
3. DAPI is a nucleic acid binding viability dye used to distinguish between dead and live cells during FACS. Other viability dyes such as propidium iodide (PI, 1 mg/mL stock use at 1:500 dilution) or 7-amino actinomycin D (7AAD, 40 μg/mL stock use at 1:20 dilution) can be alternatively used.
4. If required Trypsin-EDTA (0.25%) can be replaced by TrypLE. We have tested both and obtained comparable cell dissociation efficiencies and cell survival.
5. Collagenase type II is recommended for muscle fiber dissociation. We have tested Collagenases from different suppliers and identified Collagenase type II from Gibco™ (17101015) and Collagenase type 2 from Worthington® (LS004176) to have the best dissociation efficiencies.

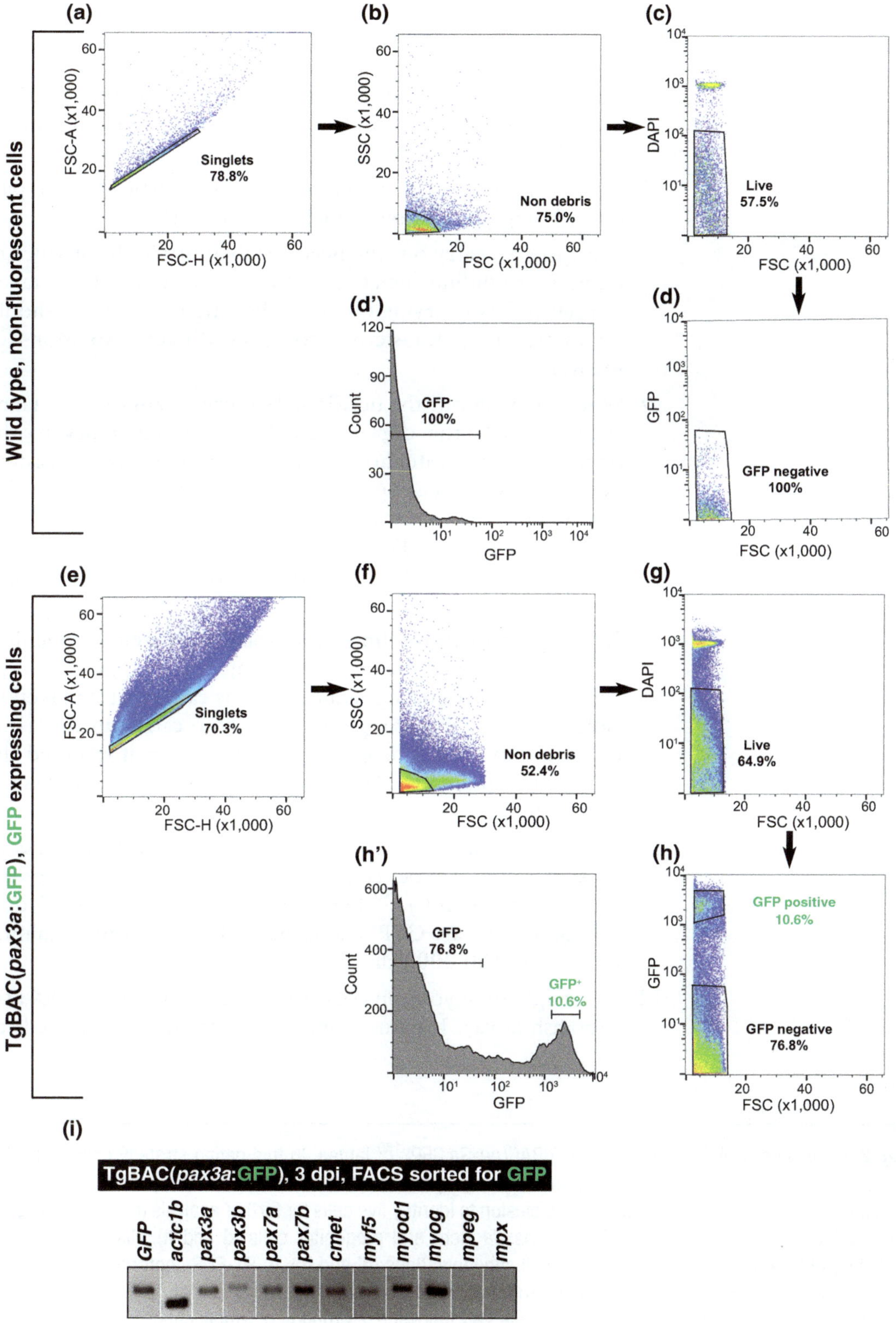

Fig. 2 FACS of GFP$^+$ *pax3a* expressing muscle stem and progenitor cells 3 days post needle stab injury. (**a–h'**) Representative flow cytometry data to illustrate the gating strategy used for FACS sorting *pax3a* cells. (**a–d'**) Profiles of the stage-matched wild-type, non-fluorescent cells. (**e–h'**) Profiles of the GFP expressing cells

6. The amount of tricaine required to anesthetize larvae varies with age, older larvae being more sensitive. Add about 8–10 drops of 0.4% tricaine and assess movement and heartbeat. If too much tricaine is added and the larval heart stops beating, quickly place the larvae in fresh Ringer's solution and wait for heart beat to resume.
7. Zebrafish larvae may sometimes stick to plastic Pasteur pipettes. Using glass Pasteur pipettes prevents this.
8. It is easier to carry out the needle stab injury in the absence of liquid surrounding the larvae. However, care should be taken to prevent larvae drying out and following injury larvae should be immediately transferred to fresh Ringer's solution for recovery.
9. When carrying out the needle stab injury care should be taken to injure only the dorsal musculature. Avoid injury to the neural tube and notochord. Try to keep injury size consistent between larvae, focusing to injure 1–2 somites on average. This can be made easier by using the same edge of the needle to carry out the injury. For example position the needle with the bevel facing upward and carry out the injury in a single puncture.
10. It is important to maximize the isolation of stem/progenitor cells responding to the needle stab injury over quiescent stem cells in the surrounding tissue. For this care must be taken to isolate only the injury region. It is easy to carry out the tissue dissection in a small drop of liquid. However, if required the liquid surrounding the larvae can be removed prior to dissecting the tissue.
11. The number of larvae required is dependent on the number of cells expected to be collected following FACS. With the *TgBAC(pax3a*:GFP)^{i150} line 3 days post injury (dpi) (7 dpf) 10% of the cells are GFP$^+$ and using 250 larvae allows obtaining on average 100,000 cells.
12. In a single tube you can collect trunk tissue of up to 300–500 zebrafish larvae. If tissue from more than 500 larvae needs to

Fig. 2 (continued) isolated from injured TgBAC(*pax3a*:GFP)^{i150} larvae. In this gating strategy cells were first gated for singlets (FSC-A vs. FSC-H, **a** and **e**) and expected cell size (SSC vs. FSC, **b** and **f**). Cells of the expected size were further gated for DAPI exclusion to identify live cells for further analysis (**c** and **g**). Live cells were gated on GFP expression to identify *pax3a* stem and progenitor cells (**d** and **h**). Histogram profiles depicting GFP expressing and GFP absent cell subpopulations (**d'** and **h'**). (**i**) RT-PCR confirming FACS sorted GFP$^+$ cells expressing known muscle stem and progenitor markers. The expression of *pax3a*, *pax3b*, *pax7a*, *pax7b*, *cmet*, *mf5*, *myod1*, and *myog* was observed following 25 rounds of amplification. The macrophage marker, *mpeg* and neutrophil marker, *mpx* were included as negative controls be collected simultaneously, separate samples into two or more microcentrifuge tubes.

13. The appropriate dissociation temperature and time are important to increase cell numbers obtained from sorting and to maintain the integrity and viability of cells. Gentle shaking during dissociation discourages cell aggregate formation.
14. Pipetting further encourages cell dissociation to optimize yield and also acts to discourage cell aggregate formation. However, care should be taken to be very gentle during pipetting as the shear force of traveling in and out of the pipette tip may damage cells.
15. The total dissociation time may need to be altered depending on the type tissue used. Optimal dissociation is achieved when you observe "fluffy" tissue. This will usually be achieved within 20–40 min.
16. Terminate trypsin digestion when most of the tissue has disappeared and the solution appears cloudy but fluid. This will usually be achieved within 15–30 min in total.
17. Adding cell dissociation solution that contains FBS inhibits trypsin activity.
18. For FACS 10^6 cells/mL is desirable. A cell count will let you calculate the volume in which the dissociated cells need to be resuspended. Generally, cells from trunk tissue of 250 injured larvae at 3 dpi (7 dpf) are resuspended in 1 mL to obtain the desired concentration.
19. Final filtration step removes all cell clumps and ensures only dissociated single cells will be sorted. Cells should be sorted immediately after this filtration step.
20. Samples should be sorted immediately after the addition of viability dye. If large sample numbers need to be sorted it may be better to add DAPI in batches.
21. Keep samples on ice until sorting and return back onto ice immediately after sorting.
22. For each experiment a cell suspension from age-matched, wild-type, non-transgenic embryos should be prepared. This sample should be used to equilibrate the FACS machine and set the sorting gates. This is particularly important when sorting larval zebrafish cells that can be autofluorescent. You can use fewer larvae for this. We routinely use 30–50 age matched wild-type TU larvae as controls.
23. Sorting conditions and parameters are variable between different machines and should be discussed with the FACS facility experts to ensure optimum sorting.

24. Sorting into cell sorting solution allows cell survival. However, if the cells will subsequently be used for RNA extraction they can also be directly sorted into lysis buffer.
25. Cells are sorted into LoBind tubes that minimize RNA binding to plastic. This is advantageous if the cells are subsequently used for RNA extraction and particularly helpful if a low concentration of RNA is expected.
26. We validated our FACS sorted GFP^{+} *pax3a* cells as muscle stem/progenitors by carrying out reverse transcription PCR (RT-PCR) and looked for the expression of known stem/progenitor markers (*see* Fig. 2).

References

1. Mauro A (1961) Satellite cell of skeletal muscle fibers. J Biophys Biochem Cytol 9(2):493–495
2. Cornelison D, Wold BJ (1997) Single-cell analysis of regulatory gene expression in quiescent and activated mouse skeletal muscle satellite cells. Dev Biol 191(2):270–283
3. Cooper R, Tajbakhsh S, Mouly V, Cossu G, Buckingham M, Butler-Browne G (1999) In vivo satellite cell activation via Myf5 and MyoD in regenerating mouse skeletal muscle. J Cell Sci 112(17):2895–2901
4. Charge SB, Rudnicki MA (2004) Cellular and molecular regulation of muscle regeneration. Physiol Rev 84(1):209–238
5. Seger C, Hargrave M, Wang X, Chai RJ, Elworthy S, Ingham PW (2011) Analysis of Pax7 expressing myogenic cells in zebrafish muscle development, injury, and models of disease. Dev Dyn 240(11):2440–2451
6. Siegel AL, Gurevich DB, Currie PD (2013) A myogenic precursor cell that could contribute to regeneration in zebrafish and its similarity to the satellite cell. FEBS J 280(17):4074–4088
7. Gurevich DB, Nguyen PD, Siegel AL, Ehrlich OV, Sonntag C, Phan JM, Berger S, Ratnayake D, Hersey L, Berger J (2016) Asymmetric division of clonal muscle stem cells coordinates muscle regeneration in vivo. Science 353(6295):aad9969
8. Pipalia TG, Koth J, Roy SD, Hammond CL, Kawakami K, Hughes SM (2016) Cellular dynamics of regeneration reveals role of two distinct Pax7 stem cell populations in larval zebrafish muscle repair. Dis Model Mech 9 (6):671–684
9. Chen Y-H, Wang Y-H, Chang M-Y, Lin C-Y, Weng C-W, Westerfield M, Tsai H-J (2007) Multiple upstream modules regulate zebrafish myf5 expression. BMC Dev Biol 7(1):1
10. Cole NJ, Hall TE, Don EK, Berger S, Boisvert CA, Neyt C, Ericsson R, Joss J, Gurevich DB, Currie PD (2011) Development and evolution of the muscles of the pelvic fin. PLoS Biol 9 (10):e1001168
11. Berger J, Currie PD (2013) 503unc, a small and muscle-specific zebrafish promoter. Genesis 51(6):443–447
12. Elworthy S, Hargrave M, Knight R, Mebus K, Ingham PW (2008) Expression of multiple slow myosin heavy chain genes reveals a diversity of zebrafish slow twitch muscle fibres with differing requirements for hedgehog and Prdm1 activity. Development 135 (12):2115–2126
13. Nord H, Burguiere A-C, Muck J, Nord C, Ahlgren U, von Hofsten J (2014) Differential regulation of myosin heavy chains defines new muscle domains in zebrafish. Mol Biol Cell 25 (8):1384–1395
14. van Impel A, Zhao Z, Hermkens DM, Roukens MG, Fischer JC, Peterson-Maduro J, Duckers H, Ober EA, Ingham PW, Schulte-Merker S (2014) Divergence of zebrafish and mouse lymphatic cell fate specification pathways. Development 141(6):1228–1238
15. Elsalini OA, Rohr KB (2003) Phenylthiourea disrupts thyroid function in developing zebrafish. Dev Genes Evol 212(12):593–598
16. Li Z, Ptak D, Zhang L, Walls EK, Zhong W, Leung YF (2012) Phenylthiourea specifically reduces zebrafish eye size. PLoS One 7(6): e40132
17. Parker MO, Brock AJ, Millington ME, Brennan CH (2013) Behavioral phenotyping of casper mutant and 1-pheny-2-thiourea treated adult zebrafish. Zebrafish 10(4):466–471

Chapter 15

Preparation of Proliferated Bovine Primary Skeletal Muscle Cells for Bottom-Up Proteomics by LC-MSMS Analysis

Eva Veiseth-Kent, Vibeke Høst, and Mona E. Pedersen

Abstract

This chapter outlines a method for sample preparation for bottom-up proteomics by LC-MSMS analysis of in vitro proliferated bovine primary skeletal muscle cells. The protocol describes the isolation of bovine primary skeletal muscle cells, extraction of proteins, proteolytic digestion of proteins, and desalting of the final peptide samples. The final peptide samples can be analyzed using various LC-MSMS systems after reconstitution in a suitable elution buffer.

Key words Proteomics, Isolation of primary muscle cells, Protein extraction, Peptide generation, LC-MSMS

1 Introduction

Bottom-up proteomics is an approach to determine the protein composition of a given tissue, cell, or organelle, with respect to protein identification and quantification and identification of post-translational modifications. This is accomplished by performing controlled proteolytic digestion of proteins prior to LC-separation, mass spectrometry, and bioinformatics analysis [1]. Bottom-up proteomics has become a popular and powerful approach to determine the protein composition of a given sample. One of the main advantages of bottom-up proteomics is its ability to unravel systematic effects on a functional level without a priori knowledge by combining LC-MSMS analysis with bioinformatics and statistics. By applying this approach to primary skeletal muscle cell cultures, more knowledge can be gained regarding various growth transitions but also the effect of different stimuli. The main procedures of this protocol are outlined in Fig. 1.

Satellite cells are located between the basal lamina and sarcolemma (plasma membrane) of the skeletal muscle fibers and are normally quiescent in the adult muscle. However upon exercise, injury, or disease they can be activated and are involved in skeletal

Sissel Beate Rønning (ed.), *Myogenesis: Methods and Protocols*, Methods in Molecular Biology, vol. 1889, https://doi.org/10.1007/978-1-4939-8897-6_15, © Springer Science+Business Media, LLC, part of Springer Nature 2019

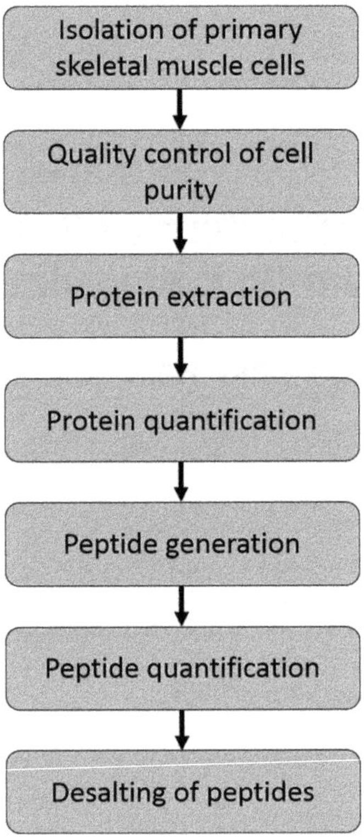

Fig. 1 Schematic workflow indicating the main steps of the protocol

muscle regeneration [2]. These cells have a remarkable ability to self-renew, and undergo myogenic differentiation to fuse and restore damaged muscle [3]. It is possible to isolate these cells from adult muscle tissue by sequential removal of connective tissue by use of collagenase and further tryptic degradation for release of the satellite cells from the myofibers. For removal of fast-adhering fibroblast cells from the primary cell cultures, isolated cells are placed in uncoated cell flasks. This allows fibroblast cells to adhere to the plastic. Then non-adhering cells (satellite cells) are further collected and seeded onto culture flasks coated with ECL (entactin-collagen-laminin). The coating affects the growth and expansion of the primary cells, and ECL is routinely used at our lab for adherence of bovine skeletal satellite cells, as it mimics an extracellular matrix consisting of several extracellular matrix proteins. The purity of the isolated primary cell is checked by staining against fibroblast and myogenic satellite markers, and less than 10% fibroblast is resident in the isolated cell fraction by our isolation procedure [4].

To perform bottom-up LC-MSMS analysis of primary skeletal muscle cells, the proteins in the cells need to be extracted and solubilized. This procedure involves the use of an extraction buffer containing detergents, reducing agents, and buffer components, in combination with both physical abruption and heat treatment.

To enable quantitative comparison between different samples, it is essential that the protein amount from each sample is standardized, and this is ensured by performing assays for protein determination following the initial extraction procedure. Proteolytic digestion of proteins prior to mass spectrometry, from here on referred to as peptide generation, is an essential step of bottom-up proteomics. The protocol described in this chapter is based on a filter aided proteome preparation (FASP) method [5], where the procedure is performed on a standard filtration device with a molecular cutoff from 10 kDa. This procedure allows a complete solubilization of proteins using SDS, which is then exchanged by urea prior to proteolytic digestion. After digestion, peptides are eluted from the filter, and a final procedure for desalting of the peptide samples is needed prior to LC-MSMS analysis. The desalting protocol is performed using STAGE (STop And Go Extraction) tips, where the peptide is purified by reversed-phase chromatography using C18-columns [6, 7]. From the point of protein extraction to final peptide samples, the protocol is also applicable to other types of cells and tissues.

2 Materials

2.1 Bovine Primary Skeletal Muscle Cell Isolation

1. Seeding medium: Dulbecco's modified Eagle's medium (DMEM) supplemented with 10% fetal bovine serum (FBS), 100 U/mL penicillin, 100 μg/mL streptomycin, and 250 μg/mL fungizone.
2. Proliferation medium: Dulbecco's modified Eagle's medium (DMEM) supplemented with 2% FBS, 2% Ultroser G, 100 U/mL penicillin, 100 μg/mL streptomycin, and 250 μg/mL fungizone (*see* **Note 1**).
3. 0.05% trypsin-EDTA solution.
4. Collagenase solution: 0.72 mg/mL solubilized in Proliferation medium.
5. 25- and 75 cm^2 tissue culture flask.
6. 15- and 50 mL Falcon plastic centrifugation tubes.
7. Water bath incubator with shaker.
8. Cell strainers, 100 μm nylon.
9. Sterile disposable bottle top filter (0.2 μm) with Polyethersulfone (PES) Membrane.
10. Biopsy needle.
11. Bench top centrifuge.
12. Cell incubator with CO_2.
13. Sterile Laminar Flow bench.

2.2 Purity of Muscle Cells

1. Coverslips.
2. Phosphate buffered saline (PBS).
3. 4% paraformaldehyde (PFA).
4. ECL commercial mixture.
5. Dulbecco's modified phosphate buffered saline (DPBS) pH 7.4.
6. Permabilizing buffer: 0.1 Triton-X100 in PBS.
7. Blocking buffer: 5% nonfat dry milk in PBS.
8. Mouse anti-TE7 (fibroblast marker) and mouse anti- NCAM (myogenic marker) and secondary Alexa 488-conjugated goat anti-mouse antibody.
9. Antibody buffer: PBS.
10. Cell fluorescent mounting medium (Dako).
11. 300 nM DAPI blue (DAPI) commercial fluorescence staining in PBS.
12. Fluorescence microscope with camera.
13. Image analysis software.

2.3 Protein Extraction

1. PBS.
2. SDT-extraction buffer: 4% sodium dodecyl sulfate (SDS), 0.1 M dithiothreitol (DTT), 100 mM Tris–HCl pH 7.6. Make fresh (*see* **Note 2**).
3. Precellys Evolution homogenizer (Bertin Corp.).
4. Precellys ®CK28 lysing kit (Bertin Corp.).
5. Water bath.
6. Police rubber.
7. Ultrasonic water bath.
8. Crushed ice.
9. Bench top centrifuge.
10. 1.5 mL tubes.

2.4 Protein Assay for Samples Containing Detergents

1. *RCDC* protein assay kit (BIO-RAD).
2. Bovine serum albumin (BSA) 2 mg/mL standard.
3. 1.5-mL tubes.
4. Bench top centrifuge.
5. 96-well microtiter plate.
6. 96-well plate reader.

2.5 Peptide Generation

1. 50 mM Tris–HCl pH 8.0.
2. Urea A solution: 8 M urea, 100 mM tris/HCl pH 8.5. Prepare fresh (*see* **Note 3**).

3. IAA solution: 0.05 M iodacetamide in Urea A solution. Prepare fresh.
4. Urea B solution: 8 M urea, 100 mM Tris–HCl pH 8.0. Prepare fresh (*see* **Note 3**).
5. 1 μg/μL trypsin/Lys-C stock solution. Keep on ice (*see* **Note 4**).
6. Microcon® YM-10 centrifugal filter device (Merck Millipore).
7. Bench top centrifuge.
8. Incubator.

2.6 Protein Assay for Peptide Samples

1. *DC* protein assay kit (BIO-RAD).
2. Bovine serum albumin (BSA) 2 mg/mL standard.
3. 96-well microtiter plate.
4. 96-well plate reader.

2.7 Desalting of Peptide Samples

1. Pipette tips, 200 μL.
2. Empore C18 membrane.
3. 18-gauge blunt-ended syringe needle.
4. Syringe.
5. 2 mL tubes.
6. Centrifuge adapters (GL Sciences).
7. Bench top centrifuge.
8. 0.1 % trifluoroacetic acid (TFA) solution.
9. 70% acetonitrile (ACN) solution.
10. Speed-vacuum system.

3 Methods

3.1 Bovine Primary Muscle Cell Isolation

Sampling of larger pieces of skeletal muscle (~250 g) at an abattoir must be performed immediately after slaughter and can be stored at room temperature (2–3 h) under transportation before the cell isolation procedure. Isolation of bovine primary muscle cells must further be prepared under sterile condition in a sterile bench and use of sterile equipment.

1. Mark 50 mL tubes and add the following content: Tube I–10 mL collagenase medium (20 tubes), Tube II–1 mL FBS (20 tubes), Tube III–2 mL FBS (20 tubes), Tube IV–2 mL FBS (20 tubes), Tube V –2 mL FBS (20 tubes) (*see* **Note 5**).
2. Coat 20 culture flasks (25 cm^2 or 75 cm^2) with 25 or 75 μL respectively, of commercial mixture with ECL with use of a police rubber and further air-drying 2–3 h.

3. Prepare small muscle pieces (~1–2 g) from the muscle collected using a biopsy needle before connective tissue digestion in 50 mL Falcon tubes. The biopsies are taken from inside of the muscle to maintain sterile conditions.
4. Perform connective tissue digestion in 10 mL DMEM with 0.72 mg/mL collagenase for 1 h and 70 rpm shaking in water bath (37 °C) (tube I).
5. Centrifuge at 550 × *g* for 10 s.
6. Filtrate the supernatant with cell strainers into new 50 mL tubes with 1 mL FBS (tube II). Pellet the cells by an additional centrifugation at 550 × *g* for 5 min, and discard the supernatant. This cell fraction is put to side.
7. Cells not removed by collagenase in the remaining tissue (sediment from **step 3**, tube I) are resuspended in 20 mL 0.05% trypsin/EDTA and incubated for 15 min and 70 rpm shaking in water bath (37 °C), mix the tube every 5 min.
8. Repeat **steps 5–7** three times, and for each time filtrate the supernatants after centrifugation and collect the cell fraction into new tubes (III, IV, and V).
9. Pool the cell fractions (tubes II–V) and resuspend in 3 mL seeding medium.
10. For the removal of fast-adhering fibroblast cells from the primary cell cultures, place the isolated cells in uncoated cell flasks for 1 h at 37 °C.
11. Then non-adhering cells are collected and further seeded onto 25 cm^2 culture flasks coated with ECL.
12. Incubate the cells for 48 h and thereafter change the medium to proliferation medium. Maintain the cells at 37 °C in a humidified atmosphere of 5% CO_2 and routinely sub-cultivate at a concentration of 5000 cells/cm^2 (usually twice a week) (*see* **Note 6**).

3.2 Determination of Purity of the Isolated Muscle Cells

1. Determine the purity of isolated muscle cell fraction by seeding an aliquot of the cells (density 5000 cells/cm^2) onto ECL-coated coverslips and culture for 1 day at 37 °C in a humidified atmosphere of 5% CO_2.
2. Wash the coverslips 3 × in PBS before fixation in 4% PFA for 15 min.
3. Wash the cells 3 × in PBS, permeabilize by using 0.1% Triton X-100 in PBS for 15 min, and incubate with 5% nonfat dry milk for 30 min.
4. Incubate the cells with primary mouse anti-TE7 (fibroblast marker) and mouse anti-NCAM (myogenic marker) for 1 h and subsequent secondary Alexa 488-conjugated goat anti-mouse antibody for 30 min (both steps at room temperature).

5. Equilibrate the coverslips in PBS before counterstaining with DAPI.
6. Add 300 nM DAPI to cover the cells (approximately 300 μL) for 5 min.
7. Mount the cell by using cell fluorescence mounting media.
8. Quantify ± NCAM and ± TE7 myoblasts cells by fluorescence microscopy and image analysis software. Choose pictures randomly from at least three regions to ensure a non-biased quantification.

3.3 Protein Extraction

1. Wash proliferated cells grown to 80% confluence 3 × in PBS.
2. Add 8 μL SDT-lysis buffer per cm^2 culture flasks and collect the cells using a police rubber to loosen the cells from culture flasks and transfer the cells to a Precellys homogenize tube.
3. Homogenize samples in Precellys Evolution Homogenizer at 6000 rpm, 2 × 20 s, with a 5 s break between cycles.
4. Heat samples for 5 min at 95 °C in a water bath.
5. Sonicate samples in ultrasonic water bath for 5 min.
6. Spin samples at 14,000 × *g* for 5 min at room temperature.
7. Transfer supernatant to new clean 1.5-mL tubes.
8. Make 2–3 aliquots (e.g., 50-μL aliquots) of samples needed for further analysis and transfer them to −80 °C freezer immediately.
9. Use the rest of the samples to measure protein content.

3.4 Protein Assay for Samples Containing Detergents

1. Prepare a BSA-protein standard curve: Perform a serial dilution of 2 mg/mL BSA in water, to achieve a standard curve containing the following protein concentrations: 2, 1, 0.5, 0.25, and 0 mg BSA/mL.
2. In triplicates, pipette 25 μL of each sample and the five BSA standard curve solutions into labeled 1.5 mL tubes.
3. Add 125 μL of the RC reagent I into each tube, vortex and incubate on the bench for 1 min.
4. Add 125 μL of the RC reagent II into each tube, vortex and centrifuge the tubes for 5 min at 16,000 × *g*, 4 °C.
5. Prepare Reagent A′ by adding 5 μL DC reagent S per 250 μL DC reagent A.
6. Following the centrifugation in **step 4**, carefully remove the supernatant by pipetting from all tubes. Allow the tubes to stand on the bench to dry completely before continuing.
7. Add 127 μL Reagent A′ to each tube, vortex and incubate at room temperature for 5 min. Repeat vortex and check that

the precipitate is completely dissolved in all tubes before continuing.

8. Add 1 mL DC Reagent B to each tube. Vortex and incubate on the bench for 15 min (*see* **Note 7**).
9. Read the absorbance at 750 nm by pipetting 200 μL of the solutions into a 96-well plate when reading the absorbance using a 96-well plate reader (*see* **Note 8**).
10. Prepare a standard curve using the serial-diluted BSA standard curve solutions and use this to calculate the protein concentration of the samples (*see* **Note 9**).

3.5 Peptide Generation

1. Insert each Microcon filter device into a clean tube (*see* **Note 10**).
2. Pipette 200 μL Urea A solution into each filter device.
3. Add 60 μg of your protein sample to the filter device (*see* **Note 11**).
4. Place assembly in a centrifuge and spin at 14,000 × *g* for 40 min, or until the total sample volume has been transferred through the filter.
5. Pipette 200 μL Urea A solution into each filter device and spin at 14,000 × *g* for 40 min.
6. Discard the flow-through.
7. Pipette 100 μL IAA solution to each filter device, place them in a table shaker, and let them shake for 1 min at ~600 rpm and then incubate for 5 min with no shaking.
8. Spin at 14,000 × *g* for 30 min.
9. Pipette 100 μL Urea B solution to each filter device and spin at 14,000 × *g* for 40 min. Repeat two more times, i.e., 3 washing steps in total, and discard the flow-through after each wash.
10. Transfer the filter units to new and clean tubes.
11. Make enzyme working solution: Dilute 2 μL of the trypsin/Lys-C stock solution into 158 μL 50 mM Tris–HCl pH 8.0 (*see* **Note 12**). Keep this solution on ice until you add this to your sample (*see* **Note 4**).
12. Add 160 μL of freshly made trypsin/Lys-C working solution and incubate overnight at 37 °C in a wet chamber (*see* **Note 13**).

3.6 Protein Assay for Peptide Samples

1. Prepare a BSA-protein standard curve, *see* **step 1** in Subheading 3.4.
2. In triplicates, pipette 5 μL of each sample and the five BSA standard curve solutions into a 96-well microtiter plate.

3. Prepare Reagent A′, by adding 5 μL DC reagent S per 250 μL DC reagent A.
4. Add 25 μL of reagent A′ into each well.
5. Add 200 μL DC reagent B to each well.
6. Mix by gentle agitation or use a 96-well plate reader with mixing function, and incubate on the bench for 15 min (*see* **Note** 7).
7. Read the absorbance at 750 nm using a 96-well plate reader.
8. Prepare a standard curve using the serial-diluted BSA standard curve solutions, and use this to calculate the protein concentration of the samples (*see* **Note 9**).

3.7 Desalting of Peptide Samples

1. Prepare STAGE tips (*see* **Note 14**): Cut one or two pieces of Empore C18 membrane with the 18-gauge blunt-ended syringe needle, and blow these into a 200 μL pipette tip with a syringe (*see* **Note 15**). A gap of several millimeters should be visible between the C18 membrane disk and the end of the pipette tip (*see* Fig. 2).

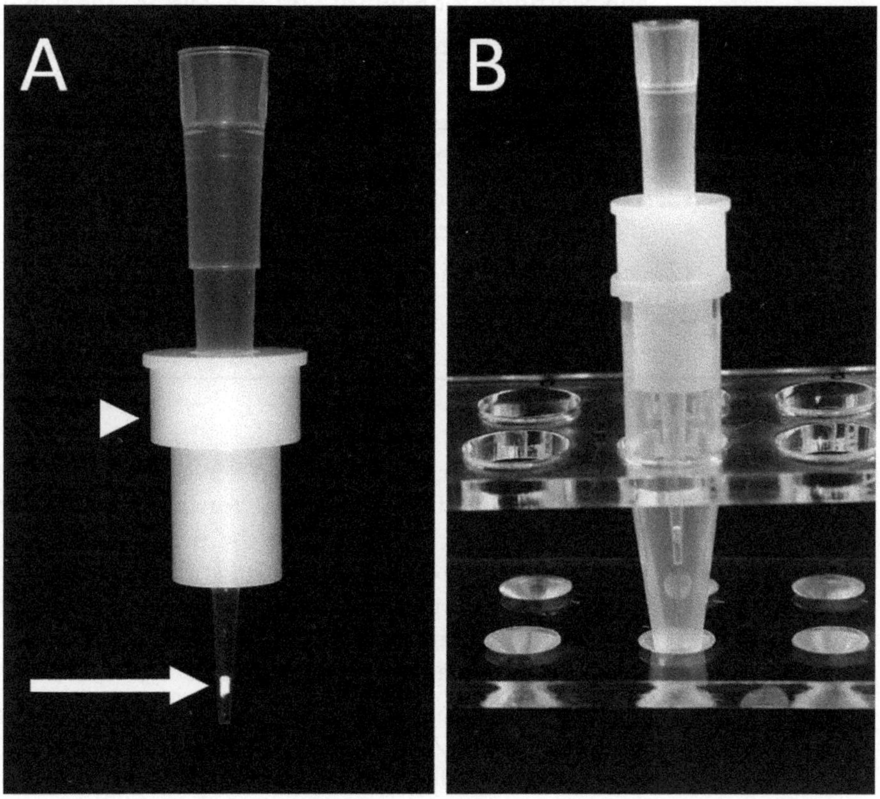

Fig. 2 Assembly of the STAGE tip. (**a**) The 200-μL pipette tip with double C18 membrane disks (arrow) mounted in the adaptor (arrow head). (**b**) The STAGE tip mounted inside a 2 mL collection tube

2. Place an adaptor on a 2 mL collection tube (with lid cut off) and mount the STAGE tip as shown in Fig. 2.
3. Wet the STAGE tip by pipetting 100 μL 0.1% TFA to the tip, and centrifuge for 5 min at 1500 × *g*. Make sure that all liquid has passed through the column—this will need to be checked after each centrifugation step in this protocol. Discard the flow-through.
4. Wash the STAGE tip by pipetting 100 μL 70% ACN to the tip, and centrifuge for 5 min at 900 × *g*. Discard the flow-through.
5. Equilibrate the STAGE tip by pipetting 100 μL 0.1% TFA to the tip, and centrifuge for 5 min at 900 × *g*. Discard the flow-through.
6. Change the collection tube to a new 2 mL tube without lid.
7. For peptide binding, transfer a sample volume equivalent to, e.g., 20 μg protein (max 100 μL) to the STAGE tip, and centrifuge for 5 min at 900 × *g*.
8. Transfer the flow-through from the collection tube to the STAGE tip, and centrifuge again for 5 min at 900 × *g*.
9. Wash the STAGE tip by pipetting 100 μL 0.1% TFA to the tip, and centrifuge for 5 min at 900 × *g*. Discard the flow-through.
10. Change the collection tube to a new and labeled 2 mL tube with lid.
11. Elute the peptides by pipetting 50 μL 70% ACN to the tip, and centrifuge for 5 min at 900 × *g*.
12. Dry the flow-through with a speed-vacuum system until samples are completely dry.
13. Store dried peptides in the tubes at −20 °C until LC-MSMS analysis.

4 Notes

1. Syntetic Ultroser G is used as a replacement for FBS in the proliferation medium. This is to reduce batch variations which often could be a problem with the use of FBS as a nutrition supplement. Ultroser G serum substitute has a concentration five times higher than fetal calf serum (i.e., 2% of reconstituted Ultroser G serum substitute is equivalent to 10% fetal calf serum in the basal medium).
2. You can make stock solutions for preparation of the extraction buffer: 1 M tris/HCl pH 7.6—store at 4 °C. 1 M DTT in water—store in aliquots at −20 °C. 10% SDS in water—store at room temperature.

3. You can make stock solutions of Tris–HCl; 1 M Tris–HCl pH 8.5 and pH 8.0—store at 4 °C. Dissolving urea in water is an endothermic reaction, so when you make the Urea A and B buffers, it might help to slightly heat the buffer in your hand to dissolve the urea.
4. It is important to keep the trypsin/Lys-C solutions on ice to avoid activation and loss of activity of the proteases prior to mixing with the protein samples.
5. We normally collect 20 biopsy samples of ~1–2 g per muscle collected at the abattoir, using one tube of collagenase medium per biopsy sample. It is possible to use less tubes than 20. The number of isolated cells is then reduced equivalent.
6. It is possible to store the proliferating cells. This is performed by washing the cells three times with DPBS before adding 0.05% trypsin-EDTA solution covering the cell layer (1 min, 37 °C) to loosen the cells. Transfer the cells to a collection tube before 5 min centrifugation at 1000 × *g*, and resuspend the cell pellet in 1 mL proliferation medium supplemented with 10% DMSO. Finally, transfer the resuspended cells to 1 mL cryo-tube and freeze the cells. The freezing should be performed sequentially at first −20 °C (30 min) and then −80 °C (overnight) before long-term storage in liquid nitrogen.
7. The absorbance at 750 nm is stabile for up to 1 h.
8. The absorbance at 750 nm can also be measured by transferring the content of the tubes to cuvettes and using a spectrophotometer.
9. If the protein concentration of the samples exceeds the standard curve, the protein assay will need to be repeated with diluted samples.
10. To avoid contaminating the samples with keratin (abundant protein in skin and hair), work surfaces should be cleaned, and gloves should be worn during all steps. To avoid contamination of detergents in the upcoming LC-MSMS analysis, it is crucial that all glassware used for preparing various solutions is cleaned either without detergents, or that they have been thoroughly rinsed. Moreover, to avoid leaching of polymers that may interfere with the LC-MSMS analysis, all plastic ware should be of high-quality polypropylene. Finally, all steps should be carried out in room temperature if nothing else is described.
11. Smaller amounts of protein can also be used, e.g., 30 μg.
12. For each 30 μg of protein you will need to add 1 μg of protease. Thus, if you added 30 μg of protein in Subheading 3.5, **step 3**, then dissolve 1 μL of the 1 μg/μL trypsin/Lys-C stock solution in 159 μL 50 mM Tris–HCl pH 8.0.

13. To avoid drying out of samples during the incubation, make sure caps are securely closed and incubate the samples in wet chamber; this can be prepared by placing the tube rack into a container with 0.5 cm layer of water in the bottom, and cover this container during the incubation.
14. To avoid contamination of detergents in the upcoming LC-MSMS analysis, it is crucial that all glassware used for preparing various solutions is cleaned either without detergents, or that they have been thoroughly rinsed. To avoid leaching of polymers that may interfere with the LC-MSMS analysis, all plastic ware should be of high-quality polypropylene. Finally, all steps should be carried out in room temperature if nothing else is described.
15. The estimated peptide-holding capacity using one or two pieces of C18 membrane is 20 and 40 μg, respectively.

Acknowledgments

This work was supported by a grant from the Foundation for Research levy on Agricultural Products. We thank Ragnhild Stenberg Berg for critical comments and suggestions regarding the book chapter.

References

1. Chait BT (2006) Chemistry. Mass spectrometry: bottom-up or top-down? Science 314 (5796):65–66. https://doi.org/10.1126/science.1133987
2. Mauro A (1961) Satellite cell of skeletal muscle fibers. J Biophys Biochem Cytol 9:493–495
3. Tsivitse S (2010) Notch and Wnt signaling, physiological stimuli and postnatal myogenesis. Int J Biol Sci 6(3):268–281
4. Ronning SB, Pedersen ME, Andersen PV, Hollung K (2013) The combination of glycosaminoglycans and fibrous proteins improves cell proliferation and early differentiation of bovine primary skeletal muscle cells. Differentiation 86 (1-2):13–22. https://doi.org/10.1016/j.diff.2013.06.006
5. Wisniewski JR, Zougman A, Nagaraj N, Mann M (2009) Universal sample preparation method for proteome analysis. Nat Methods 6(5):359–362. https://doi.org/10.1038/nmeth.1322
6. Rappsilber J, Ishihama Y, Mann M (2003) Stop and go extraction tips for matrix-assisted laser desorption/ionization, nanoelectrospray, and LC/MS sample pretreatment in proteomics. Anal Chem 75(3):663–670
7. Yu Y, Smith M, Pieper R (2014) A spinnable and automatable StageTip for high throughput peptide desalting and proteomics

Chapter 16

Myogenesis in *Drosophila melanogaster*: Dissection of Distinct Muscle Types for Molecular Analysis

Anton L. Bryantsev, Lizzet Castillo, Sandy T. Oas, Maria B. Chechenova, Tracy E. Dohn, and TyAnna L. Lovato

Abstract

Drosophila is a useful model organism for studying the molecular signatures that define specific muscle types during myogenesis. It possesses significant genetic conservation with humans for muscle disease causing genes and a lack of redundancy that simplifies functional analysis. Traditional molecular methods can be utilized to understand muscle developmental processes such as Western blots, in situ hybridizations, RT-PCR and RNAseq, to name a few. However, one challenge for these molecular methods is the ability to dissect different muscle types. In this protocol we describe some useful techniques for extracting muscles from the pupal and adult stages of development using flight and jump muscles as an example.

Key words *Drosophila*, Indirect flight muscles, Tergal depressor of trochanter, Muscular dystrophy, Myopathy, Myogenesis

1 Introduction

Muscle development and remodeling are complex processes that span from embryogenesis to the aging adult and can be dysregulated leading to disease. Degenerative diseases such as muscular dystrophies and myopathies are frequently a consequence of genetic mutations that affect the structural and contractile components of muscles [1, 2]. Striated muscle tissue in many species across the animal kingdom is typically made of different fiber types that can be identified by their morphological, biochemical, and molecular features [3–6]. The typical parameters used to characterize mammalian muscle fiber types are the expression of structural proteins (e.g., myosin heavy chain isoforms), mitochondrial content, and contractile properties (e.g., contraction speed and force) [7–9]. Muscle fibers can be loosely classified into Type I and Type II. Type I fibers have high oxidative capacity and utilize oxygen to produce continuous muscle contractions over long periods of time while type II fibers have high glycolytic activity and use

Sissel Beate Rønning (ed.), *Myogenesis: Methods and Protocols*, Methods in Molecular Biology, vol. 1889,
https://doi.org/10.1007/978-1-4939-8897-6_16, © Springer Science+Business Media, LLC, part of Springer Nature 2019

anaerobic metabolism to produce short bursts of energy, but cannot maintain continuous muscle contractions for long [10]. With the advances in molecular diagnostics, fiber classification becomes even more complex and includes highly specialized fibers (e.g., periocular muscles) and transitional forms between fiber types [11, 12]. Fiber type composition of a muscle is an important parameter, as it determines overall functional performance of muscles [13] and influences systemic metabolism [14]. Mammalian muscle fibers can change their fiber-type during normal development [15], or in response to denervation and disuse [16]. Fibers of different types demonstrate different susceptibility to pathological conditions (e.g., diabetes, cancer) and aging [17–19]. Nevertheless, the exact genetic mechanisms that specify muscle diseases are currently not well understood. Given that in many cases muscle diseases can impact one fiber type to a greater extent than another, there is a need to gain a better understanding of normal human muscle development, to establish the basic knowledge required for potential treatments and therapies.

Drosophila provides a convenient experimental platform to study the heterogeneity of muscle fiber types. Most somatic muscles in adult *Drosophila* flies are made of multiple, but homogenous fibers. However, muscle fibers belonging to different muscles can differ substantially. From a practical standpoint, the two biggest muscle groups in *Drosophila*, namely indirect flight muscles (IFMs, or flight muscles) and tergal depressors of the trochanter (TDTs, or jump muscles), are a pair of muscles with highly distinct properties. Flight muscles are formed by 26 large, mitochondria-rich fibers that occupy the center of the thoracic cavity (Fig. 1). Flight muscles are highly fatigue-resistant; they power flight by slightly warping the exoskeleton of the thorax to make the wings flap, which can be done at a rate of hundreds of contractions per second. In contrast, two jump muscles are located at the opposing lateral sides of the thorax, each muscle comprising 25–30 individual fibers with low mitochondrial content. Jump muscles connect the top of the thorax with the base of the second pair of legs and, upon contraction, exclusively provide power for jumping. Flight and jump muscles demonstrate a striking difference in their morphological organization, expression of muscle genes, and alternative transcript splicing [20–22] analogous to vertebrate slow twitch (Type I) or fast twitch (Type II) fibers [12]. Thus, the conservation of muscle fiber diversity between *Drosophila* and vertebrates allows us to investigate the molecular mechanisms underlying muscle fiber diversity in an ideally simplified model.

Here, we present dissection protocols for the isolation of individual flight (IFMs) and jump (TDT) muscles at multiple developmental time points. These sampling approaches have been instrumental in identifying fiber-specific muscle proteins [23] and deciphering the genetic program regulating fiber-specific differentiation and pre-mRNA alternative splicing [21, 24].

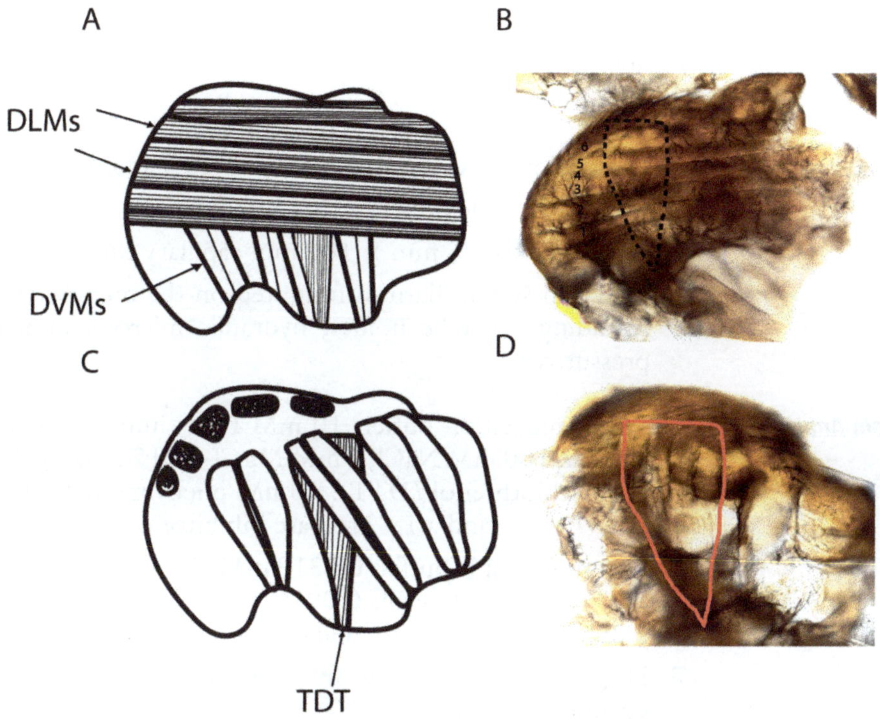

Fig. 1 Schematic overview of the muscles developing in the adult fly thorax. (**a**, **c**) Illustrated image depicting the major muscles found in the adult fly thorax: DVMs, DLMs, and TDT (**b**) Control adult fly showing live image of the six DLM fibers in a bisected thorax. Notice the fibers are numbered and the TDT is outlined which sits behind the DLM fibers. (**d**) The DLMs are removed to reveal the triangle shape fibers of the TDT which is outlined in red

2 Materials

2.1 Fly Microdissection

1. Binocular dissecting microscope.
2. Disposable plastic 100 mm petri dish lid.
3. 22-gauge syringe needle.
4. Micro dissecting needle probe and needle probe holder.
5. 1× PBS.
6. Rotator for microcentrifuge tubes.
7. 1 M sucrose solution prepared in 1× PBS.

2.2 Freezing Flies and Cryosectioning

1. Large Petri dish lid 92 × 16 mm.
2. Double-sided sticky tape.
3. 22-gauge syringe needle.
4. Tissue freezing medium.
5. Metal spatula.

6. Liquid Nitrogen.
7. Standard glass microscope slides 25 × 75 × 1 mm.
8. Cryotome.

2.3 Microsampling

1. Inverted microscope equipped with 10× and 20× objectives.
2. Capillary needle puller.
3. 1.00 mm × 0.75 mm × 100 mm capillary tubes.
4. Injection system harness mounted on the microscope's stage, including a needle holder, hydraulic micromanipulator, and pressurizer.

2.4 Protein Analysis

1. Homogenization buffer: 10 mM Potassium phosphate buffer pH 7.0, 100 mM NaCl, 2 mM $MgCl_2$ and 2 mM EGTA, 1 mM 1,4-Dithiothreitol (DTT), 10 μM phenylmethylsulfonyl fluoride (PMSF) and a 1× protease inhibitor cocktail.
2. Laemmli sample buffer: 0.0313 mM Tris–HCl pH 6.8, 1% w/v SDS, 5% v/v glycerol, 0.001% w/v bromophenol blue, and 1.25% v/v 2-Mercaptoethanol.
3. Disposable pestles that fit into the bottom of a 1.5 mL eppendorf tube.
4. 4–20% gradient precast polyacrylamide gel.

3 Methods

3.1 Freezing Samples for Cryosectioning and Microsampling

1. Obtain time 0 h pupae. 2-h larva will crawl up the side of the vial or bottle and eventually choose a site for pupariation. Once they choose their site, the larvae become shorter and extend spiracles from the head while still remaining the light white color of wandering larva. It is at this point that you can circle them and mark time zero for subsequent pupal aging.
2. Stage pupae appropriately to the selected hour [APF (after puparium formation)—16 h, 24 h, 48 h, 72 h, and 96 h pupae] (*see* **Note 1**).
3. Attach approximately 5 cm of double-sided tape to a large Petri dish lid. Using a paintbrush, gently adhere pupae to the tape in a row, positioning them side by side.
4. Begin removing the pupal case from the pupae by first removing the anterior/top of the pupa, then make small incisions along a line at the lateral side along the body, lift up the pupal case to reveal the thorax, and finally remove the flap from the thorax (Fig. 2a–c).
5. Cover pupae or anesthetized adult flies with tissue freezing medium (Fig. 2e, f).

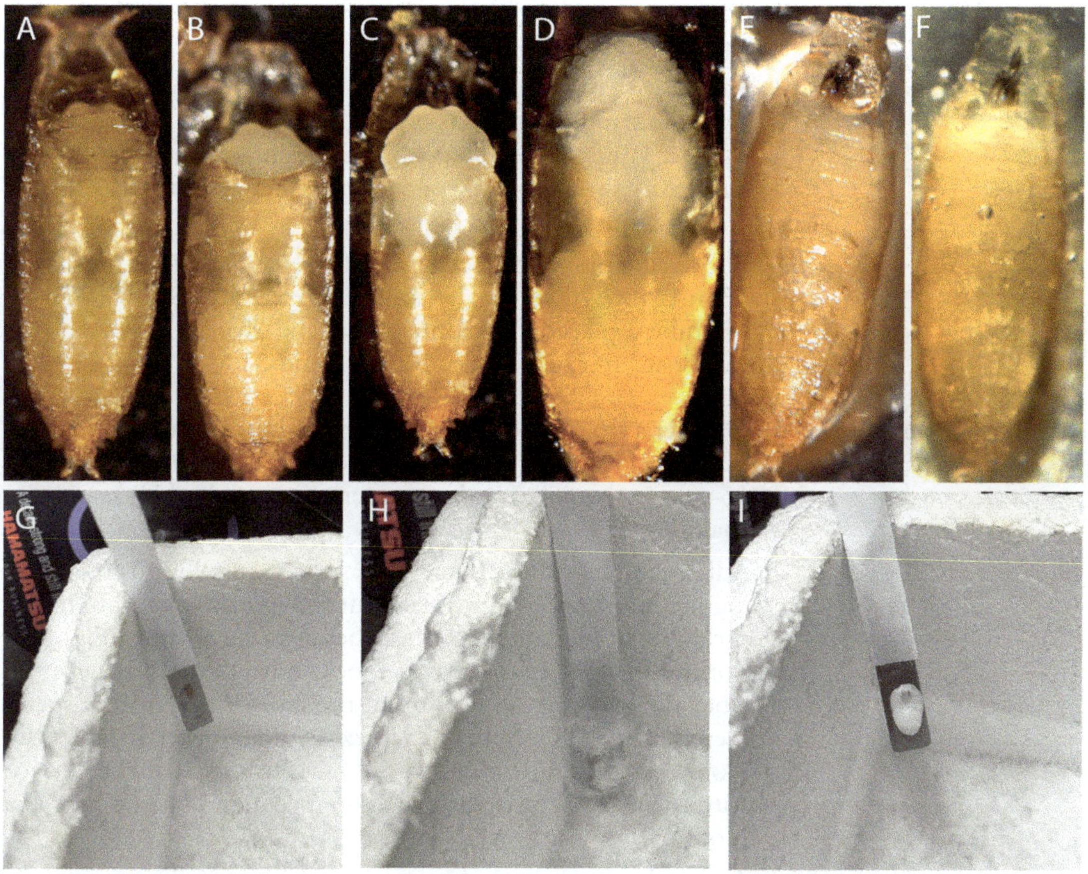

Fig. 2 Pupal sample preparation for cryosectioning. (**a**) Placed timed pupae on double-sided tape. Ensure the pupa is secure and dorsal facing upward. (**b**) 16-h pupa with the first step of the casing removed. (**c**) Continuing the removal of the case; notice the extra casing removed on either side. (**d**) The 16-h pupa that is finally prepared is shown where the removal is only necessary around the thorax (**e**) Samples are aligned onto the spatula, notice that some pupae may roll. Ensure that the pupa has not rolled with the ventral side up (**f**) A liberal amount of tissue freezing medium immerses the pupa in order to properly stabilize the tissue. Add a sufficient amount of tissue freezing medium to cover the samples, but not too large as a large piece may break when freezing in liquid nitrogen. (**g**) Prepared sample is placed in deepest corner of Styrofoam box filled with liquid nitrogen. (**h**) Samples are completely submerged in liquid nitrogen and removed when bubbling is minimal. (**i**) Samples can now be stored in a 15 mL centrifuge tube in the −80 °C freezer

6. Place a small piece of double-sided tape at the end of a metal spatula.
7. Place a thin layer of tissue freezing medium onto the double-sided sticky tape. Add a label to the end away from the tip of spatula.
8. Lay pupae/adults on their backs onto the spatula, ventral sides with legs facing up.
9. Arrange multiple pupae/adults closely side by side removing any large air bubbles found near or between the flies.

10. Take liquid nitrogen into a Styrofoam box, and tip the box so the liquid nitrogen accumulates in one corner.
11. Slowly lower the spatula to the liquid nitrogen surface, stop until the sample begins to whiten, and then fully submerge the sample. Wait until sample stops bubbling; discard the sample back into the liquid nitrogen with label attached. Continue freezing more samples (Fig. 2g–i).
12. For long storage, collect several frozen blocks in a prechilled 15 mL tube that can be stored frozen at −80 °C. Up to 15 tubes, placed horizontally, can be stored in a standard cardboard freezer box (*see* **Notes 2** and **3**).

3.2 Cryosectioning

1. Transfer a frozen block containing flies into the working chamber of a cryotome (preset to approximately −20 °C). Wait for 15 min to equilibrate the temperatures of the block and the chamber (Fig. 3a).
2. Adhere the block to a detachable specimen holder using freezing medium. Apply a liberal amount of freezing medium to the specimen holder and allow it to begin to turn white (Fig. 3b). For mounting, flip the frozen block to bring the side with the label and the tape on top; bring the block's base to the specimen holder with semi-frozen freezing medium on it and let the block adhere for 10 min (Fig. 3c). Make sure that the surface with the label is in parallel with the specimen holder's base (*see* **Note 4**).
3. Attach the specimen holder to the cryotome and adjust the tilt so the plane of sectioning is parallel to the surface of the block. Briefly section the block until the flies inside of the block are beginning to be exposed (Fig. 3d–f).
4. Section through the frozen fly samples, producing sections of $>14\ \mu m$ thickness.
5. Pick up good, even sections coming out from under the cryotome blade with a standard microscopy slide. Apply sequential sections contiguously along the slide (*see* **Note 5**).
6. Let cryosections dry on the slide for 15 min at room temperature (*see* **Note 6**).

3.3 Microsampling Cryosections

1. Prepare a needle from a glass capillary tube using a needle puller.
2. Wear protective eyewear and trim the needle by rapidly running forceps or a metal syringe needle across the prepared glass needle tip. Some practice will be necessary to determine how far from the needle's end the breakage has to occur. The goal here is to produce an opening of approximately 50–100 μm in diameter.

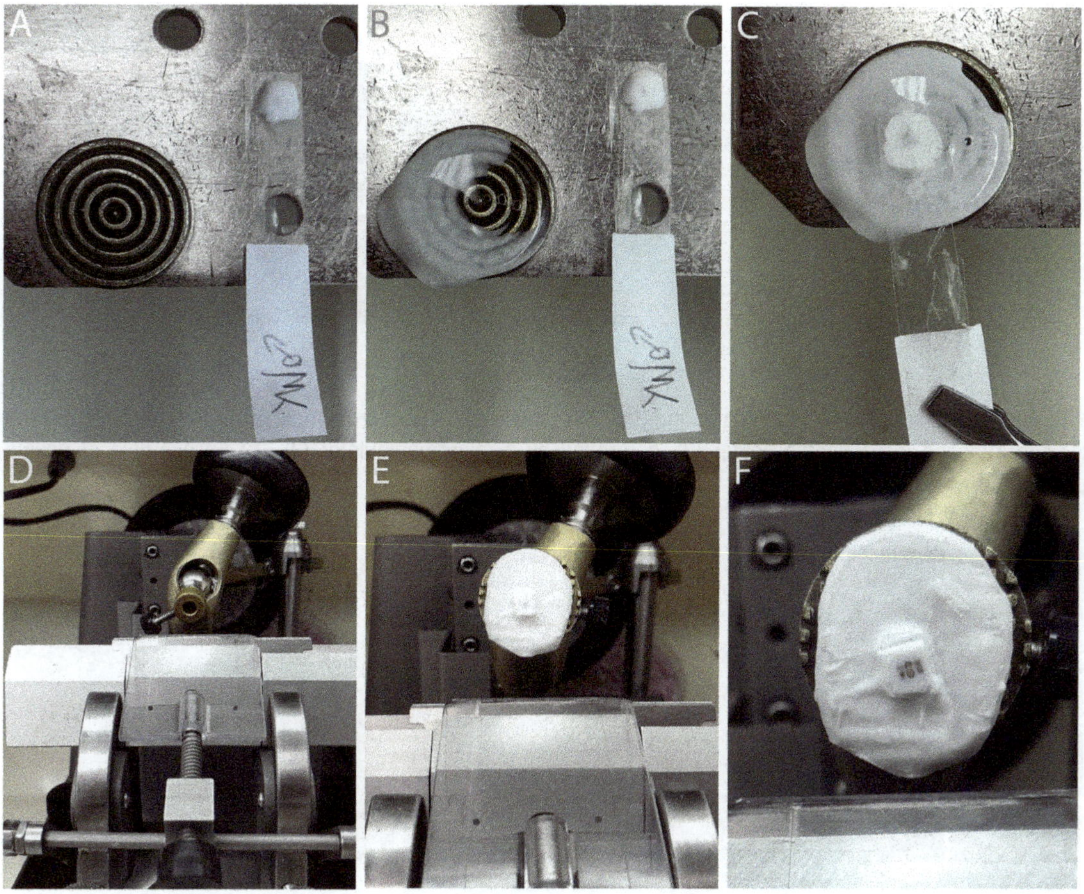

Fig. 3 Sample preparation for cryosectioning. (**a**) Sample and cryosection specimen holder equilibrating to the temperature of the cryostat. (**b**) Freezing medium added to the specimen holder is beginning to freeze. (**c**) Once the freezing medium has turned white, the sample is pressed into it. (**d**) Cryostat mount before the specimen holder is positioned. Take note of the specimen orientation ball assembly and the specimen orientation handle in black. (**e**) Mounted specimen holder. (**f**) The sample in (**e**) was sectioned through until the flies became visible. This is the point to start collecting sections on a slide

3. Secure the trimmed needle within the needle holder mounted on the microscope's stage, bring it into focus, and rotate the needle to make the working edge face downward. The ideal working edge should be perpendicular to the needle's axis (i.e., not angled) and contain no ragged surfaces (Fig. 4a, b).
4. Place a slide containing dried cryosections on the microscope stage, so the sections are facing upward; bring the sample into focus and move the needle tip to the center of the view, keeping it above the slide and out of focus.
5. Find the area of interest (e.g., cross-sectioned jump muscle, Fig. 5a–f) and lower the needle so the working edge is in a flush contact with the surface of the slide. Using micromanipulators

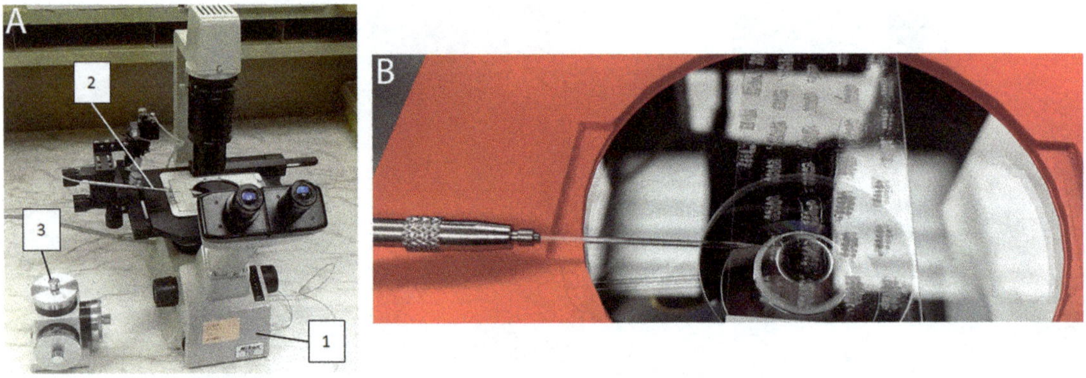

Fig. 4 Microscope setup for microsampling. (**a**) Hardware setup for microsampling: an inverted microscope (1) has attached glass needle holder (2), controlled by a micromanipulator (3). (**b**) An example of needle placement over thick pupal sections

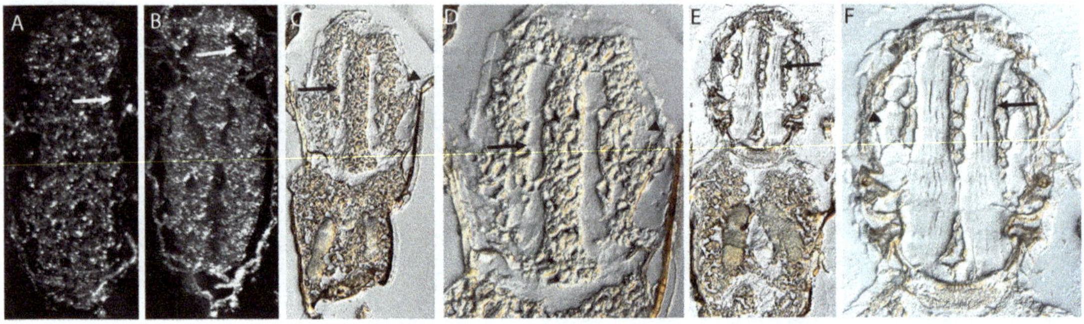

Fig. 5 Identification of specific muscles for microsampling. (**a**) Sectioned pupa at 16 h APF, notice the white arrows indicate the area of muscle precursors of the flight muscles. (**b**) Sectioned pupa at 24 h APF, notice the white arrows indicate the area of muscle precursors of the flight muscles. (**c, d)** 48 h APF sample is sectioned, notice the black arrows indicate the myofibers of the DLMs and the arrowheads indicate the TDT. (**e**, **f**) The sample is 72 h APF, and notice the black arrows indicate the myofibers of the DLMs and the arrowheads indicate the TDT

and microscope stage controls, produce a series of successive grooves across the area of interest, converting the tissue within that area into a pile of loose material (scraps). For better precision, scraping should be performed under a 20× objective.

6. Switch to a smaller power 10× objective and pick up the piled tissue with a fine dissecting pin attached to a paintbrush handle (*see* **Note 7**).
7. Transfer amassed material at the end of the pin into an Eppendorf tube for RNA extraction. The sample can be frozen at −20 °C for long-term storage or used immediately.

3.4 Adult Muscle Microdissection

1. Obtain adult flies of desired genotypes and ages (*see* **Note 8**).
2. Apply a large drop of tissue freezing medium at the center of an overturned (placed upside-down) Petri dish. Anesthetize flies with CO_2 (Fig. 6a), transfer and submerge the fly into tissue freezing medium (Fig. 6b–e, *see* **Note 9**).
3. Under a dissection microscope, cut off and discard the abdomen, the legs, the wings, and the head using the sharp edges of a 22-gauge syringe needle. Keep the liberated thorax for further work (Fig. 6e).
4. Using either a 22-gauge syringe needle or a microdissection probe, retrieve the thorax from tissue freezing medium, taking care to remove excess medium. Transfer the thorax into a standard microcentrifuge tube filled with 1 mL of 1× PBS. Do not load more than 10 thoraces per 1 mL of PBS; for more flies use additional tubes. Mix gently by inverting the tube for 2 min on a rotator. Let the tube stand vertically to collect thoraces at the bottom and then remove the buffer by pipette. Fill with 1 mL of fresh PBS. Repeat **step 3** two more times, leave thoraces in the tube with 200 μL of PBS, place and keep on ice for the entirety of the dissection procedure (*see* **Note 10**).
5. Transfer a fly thorax into a 30 μL droplet of 1 M sucrose placed on the surface of an overturned Petri dish (Fig. 6f).
6. Split the thorax with a 22-gauge needle along the vertical midline to produce two equal hemisegments: first, insert the needle into the bigger opening on the posterior part of the thorax, where it was connected to the abdomen; next, push the needle down to make an incision through the ventral side separating the left and right legs; make a symmetrical second incision through the dorsal side to completely separate thoracic hemisegments (Fig. 6g–j).
7. Locate and detach an array of dorsal longitudinal muscles (DLMs)—a group of muscle fibers belonging to IFMs—inside of the dorsal part of each hemisegment. Working with a 22-gauge needle, make an incision underneath the DLM array to release a patch of cuticle with attached DLM fibers (Fig. 6k, l).
8. Carefully separate DLM fibers from large tracheal ducts (the latter appear as silvery/white lines running along muscle fibers) and detach the muscles from the cuticle (Fig. 6m, n).
9. Immediately transfer separated DLMs into a new standard microcentrifuge tube for RNA extraction or freeze at −20 °C for long storage.
10. After DLM removal, large vertical muscle fibers can be seen in the remaining part of the hemithorax. These are dorsoventral

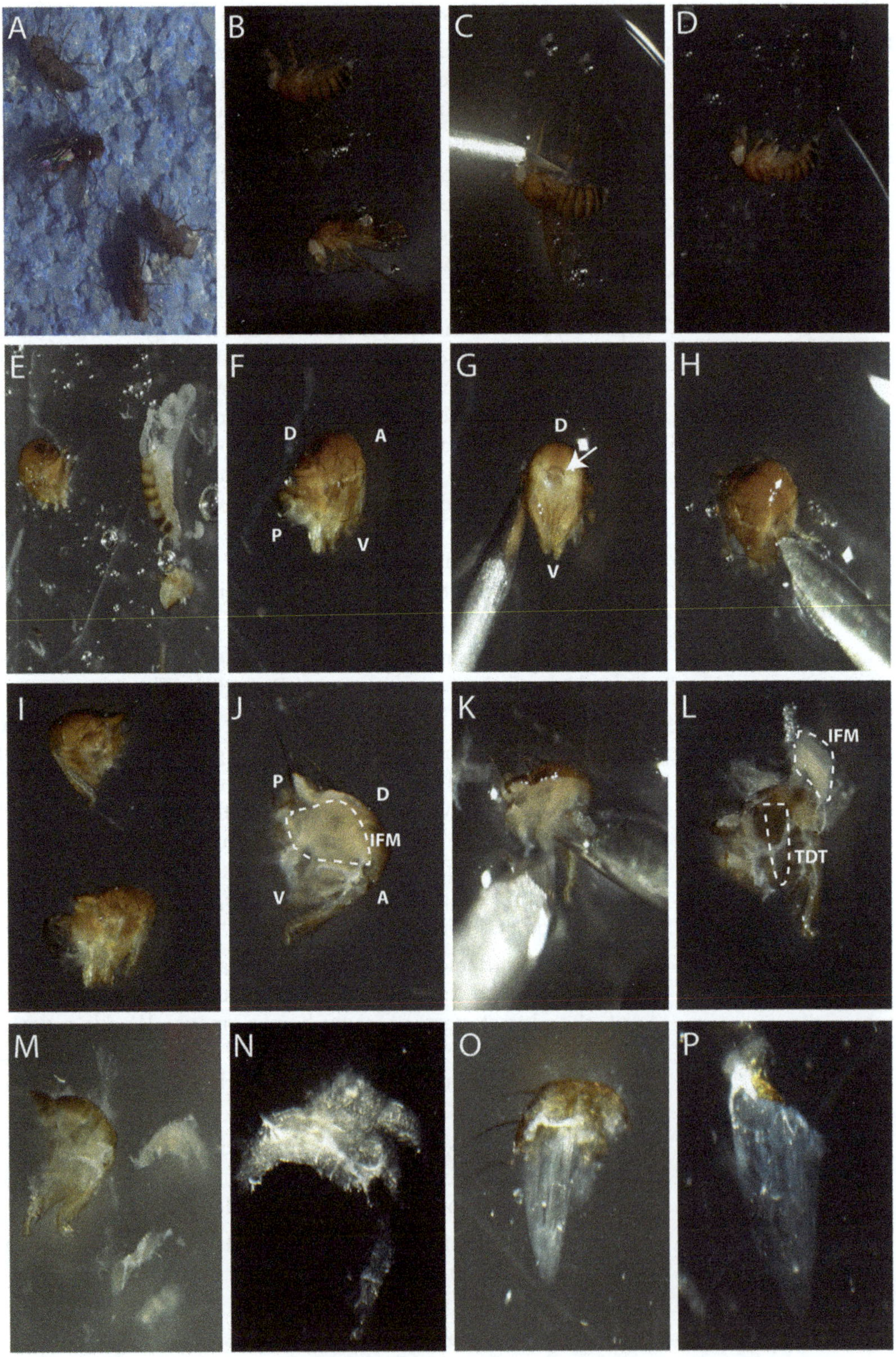

Fig. 6 IFM and TDT dissections. (**a**) Anesthetized flies on CO_2 pad. (**b**) Anesthetized flies in OCT media. (**c**) Removal of legs using sharp edge of the needle. (**d**) Complete removal of legs and wings. (**e**) Removal of head

muscles, DVMs, that also belong to the IFMs (Fig. 1a). They are typically not collected, but they need to be removed to get access to a TDT muscle that underlies them. The jump muscle can be morphologically identified by a clearer appearance than IFM fibers and its characteristic triangular shape. The base of the triangle is attached to the dorsal cuticle while the narrow tip is connected at the base of the second leg (Figs. 1c, d and 6k, m, o, p).

11. Subsequently, using a 22-gauge needle, cut off the cuticle with DVM attachment sites and the bases of the first and third legs, reducing the hemithorax into a smaller, roughly triangular piece with two longer sides converging upon the base of the second leg. Locate the TDT muscle that now should be accessible in the middle of this tissue chunk and free it by cutting and removing surrounding tissues (Fig. 6l, o, p).
12. When the TDT muscle is free, cut away the attached cuticle and transfer the muscle into an Eppendorf tube for subsequent RNA extraction or freeze at −20 °C for longer storage (Fig. 6p). The amount of material used in this procedure is sufficient to carry out RT-PCR to identify expression of alternative transcript isoforms between muscle types (Fig. 7).

3.5 Preparation of Dissected Samples for Protein Analysis

1. Start with 10 or 15 dissected muscle fibers from flight or jump muscles, respectively.
2. Wash dissected muscles with 100 μL of homogenization buffer to remove traces of the tissue freezing medium (*see* **Note 11**).
3. Centrifuge muscles at 14,000 × *g* and carefully remove the supernatant with a narrow tip of an automatic pipette.
4. Add 30 μL of 1× Laemmli sample buffer and homogenize samples in a standard microcentrifuge tube using a disposable pestle.
5. Heat samples at 100 °C for 3 min, briefly spin down, mix by vortexing, and centrifuge at 14,000 × *g* for 15 min.
6. Dilute a sample for protein assay analysis: take 5 μL of the lysate and mix it with 20 μL of 1× Laemmli sample buffer.

Fig. 6 (continued) and abdomen in tissue freezing media using the sharp edge of a needle. (**f**) Thorax in 1 M Sucrose. (**g**) Anterior view of the thorax, arrow shows hole where head was attached. (**h**) Inserting the needle on the anterior side of the thorax to make vertical cut. (**i**) Bisected thorax. (**j**) Visible IFMs on bisected thorax. (**k**) Detaching the IFMs using tip of the needle. (**l**) IFMs lifted out but still attached to thorax. TDT visible underneath IFMs. (**m**) Extracted IFMs next to one half of the thorax. (**n**) Bundle of IFMs. (**o**) Extracted TDT with dorsal cuticle attached. (**p**) TDT in 1 M sucrose media. *D* dorsal, *V* ventral, *A* anterior, *P* posterior

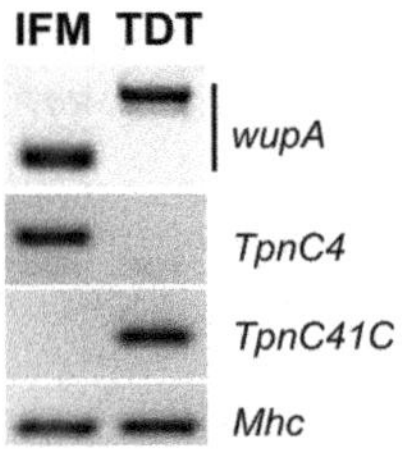

Fig. 7 Molecular diversity of *Drosophila* flight and jump muscles revealed with the microsampling technique. Frozen young adults were sectioned and, under microscope control, flight (IFM) and jump (TDT) muscles were scraped and collected for RNA isolation, cDNA was made with random hexamer oligos and equalized cDNA samples were used for end-point amplification of specific gene products using sequence-specific primers. Note that this technique reveals fiber-specific preferences in alternative mRNA splicing of Troponin I transcripts (wupA), as well as mutually exclusive expression of Troponin C genes (TpnC4 and TpnC41C); amplification of pan-specific Myosin Heavy Chain (Mhc) transcripts was used as loading control

7. Measure the concentration of protein and adjust to 2 mg/mL. Load 5 μL (i.e., 10 μg of total protein) per lane of a 4–20% gradient polyacrylamide gel.
8. Run protein gels in accordance with the general procedure for protein electrophoresis.
9. Some applications may require analysis of muscle structural proteins only. Preparations of "skinned" muscle lysates, enriched in structural proteins, can be obtained by additional treatment of dissected muscles with Triton X-100. In this case, homogenize 16 IFM or 30 TDT muscles in 100 μL of ice-cold homogenization buffer containing 0.5% Triton X-100. Centrifuge muscles at 14,000 × *g*, 4 °C for 5 min and carefully remove the supernatant. Wash skinned muscles three times with 100 μL of cold homogenization buffer to remove traces of Triton X-100. After centrifugation and removal of supernatant, resuspend muscles in 60 μL of 1× Laemmli sample buffer and proceed with **steps 5–8** above [25].
10. After electrophoresis gels can be used for Coomassie staining, silver staining, or Western blotting. The amount of material (*see* **step 1**) is sufficient to visualize multiple bands in Coomassie-stained gels (Fig. 8).

4 Notes

1. Developmental time points need to be adjusted depending on incubation temperature.
2. Great care should be exercised to avoid an accidental transfer of liquid nitrogen into a tube with samples. Do not tighten tube's

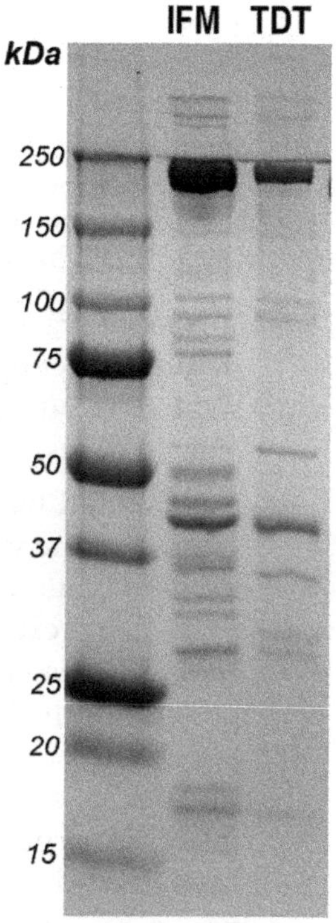

Fig. 8 Analysis of the diversity of IFM and TDT structural proteins. IFM and TDT muscles of wild-type flies were dissected, and skinned muscle lysates were prepared according to the described protocol and procedure [25]. 10 μg of proteins from each sample were resolved in 4–20% gradient polyacrylamide gel and stained with Coomassie blue

cap to avoid pressure building from liquid nitrogen vapors that can potentially result in dangerous tube bursting.

3. Frozen blocks can be stored in a non-auto defrost −20 °C freezer, but a −80 °C freezer is preferential. During sample storage and handling, avoid warming of frozen blocks, as it will result in crystallized water formation and subsequent damage of tissues.
4. Applying tissue freezing medium to a cold specimen holder will lead to inadequate adhesion and may result in block separation during sectioning. Therefore, keep unused cryotome specimen holders at ambient temperature, apply freezing medium first and then place the specimen holder into the cryotome chamber for sample mounting.
5. We apply sections in a row following along the longer axis of the slide; when reaching slide's edge turn the direction and produce a second row of sections. With sufficient training, a standard slide can hold six rows of sequential sections, fitting the block of five to six flies.

6. If necessary, air-dried slides with sections can be stored in a slide box at +4 °C, although it should be avoided for samples used for microsampling (which should always be carried out on freshly sectioned samples). Always let a box with slides sit at room temperature to equilibrate to ambient temperature before opening it. This prevents moisture from the air precipitating on sections and damaging them.
7. It is more practical to scrape the entire row of cryosections first, before collecting the scraped material. Lean on the microscope stage to steady the hand; use stage controls to keep the field of view in focus.
8. Care should be taken to ensure that control and experimental flies are of the same age, since expression of numerous genes in muscles is age-dependent. A dramatic decline in muscle structural gene expression occurs within the first 2 days post-eclosion. Dissection of flight and jump muscles from pupae is possible at the late stages of pupal development (72–96 h after puparium formation, apf), but it is difficult due to the small size of muscles and a high chance of sample contamination with histolysis products—the microsampling protocol should be used instead.
9. The fly's cuticle is hydrophobic; bathing it with viscous freezing medium will prime fly surfaces for 1 M sucrose solution used in downstream dissecting procedures.
10. We discovered that some components of the tissue freezing medium interfere with downstream RNA purification procedures and result in lower yields of RNA. This washing step will remove freezing medium from thoraces.
11. It is important to remove freezing medium completely, as it may cause smudging of protein lanes during gel electrophoresis.

Acknowledgments

Research at UNM is supported by grants from NIH/NIGMS: COBRE Center for Evolutionary and Theoretical Immunology (P30GM110907), the NIH (GM061738 and GM124498) and NSF (1518073) to R.M. Cripps.

References

1. Bönnemann CG, Laing NG (2004) Myopathies resulting from mutations in sarcomeric proteins. Curr Opin Neurol 17(5):529–537
2. D'Amico A, Bertini E (2008) Congenital myopathies. Curr Neurol Neurosci Rep 8:73–79
3. Schiaffino S, Reggiani C (2011) Fiber types in mammalian skeletal muscles. Physiol Rev 91 (4):1447–1531
4. Ariano MA, Armstrong RB, Edgerton VR (1973) Hindlimb muscle fiber populations of

five mammals. J Histochem Cytochem 21 (1):51–55

5. Smith RS, Ovalle WK Jr (1973) Varieties of fast and slow extrafusal muscle fibres in amphibian hind limb muscles. J Anat 116(Pt 1):1–24
6. Gleeson TT, Putnam RW, Bennett AF (1980) Histochemical, enzymatic, and contractile properties of skeletal muscle fibers in the lizard Dipsosaurus dorsalis. J Exp Zool 214 (3):293–302
7. Thorstensson A, Grimby G, Karlsson J (1976) Force-velocity relations and fiber composition in human knee extensor muscles. J Appl Physiol 40(1):12–16
8. Harridge SD et al (1996) Whole-muscle and single-fibre contractile properties and myosin heavy chain isoforms in humans. Pflugers Arch 432(5):913–920
9. Hoppeler H et al (1973) The ultrastructure of the normal human skeletal muscle. A morphometric analysis on untrained men, women and well-trained orienteers. Pflugers Arch 344 (3):217–232
10. Herbison GJ, Jaweed MM, Ditunno JF (1982) Muscle fiber types. Arch Phys Med Rehabil 63 (5):227–230
11. Rossi AC et al (2010) Two novel/ancient myosins in mammalian skeletal muscles: MYH14/ 7b and MYH15 are expressed in extraocular muscles and muscle spindles. J Physiol 588 (Pt 2):353–364
12. Korfage JA, Van Eijden TM (2003) Myosin heavy-chain isoform composition of human single jaw-muscle fibers. J Dent Res 82 (6):481–485
13. Costill DL et al (1976) Skeletal muscle enzymes and fiber composition in male and female track athletes. J Appl Physiol 40 (2):149–154
14. Agudelo LZ et al (2014) Skeletal muscle PGC-1alpha1 modulates kynurenine metabolism and mediates resilience to stress-induced depression. Cell 159(1):33–45
15. Butler-Browne GS, Whalen RG (1984) Myosin isozyme transitions occurring during the postnatal development of the rat soleus muscle. Dev Biol 102(2):324–334
16. Ciciliot S et al (2013) Muscle type and fiber type specificity in muscle wasting. Int J Biochem Cell Biol 45(10):2191–2199
17. Lexell J (1995) Human aging, muscle mass, and fiber type composition. J Gerontol A Biol Sci Med Sci 50:11–16
18. Oberbach A et al (2006) Altered fiber distribution and fiber-specific glycolytic and oxidative enzyme activity in skeletal muscle of patients with type 2 diabetes. Diabetes Care 29 (4):895–900
19. Tanner CJ et al (2002) Muscle fiber type is associated with obesity and weight loss. Am J Physiol Endocrinol Metab 282(6): E1191–E1196
20. Bryantsev AL et al (2012) Differential requirements for myocyte enhancer factor-2 during adult myogenesis in *Drosophila*. Dev Biol 361 (2):191–207
21. Oas ST, Bryantsev AL, Cripps RM (2014) Arrest is a regulator of fiber-specific alternative splicing in the indirect flight muscles of *Drosophila*. J Cell Biol 206(7):895–908
22. Chechenova MB et al (2017) Functional redundancy and non-redundancy between two troponin C isoforms in *Drosophila* adult muscles. Mol Biol Cell 28:760
23. Chechenova MB, Bryantsev AL, Cripps RM (2013) The *Drosophila* Z-disc protein Z(210) is an adult muscle isoform of Zasp52, which is required for normal myofibril organization in indirect flight muscles. J Biol Chem 288 (6):3718–3726
24. Bryantsev AL et al (2012) Extradenticle and homothorax control adult muscle fiber identity in *Drosophila*. Dev Cell 23(3):664–673
25. Cripps RM, Sparrow JC (1992) Polymorphism in a *Drosophila* indirect flight muscle-specific tropomyosin isozyme does not affect flight ability. Biochem Genet 30:159

Chapter 17

Measuring Both Glucose Uptake and Myosin Heavy Chain Isoform Expression in Single Rat Skeletal Muscle Fibers

Mark W. Pataky, Edward B. Arias, and Gregory D. Cartee

Abstract

Glucose uptake by skeletal muscle is important for metabolic health. Because skeletal muscle is composed of multiple fiber types that have differing metabolic and contractile properties, studying glucose uptake in whole muscle tissue does not elucidate differences at the cellular level. Here, we describe a procedure that enables the measurement of both glucose uptake and fiber type (by myosin heavy chain isoform expression) in individual rat epitrochlearis muscle fibers.

Key words Muscle fiber type, Myosin heavy chain, Glucose transport, Glucose uptake, SDS-PAGE

1 Introduction

Skeletal muscle is a heterogeneous tissue composed of hundreds to many thousands of individual muscle cells (muscle fibers). Different muscle fibers can have strikingly different metabolic and contractile properties [1]. Measuring the myosin heavy chain (MHC) expression via SDS-PAGE is the gold standard for classifying muscle fiber type at both the tissue and single fiber levels [2]. Four MHC isoforms are expressed in adult rat or mouse skeletal muscle (MHC type I, IIA, IIX, and IIB) and three MHC isoforms are expressed in adult human skeletal muscle (MHC type I, IIA, and IIX) [1]. Additionally, muscle fibers that express more than one MHC isoform are classified as "hybrid" fibers which typically possess characteristics that are intermediate to the values of the individual MHC isoforms that they express. There is evidence that certain conditions such as chronic exercise training [3, 4], and old age [5, 6] may alter fiber type composition, and this may have an impact on metabolic health.

In skeletal muscle, glucose transport into the cell is a rate-limiting step for glucose disposal. Glucose uptake in whole skeletal muscle tissue is often measured using radiolabeled glucose analogs such as 2-deoxyglucose (2-DG; which can be phosphorylated by

Sissel Beate Rønning (ed.), *Myogenesis: Methods and Protocols*, Methods in Molecular Biology, vol. 1889,
https://doi.org/10.1007/978-1-4939-8897-6_17,

hexokinase with limited subsequent metabolism) or 3-O-methyl-glucose (3-MG; which is not metabolized after cell entry) in response to insulin [7, 8], exercise [9, 10], contraction [11, 12], or various other stimuli [8, 12]. Conventionally, researchers have attempted to evaluate fiber type-related differences for glucose uptake by comparing whole muscle or regions of whole muscle which have enriched expression of a given fiber type. However, comparing glucose uptake at the whole muscle level is not ideal from the perspective of understanding fiber type-specific differences because: (1) no whole muscle is only comprised of one fiber type, (2) hybrid fibers cannot be evaluated using whole muscle tissue analysis, (3) whole muscles contain multiple cell types (including neural, vascular, connective, and adipose) which contribute to tissue glucose uptake, and (4) no whole muscle in rat has been identified that is primarily composed of type IIX fibers.

In 2012, MacKrell and Cartee developed and validated a novel method for measuring glucose uptake and fiber type in the same single fiber from rat skeletal muscle [13]. 2-DG was the glucose analog used because 2-DG that enters myofibers is rapidly phosphorylated by hexokinase, and the resultant 2-DG-6P is trapped in the fiber. A fundamental feature of measuring glucose uptake in isolated whole muscles involves the calculation of [^{3}H]-2-DG within the muscle fibers with correction of extracellular [^{3}H]-2-DG, e.g., using [^{14}C] mannitol to estimate the extracellular space. Because isolated muscle fibers do not have extracellular space, extracellular [^{14}C] mannitol is not required for single fiber glucose uptake. Therefore, the conventional whole muscle glucose uptake method (with [^{3}H]-2-DG and extracellular [^{14}C] mannitol) was compared with incubation of whole muscle in [^{3}H]-2-DG followed by three washing steps to remove extracellular [^{3}H]-2-DG that was not trapped in the muscle cells, and no differences were detected [13]. Because the procedure to measure single fiber glucose uptake requires enzymatic digestion of the collagen surrounding muscle fibers by collagenase treatment, a control experiment was performed which demonstrated that collagenase did not alter glucose uptake by isolated muscle tissue. Finally, incubation in cytochalasin B, a fungal metabolite that inhibits glucose transporter-mediated glucose transport [14], completely blocked insulin-stimulated glucose uptake in all fiber types, indicating that the glucose uptake measured in single fibers was attributable to glucose transporter-mediated transport.

After validating this single fiber method, our laboratory performed a series of experiments to characterize the fiber type-specific effects of various conditions and interventions on glucose uptake. The seminal publication by MacKrell and Cartee which evaluated type IIA, IIB, IIX, and IIBX fibers demonstrated that glucose uptake was increased by insulin in a fiber type-specific manner, where type IIA fibers displayed the greatest insulin-stimulated glucose uptake and type IIX and IIB fibers had the lowest insulin-stimulated glucose uptake [13]. In the original study, we failed to isolate any type I fibers, but we subsequently modified the protocol and were successful in isolating type I fibers from the epitrochlearis muscle. With the modified protocol, we recently found that insulin-stimulated glucose uptake was similar for type I fibers versus type IIA fibers, and values for each of these fiber types significantly exceeded the values for other fiber types (IIB, IIX, and IIBX) [15]. Using the single fiber method we have discovered that some conditions or interventions do not result in fiber type-selective effects on glucose uptake. For example, in a comparison of obese versus lean Zucker rats, insulin-stimulated glucose uptake was substantially reduced to a similar extent in each fiber type studied (types IIA, IIB, IIX, and IIBX) [13]. We also found that old rats eating a calorie restricted diet (40% below ad libitum, AL, intake) compared to AL controls had greater insulin-stimulated glucose uptake in each fiber type studied (types I, IIA, IIB, IIX, IIAX, and IIBX) [16]. However, we have also discovered fiber type-selective differences on insulin-stimulated glucose uptake induced by either acute exercise (increased for rats at 3.5 h post-exercise versus sedentary controls in all fiber types except IIX) [15], or a high-fat diet (values for rats consuming a high-fat diet were reduced versus low-fat diet-fed controls for all fiber types except type I) [17].

Insulin is the most important physiologic stimulator of glucose transport, but there are also insulin-independent mechanisms to increase glucose transport in skeletal muscle. Using muscle incubated with AMP-activated kinase stimulating compound, AICAR, we found a relatively greater increase in glucose uptake in type IIB fibers compared to type IIA fibers [13]. In response to electrical stimulation-induced muscle contraction, we observed similar increases in glucose uptake in each fiber type studied (types IIA, IIB, IIX, and IIBX) [18]. Whether or not the method has revealed fiber type-selective responses to a given intervention, the ability to interrogate muscle at the single fiber level has offered valuable insights at the cellular and fiber type-specific level that would have otherwise been impossible to determine based solely on tissue-level analysis.

For several reasons, we have exclusively used rat epitrochlearis muscles in all of our publications that evaluated glucose uptake and fiber type in single muscle fibers. We initially used the rat epitrochlearis for single fiber glucose uptake because we and many other researchers have frequently used the epitrochlearis for analysis of whole muscle glucose uptake, resulting in and extensive foundation of information regarding this muscle. The epitrochlearis is a thin forelimb muscle that originates at the distal tendon of the latissimus dorsi and inserts at the medial epicondyle of the humerus [19]. Because the epitrochlearis is only ~25 fibers thick, it has a relatively thin diffusion distance that facilitates the access of extracellular nutrients to the fibers. The epitrochlearis expresses each of the MHC isoforms that are expressed in adult rat skeletal muscle, and the fiber type proportion of the epitrochlearis muscle is comparable to the fiber type profile that has been reported for 76 skeletal muscles or muscle regions that were sampled from the rat neck, trunk, forelimbs, and hindlimbs [20], making it a representative muscle from the perspective of fiber type. When we initially developed this method, we made very limited attempts to isolate single fibers from rat soleus and extensor digitorum (EDL) muscles. These limited efforts suggested that the epitrochlearis was superior to these muscles with regard to the ease of fiber isolation using our approach, so we discontinued further attempts using these muscles. It remains possible that others will identify modifications of our methods or other muscles that are suitable for single fiber glucose uptake analysis.

Here, we describe the procedure that enables the measurement of both glucose uptake and fiber type in the same individual rat epitrochlearis muscle fiber. The following description incorporates improvements to the original method that have been noted in several subsequent publications, as well as providing more details than have been previously documented [15–18, 21].

2 Materials

2.1 Experimental Setup

1. Temperature controlled shaking water bath capable of 35 °C and 45 revolutions per minute (rpm).
2. 20 mL glass scintillation vials (used for both incubating muscles and for scintillation counting).
3. Krebs-Henseleit Buffer (KHB): Prepare 10× KHB Stock #1 (1 L deionized H_2O, 67.8 g NaCl, 3.4 g KCl, 1.58 g KH_2PO_4, and 21.5 g NaH_3CO_3) and 10× KHB Stock #2 (1 L deionized H_2O, 2.82 g $CaCl_2$, and 1.4 g $MgSO_4$). 10× KHB Stock #1 and #2 can be stored at 4 °C for at least 1 month. To make KHB add 50 mL of 10× KHB Stock #1 to 400 mL deionized ultrapure H_2O and gas with 95% O_2/5% CO_2 with stainless steel aeration stone for 15 min on ice, then add 50 mL of 10× KHB Stock #2 and mix.
4. Incubation Medium #1: KHB supplemented with 0.1% RIA-grade bovine serum albumin (BSA), 2 mM sodium pyruvate, and 6 mM mannitol (*see* **Note 1**).
5. Incubation Medium #2: KHB supplemented with 0.1% BSA, 2 mM sodium pyruvate, 6 mM mannitol, and 0.1 mM 2-deoxy-D-glucose (2-DG) [specific activity of 13.5 mCi/mmol [^{3}H]-2-DG] (*see* **Note 2**).

6. Wash Medium: Ca^{2+}-free KHB (*see* **Note 3**) supplemented with 0.1% BSA and 8 mM glucose.
7. 2.5% type II Collagenase Medium: Ca^{2+}-free KHB supplemented with 0.1% BSA, 8 mM glucose, and 2.5% type II collagenase (*see* **Note 4**).
8. Gas washing bottle with gas dispersion disc.
9. 95% O_2/5% CO_2 compressed gas cylinder.
10. Gas regulator appropriate for 95% O_2/5% CO_2.
11. Gas manifold apparatus (with sufficient number of ports to accommodate the number of incubation vials used in the experiment).
12. Support stand setups (base and rods) with appropriate clamps to secure each manifold apparatus.
13. 3/8″ × 1/16″ wall flexible polyethylene tubing (enough to reach from the compressed gas cylinder to the gas washing bottle and manifold apparatus).
14. Flexible polyethylene microbore tubing (size 0.04 × 0.07; 1 ft. per vial to reach from manifold to incubation vials).
15. 1.5 in. blunt tip 18-gauge needles (sufficient number to accommodate for the number of incubation vials used in the experiment).
16. Rubber stoppers for vials (size 0; sufficient number to accommodate the number of incubation vials used in the experiment).

2.2 Muscle Dissection and Incubation

1. Surgical instruments for rat epitrochlearis dissection.
2. Stainless steel Gerald forceps.
3. Stainless steel laboratory spatula.
4. Orbital Shaker capable of 90 rpm.
5. Glass petri dishes.
6. 30 mm culture dish.

2.3 Single Fiber Isolation and Processing

1. Camera enabled dissecting microscope capable of 8–20× magnification.
2. Fine surgical forceps (*see* **Note 5**).
3. 200 μL microcentrifuge tubes.
4. Lysis Buffer: tissue protein extraction reagent (T-PER; *see* **Note 6**) supplemented with 1 mM Na_3VO_4, 1 mM EGTA, 2.5 mM sodium pyrophosphate tetrabasic decahydrate, 1 mM β-glycerophosphate, 1 μg/mL leupeptin, and 1 mM phenylmethylsulfonyl fluoride.
5. 2× Laemmli Buffer.

6. Vortex mixer.
7. Heating block.
8. Benchtop mini-centrifuge.

2.4 Measuring Glucose Uptake

1. Liquid scintillation cocktail.
2. Scintillation counter.
3. Computer with ImageJ software (NIH, Bethesda, MD).

2.5 MHC Isoform Identification

1. 6.5% Polyacrylamide Resolving Gel (volumes for 16 gels): 14.7 mL 30% acrylamide/bis-acrylamide (50:1 ratio), 20.25 mL glycerol, 9.45 mL 1.5 M Tris–HCl (pH 8.8), 7.08 mL 1 M glycine, 2.7 mL 10% sodium dodecyl sulfate (SDS), 12.612 mL deionized H_2O, 0.675 mL 10% ammonium persulfate, and 0.04 mL TEMED.
2. 4% Polyacrylamide Stacking Gel (volumes for 16 gels): 5.332 mL 30% acrylamide/bis-acrylamide (37.5:1 ratio), 12 mL glycerol, 5.6 mL 0.5 M Tris–HCl (pH 6.7), 1.6 mL 100 mM EDTA (pH 7.0), 1.6 mL 10% SDS, 13.448 mL deionized H_2O, 0.4 mL 10% ammonium persulfate, and 0.025 mL TEMED.
3. 10× Running Buffer: 0.25 M Tris base, 1.92 M glycine, 34.7 mM SDS.
4. 1× Running Buffer: 900 mL deionized H_2O and 100 mL of 10× Running Buffer.
5. Rat extensor digitorum longus (EDL) muscle (for positive MHC control).
6. Rat soleus muscle (for positive MHC control).
7. High molecular weight (HMW) protein ladder standard.
8. Equipment used for SDS-PAGE.
9. Plastic tupperware to be filled with ice that can also hold SDS-PAGE boxes.
10. SimplyBlue™ SafeStain.
11. Bicinchoninic acid (BCA) Protein Assay Kit.
12. Microplate reader with absorbance detection capability and appropriate filters.

3 Methods

3.1 Experimental Setup

1. On the day of the experiment prepare KHB solution as described in Subheading 2.1, **item 3**.
2. Prepare Incubation Medium #1 and Incubation Medium #2 fresh on the day of the experiment (*see* **items 4** and **5** in

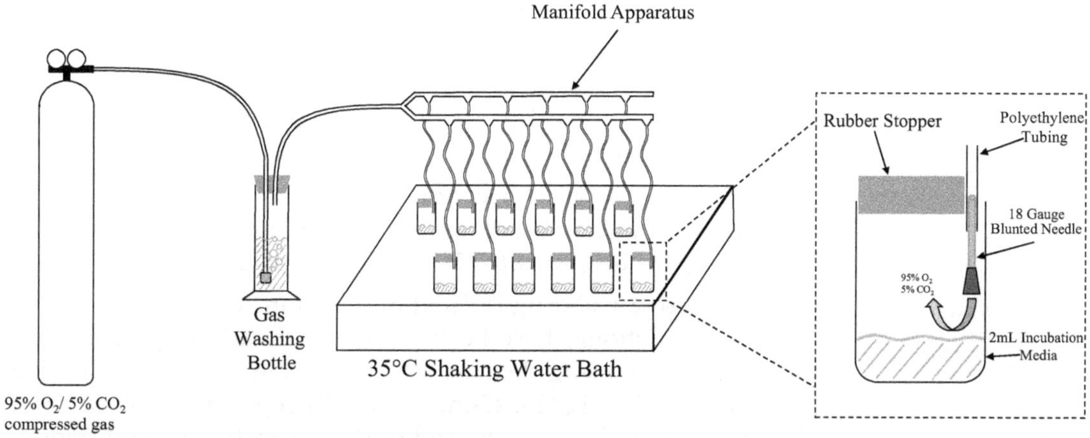

Fig. 1 Schematic diagram illustrating the system used to gas the vials during isolated muscle incubation steps. The 95% O_2/ 5% CO_2 gas cylinder with a gas regulator is attached by polyethylene tubing to the inlet tube of a gas washing bottle including deionized water. The gas washing bottle's outlet tube is attached via polyethylene tubing to a manifold that has multiple gassing ports, each of which is connected to individual segments of polyethylene tubing. The distal end of the tubing is connected to stainless steel 18-gauge blunt tip needles that are inserted into the mouth of glass scintillation vials. A size 0 rubber stopper is inserted into the vial's opening and holds each needle in place above the surface of the media that has been pipetted into the vial, which is placed in a shaking, heated water bath. During muscle incubation, individual isolated muscles are submerged in the media in the vials, and the vials are returned water bath. *See* Subheading 2.1 and **Note 8** for details

Subheading 2.1). When the incubation medium #2 is prepared, an aliquot of ~200–500 μL should be removed and stored at −20 °C until it is thawed and used for liquid scintillation counting and the calculation of glucose uptake measurement (*see* **step 3** in Subheading 3.4).

3. Pipette 2 mL of Incubation Medium #1 into an incubation vial for each muscle that will be isolated (*see* **Note** 7). Pre-gas and pre-warm the vials containing Incubation Medium #1 for at least 5 min in the shaking, heated water bath as shown in Fig. 1 (*see* **Note 8**).
4. Pipette 2 mL of Incubation Medium #2 into a separate incubation vial for each muscle, and pre-gas and pre-warm the medium.
5. Prepare Wash Medium and 2.5% type II Collagenase Medium (*see* **items 6** and 7 in Subheading 2.1) and add to appropriate vials. Wash Medium and 2.5% type II Collagenase Medium should be stored at 4 °C until just prior to use. Incubations in Wash Medium will be performed on ice. For incubation steps using 2.5% type II Collagenase Medium, vials should be placed in the shaking, heated water bath ~5 min prior to muscle incubation (*see* **Note 9**).

3.2 Muscle Dissection and Incubation

1. Remove skin from the forelimb of a deeply anesthetized rat.
2. Carefully dissect the whole epitrochlearis muscle from the rat. Take care to avoid damaging the epitrochlearis fibers. Retain the connective tissue at the proximal and distal ends of the epitrochlearis.
3. Briefly (2–3 s) wash excess blood from the isolated epitrochlearis muscle (handling only by the connective tissue) in a glass petri dish with KHB at room temperature.
4. Prior to transferring isolated muscles into vials for incubation, the vials should have been pre-warmed and pre-gassed for at least 5 min using the gassing apparatus in the heated, shaking water bath (Fig. 1). Using Gerald forceps, grasp the adjacent connective tissue on the end of the epitrochlearis, and transfer the muscle to a glass vial containing 2 mL of Incubation Medium #1 (*see* **Note 10**). During this incubation step each vial should be continually gassed with 95% O_2/5% CO_2 from above with an individual line. Additionally, vials should be shaken at 45 rpm (*see* **Note 11**) in a heated (35 °C) water bath for 10–30 min (*see* **Note 12**).
5. Using Gerald forceps, gently transfer the muscle into a pre-gassed and pre-warmed glass vial containing 2 mL of Incubation Medium #2. Vials containing isolated muscles should be continually gassed with 95% O_2/5% CO_2 while being shaken at 45 rpm in a heated (35 °C) water bath for 30–60 min (*see* **Note 13**).
6. To clear extracellular space of 2-DG, rinse the muscle 3 × 5 min in 5 mL of ice-cold Wash Medium in glass vials with shaking at 90 rpm.
7. After washing, place the muscle in a continually gassed (95% O_2/5% CO_2) glass vial containing 2 mL of 2.5% type II Collagenase Medium for 60 min (*see* **Note 14**). The vials should be placed in a water bath at 35 °C with shaking at 45 rpm. The enzymatic digestion of collagen results in loosely connected fibers, which are hereafter referred to as fiber bundles.
8. Transfer Collagenase Medium and fiber bundle by carefully pouring into a 30 mm culture dish. Under a dissecting microscope, trim the proximal and distal connective tissue from the fiber bundle using a scalpel (*see* **Note 15**). Gently remove fiber bundle from collagenase medium with a stainless steel laboratory spatula and place in a glass petri dish containing Wash Medium at room temperature (*see* **Note 16**). Fibers will be isolated from the fiber bundle in this dish.

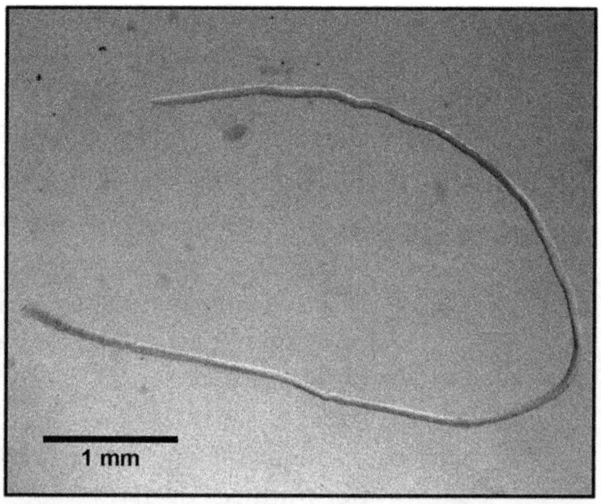

Fig. 2 Representative image of an isolated single rat epitrochlearis muscle fiber (16× magnification). The scale bar in the lower left of image is used for calibration when measuring fiber area

3.3 Single Fiber Isolation and Processing

1. Using fine surgical forceps, gently tease apart individual fibers from the fiber bundle by grasping a fiber from one end and gently pulling away it from the fiber bundle. Then release the fiber in an area of the petri dish free from debris.
2. After isolating each fiber under the microscope, capture an image of the fiber to be analyzed for fiber area (*see* **step 8** in Subheading 3.4). Include a scale bar in the image based on the magnification being used (typically 16× or 20× magnification). A representative image of an isolated fiber is shown in Fig. 2.
3. After image capture, transfer each fiber by pipette, along with 20 μL of the Wash Medium that the fibers are being isolated in, to a microcentrifuge tube (*see* **Note 17**).
4. After completing the imaging of all of the fibers that were isolated from a given muscle (typically 50–60 fibers for our experiments), and after each isolated fiber has been transferred into an individual microcentrifuge tube, add 30 μL of Lysis Buffer and 50 μL of 2× Laemmli Buffer to each microcentrifuge tube. The final volume of each tube will now be 100 μL.
5. Vortex and then lyse the fibers by heating tubes for 10 min at 95–100 °C in a heating block (*see* **Note 18**). Vortex tubes again, flash spin with mini-centrifuge, and store lysed fibers at −20 °C until processing for glucose uptake and MHC expression.

3.4 Measuring Glucose Uptake

1. Aliquot 40 μL of each fiber lysate into separate glass scintillation vials.
2. Add 8 mL of liquid scintillation cocktail to each vial, and tightly cap and vortex the vials.

3. Dilute 100 μL of Incubation Medium #2 (*saved from* **step 2** in Subheading 3.1) in 900 μL of Lysis Buffer. Aliquot 200 μL of the diluted Incubation Medium #2 into 4 separate scintillation vials, followed with 8 mL of liquid scintillation cocktail.
4. Aliquot 200 μL of non-radioactive KHB into four separate scintillation vials, along with 8 mL of liquid scintillation cocktail.
5. Place the individual scintillation vials containing lysate from each single fiber sample (after **steps 1** and **2**), quadruplicate incubation medium (**step 3**), and quadruplicate KHB background samples (**step 4**) into carrier racks that are placed in a scintillation counter to determine 2-[^{3}H]-DG disintegrations per minute (dpm).
6. The total ^{3}H dpm per fiber can then be used to calculate 2-DG accumulation in individual fibers using the following equation (*see* **Note 19**):

$$\text{Total}^3\text{H dpm per fiber} \times \frac{\mu\text{mol 2-DG per mL incubation medium}}{^3\text{H dpm per mL incubation medium}}$$

$$= \text{2-DG accumulation per fiber}$$

7. 2-DG accumulation should be expressed relative to fiber area (pmol × mm^2) which is determined based on the captured image from individual fibers.
8. Images of each fiber (*captured during* **step 2** of Subheading 3.3) can be manually traced in ImageJ using the "Polygon Selections" or "Freehand Selections" tools for an estimate of fiber area (*see* **Notes 20–22**).

3.5 MHC Isoform Identification

1. Prepare 15-well self-cast 6.5% polyacrylamide gels (*see* **items 1** and **2** in Subheading 2.4) and 1× running buffer.
2. Preparation of MHC Standards: Homogenize ~30 g of rat EDL and ~20 g of rat soleus muscles together in 1 mL of Lysis Buffer. Do not centrifuge to remove insoluble protein which include the MHC isoforms. Perform a BCA protein assay in quadruplicate for protein concentration in the lysate and then dilute the lysate to a concentration of 1 μg/μL with lysis buffer. Add equal volume of 2× Laemmli buffer to bring the final protein concentration to 0.5 μg/μL and then boil heat sample for 10 min at 95–100 °C and store at 20 °C. 12 μL (6 μg) of this 3:2 EDL:Soleus standard will be used for each MHC standard lane (*see* **Note 23**).
3. When ready to run the MHC fiber-typing gel, heat the 3:2 EDL:Soleus standard and single fiber lysates for 10 min at 95–100° prior to loading on gel (*see* **Note 24**).
4. Add 1× Running Buffer to SDS-PAGE boxes.

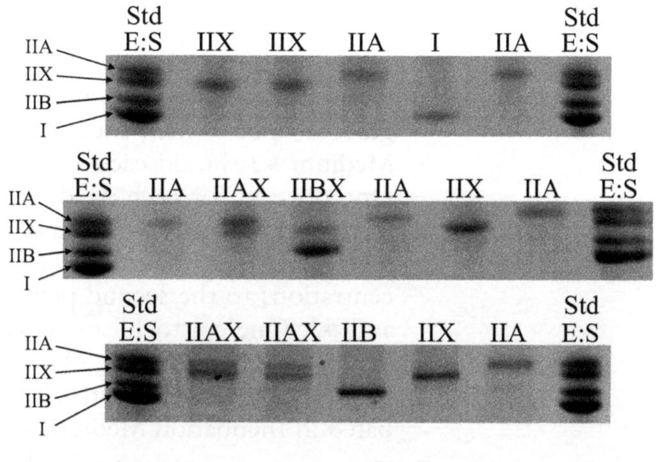

Fig. 3 Representative myosin heavy chain (MHC) single fiber gels. The 3:2 EDL: Soleus standard (Std E:S) is located on either side of all fibers. The migration of the Std E:S MHC allows for a guide to determine the MHC isoform expression of each single fiber which has been loaded in all other lanes. Fibers of all isolated fiber types (types I, IIA, IIAX, IIX, IIBX, and IIB) are displayed in the representative gels

5. In well #1 load 6–10 μL of a HMW protein ladder standard. In wells #2, #9, and #15 load 12 μL of heated 3:2 EDL:Soleus standard. In the remaining 11 wells load 18 μL of the heated fiber lysates (*see* **Note 25**).
6. Pack SDS-PAGE boxes in a container with ice and run for 1 h at 50 V in a 4 °C refrigerator.
7. After 1 h, reset the voltage to ~40 V and run for 22–24 h, or until the 200 kDa marker is almost at the bottom of the gel (*see* **Note 26**).
8. After the gel is finished running, remove the gels from the glass plates and wash gels in a shaking container with deionized water at room temperature for 4–5 min to rinse off excess glycerol (*see* **Note 27**).
9. Remove gels from deionized water and place in a shaking container with SimplyBlue™ SafeStain for 1–2 h (*see* **Note 28**).
10. Remove gels from SimplyBlue™ SafeStain and place in a shaking container with deionized water for ≥2 h (*see* **Note 29**) for destaining.
11. Gels can now be imaged and MHC expression determined by the migration of MHC isoforms in each lane. Images of representative gels with labeled MHC isoforms are shown in Fig. 3.
12. MHC isoform and 2-DG uptake (*from* Subheading 3.4) can be used to determine fiber type specific glucose uptake.

4 Notes

1. If the experiment requires evaluating the effect of insulin on glucose uptake, then Incubation Medium #1 and Incubation Medium #2 should each be divided into two equal portions. One portion for each Medium will be used for the muscles incubated without insulin, and the second portion will be used for insulin-treated muscles. Add insulin (at the desired concentration) to the second portions of Incubation Medium #1 and #2. One epitrochlearis muscle from each rat can then be incubated in Incubation Medium #1 and #2 without insulin, and the contralateral epitrochlearis from each rat can be incubated in Incubation Medium #1 and #2 with insulin.
2. The greater specific activity of [^{3}H]-2-DG for these single fiber experiments than is used for whole muscle experiments is necessary for the detectable accumulation of [^{3}H]-2-DG in single fibers. Additionally, the initial single fiber experiments from our group used 1 mM 2-DG in the incubation medium. Although we have no evidence that 1 mM 2-DG causes problems, we have since reduced the 2-DG in the incubation medium to 0.1 mM to minimize the possibility of metabolic stress related to using a higher 2-DG concentration.
3. To make Ca^{2+}-free KHB do not include $CaCl_2$ in the KHB Stock #2 (*see* **item 3**, Subheading 2.1).
4. The Type II collagenase used for previously performed experiments was from Worthington Biochemical Corporation (CAT#: LS004177). For our experiments, the activity of the type II collagenase was 295 u/mg dw. Because the activity of collagenase can vary from lot-to-lot, researchers should be aware of the collagenase activity that they are using, and they should perform some preliminary experiments to confirm it is optimal for their experiments.
5. Multiple types of fine surgical forceps may be used. We recommend Dumont #5 Forceps Inox Tip Size 0.10 × 0.06 mm.
6. Tissue protein extraction reagent (T-PER) is a proprietary detergent from ThermoFisher, Pittsburgh, PA (PI78510). Because we have not tested other detergents, we do not know which other detergents may also be suitable.
7. The volume of medium that should be added to each vial will depend on the size of vial being used. There should be enough medium to fully cover the muscle. We use 2 mL of medium in 20 mL glass scintillation vials. The size of these vials is relatively small and allows for a small volume of medium (2 mL), which reduces the cost of reagents. Additionally, when small vials are

used then a larger number of vials can be placed in the shaking water bath, which makes larger experiments possible.

8. A gas washing bottle including deionized water is used to ensure humidification of gasses, and thereby minimize evaporation of the incubation medium during gassing. The length of microbore tubing which connects the manifold to each vial is equal length (12 in.) to ensure equal gas flow to all vials. The needle that is attached to the microbore tubing and placed in each vial serves as a rigid tube that will not be constricted by the rubber stopper, allowing gas to flow into the vial. When positioned inside the vial, and held in place by inserting the rubber stopper, the hub of the needle should remain above the surface of the incubation medium in the vial, to maintain an oxygen-enriched atmosphere above the medium's surface. Bubbling the gas directly into medium that contains bovine serum albumin would cause problems, including foaming of the medium as well as cross-contamination between the needle and the medium.
9. Placing the vials which contain 2.5% type II Collagenase Medium into the 35 °C heated water bath at least 5 min prior to the transfer of muscles into these vials ensures that the muscle will be incubated at the temperature for appropriate collagenase activity.
10. It is important to transfer the muscles from one vial to another vial by gently grasping the muscle with Gerald forceps by the connective tissue on either end of the muscle to minimize damage to the muscle fibers.
11. Shaking vials during muscle incubations is important for maintaining a more homogenous oxygen gradient and for helping to dissociate/ dilute metabolic waste products away from the incubating muscle.
12. For incubation **step 1** we have had successful results using incubation times ranging from 10 to 30 min. For experiments that have evaluated 2-DG uptake immediately following exercise the **step 1** incubation was shortened to 10 min to measure [^{3}H]-2-DG uptake (incubation **step 2**) as soon after exercise as possible (because the insulin-independent effect of 2-DG uptake is relatively transient). When evaluating insulin-stimulated 2-DG uptake, the duration of **step 1** in other experiments has typically been 20–30 min.
13. Similar to incubation **step 1**, the duration of incubation **step 2** may be altered depending on the experiment. We have successfully used incubations of 30–60 min in duration. Longer duration will enable greater accumulation of [^{3}H]-2-DG, and we have used 30 or 60 min for both basal and insulin-treated conditions. However, we have used 30 min duration for the

insulin-independent effect immediately after ex vivo contractions or in vivo exercise because of the relatively transient nature of the effects of these treatments.

14. We have found that 2.5% type II collagenase allows for effective enzymatic digestion of collagen using the rat epitrochlearis muscle. In our initial experiments using 1.5% collagenase, MHC type I fibers were rarely isolated. It is possible that greater collagen levels surrounding type I fibers require a greater concentration of collagenase for proper digestion. In subsequent experiments using 2.5% collagenase, we have succeeded in increasing the number of type I fibers isolated from the epitrochlearis muscle. If future researchers aim to use other rat skeletal muscles, they may need to perform exploratory analysis to identify the optimal collagenase concentration.

15. It is important to trim the adjacent connective tissue on the proximal and distal ends of the epitrochlearis as closely to the end of the epitrochlearis fibers as possible to minimize damage to the fibers and loss of intracellular [^{3}H]-2-DG.

16. It is not recommended to grasp fiber bundles by the connective tissue. The fiber bundles are very fragile and fibers will easily tear if grasped by the proximal or distal connective tissue. We pour Collagenase Medium into a 30 mm culture dish to easily remove fiber bundles with a stainless steel laboratory spatula. This approach also minimizes the debris that is transferred with the fiber bundles.

17. Fibers can easily adhere to pipette tips or the surface tension of the meniscus. To avoid this problem, flush the pipette tip once with Ca^{2+}-free KHB solution and then draw up ~5 μL into the pipette tip before drawing up the remaining ~15 μL with the muscle fiber into the pipette tip. Eject the fiber and Ca^{2+}-free KHB into the microcentrifuge tube quickly or the fiber will adhere to the side of the pipette tip. Not gassing KHB (which will result in a high pH) or not adding 0.1% BSA will also cause issues with fibers adhering to forceps and pipette tips.

18. Because glucose uptake and MHC isoform identification require 40 and 18 μL of sample respectively, there will be 42 μL of sample remaining (42% of the lysed fiber) following these procedures. The remaining sample can be used for blotting for the abundance of proteins of interest. However, we have found that some proteins (GLUT4, MTCO1, and COXIV) are affected by heating. If western blotting is to be performed on proteins affected by heating, then it is recommended that an aliquot of the lysed fiber sample is removed before heating.

19. The "total ^{3}H dpm per fiber" value is determined by taking the dpm measured in each fiber lysate (**step 5** of Subheading 3.4)

and subtracting the average background dpm (**step 4** of Subheading 3.4). The resulting value should be divided by 0.4 to account for the 40% (or 40 μL of sample, **step 1** of Subheading 3.4) of the fiber that was used for 2-DG uptake assay. The "μmol 2-DG per mL incubation medium" represents the 0.1 μmol of 2-DG used in the incubation medium (from **item 5** of Subheading 2.1). The "^{3}H dpm per mL incubation medium" represents the average dpm taken from the quadruplicate incubation medium samples in **step 5** of Subheading 3.4, and this value should be multiplied by 50 due to the dilutions performed in **step 3** of Subheading 3.4.

20. In the original publication, we expressed glucose uptake relative to the estimated value for fiber volume ($V = \pi r^2 l$) using the mean width (mean value of the width measured at three locations per fiber) and the length of each fiber [13]. However, reproducibility in the measurement of muscle width was challenging. When two experienced researchers analyzed an identical set of fiber images using this procedure the coefficient of determination of estimated fiber volume was $r^2 = 0.9317$. Subsequently, we modified and improved our method for determining fiber size using ImageJ by calculating fiber area based on the traced perimeter of each fiber [15]. When two experienced researchers analyzed an identical set of fiber images to determine fiber area the coefficient of determination was $r^2 = 0.9885$. Because the fiber area is a more reliable measure of fiber size than estimate fiber volume, we recommend that glucose uptake be expressed relative to fiber area rather than estimated fiber volume.

21. Because of the reflective glare from the microscope lighting and the uneven brightness of captured images across the length of the fiber, we have been unable to create an automated method for measuring fiber area that is as reproducible and accurate as manual tracing. Development of an accurate automated thresholding and fiber area analysis method would save time and eliminate the tedious, but crucial step of manual fiber tracing.

22. Accuracy and reliability of fiber tracing is essential for meaningful quantification, and the training and quality control of fiber tracing is crucial. When a new researcher performs these techniques it is important to confirm the reproducibility of fiber area measurements within each researcher. We have a set of imaged fibers that each new researcher will trace multiple times to confirm internal reproducibility. Furthermore, after the trainee establishes his/her internal consistency, his/her results are compared to the values obtained by experienced researchers using the same images. We also recommend spot checking fiber tracing accuracy by having two experienced

researchers independently trace the same fibers. Researchers should be blinded to the group and treatment identities of the fibers being traced.

23. EDL:Soleus standard can be made in large volumes (~10–15 mL) and stored at −20 °C in aliquots. We have freeze-thawed and re-frozen these MHC standards up to four times without issue. We have not tested if more freeze-thaw cycles cause problems for the MHC standard.
24. The EDL:Soleus standard is used as a MHC isoform marker guide to identify fiber type of lysed fibers in adjacent lanes. The MHC isoform expression in rat EDL is <5% type I, ~10% type IIA, ~30% type IIX, and ~55% type IIB; and in rat soleus is ~90% type I, ~10% type IIA, and <1% types IIX and IIB [22]. By loading a sample containing EDL:Soleus standard beside wells containing single fiber lysates, the fiber type can be determined based on the MHC migration in the EDL: Soleus well (Fig. 3 *above*).
25. The EDL:Soleus standard is loaded into three different lanes with spacing between the standard lanes to ensure accurate fiber typing of single fiber lysates. Some MHC isoforms migrate relatively close to one another (for example: type IIA and type IIX). It can be difficult to determine the fiber type if an EDL:Soleus standard is loaded in a lane that is a number of lanes from the sample. For this reason we recommend that the EDL:Soleus standard is loaded in multiple lanes.
26. The distinct separation of MHC isoforms in the EDL:Soleus standard is crucial for determining fiber type. We have found that running the MHC (~220 kDa) approximately 2/3 of the way down the gel allows for optimal separation of these bands.
27. After radioactive samples are loaded in the gels, the 1× running buffer will now be contaminated. It is essential to dispose of this buffer in the appropriate radioactive waste container. Additionally, sufficiently rinse the SDS-PAGE apparatus with water for decontamination and dispose of rinsing water in the appropriate radioactive waste container.
28. Typically MHC bands are stained by the SimplyBlue™ SafeStain within 1 h, but if less protein is added then staining may require up to 2 h. Coomassie Stain can also be used for this procedure, but requires a longer duration of staining with more steps.
29. Typically destaining in water takes 2 h, but gels can be left in shaking water for >48 h and the MHC protein bands will remain stained.

Acknowledgments

This work was supported by grants from the National Institutes of Health (R01 DK71771 and R01 AG10026). We thank Jim MacKrell and Haiyan Wang for providing valuable suggestions about the manuscript.

References

1. Pette D, Staron RS (1990) Cellular and molecular diversities of mammalian skeletal muscle fibers. Rev Physiol Biochem Pharmacol 116:1–76
2. Pandorf CE, Garland T, Aoi W, Handschin C, Gorza L, Garcia-Roves PM, Copp SW, Tipton CM, Caiozzo VJ, Haddad F (2010) A rationale for SDS-PAGE of MHC isoforms as a gold standard for determining contractile phenotype. J Appl Physiol 108(1):222–225
3. Fitzsimons DP, Diffee GM, Herrick RE, Baldwin KM (1990) Effects of endurance exercise on isomyosin patterns in fast-and slow-twitch skeletal muscles. J Appl Physiol 68 (5):1950–1955
4. Röckl KS, Hirshman MF, Brandauer J, Fujii N, Witters LA, Goodyear LJ (2007) Skeletal muscle adaptation to exercise training: AMP-activated protein kinase mediates muscle fiber type shift. Diabetes 56(8):2062–2069
5. Korhonen MT, Cristea A, Alén M, Häkkinen K, Sipilä S, Mero A, Viitasalo JT, Larsson L, Suominen H (2006) Aging, muscle fiber type, and contractile function in sprint-trained athletes. J Appl Physiol 101 (3):906–917
6. Nilwik R, Snijders T, Leenders M, Groen BB, van Kranenburg J, Verdijk LB, van Loon LJ (2013) The decline in skeletal muscle mass with aging is mainly attributed to a reduction in type II muscle fiber size. Exp Gerontol 48 (5):492–498
7. Hansen PA, Nolte LA, Chen MM, Holloszy JO (1998) Increased GLUT-4 translocation mediates enhanced insulin sensitivity of muscle glucose transport after exercise. J Appl Physiol 85(4):1218–1222
8. Lee AD, Hansen PA, Holloszy JO (1995) Wortmannin inhibits insulin-stimulated but not contraction-stimulated glucose transport activity in skeletal muscle. FEBS Lett 361 (1):51–54
9. Wallberg-Henriksson H, Constable S, Young D, Holloszy J (1988) Glucose transport into rat skeletal muscle: interaction between exercise and insulin. J Appl Physiol 65 (2):909–913
10. Wojtaszewski JF, Higaki Y, Hirshman MF, Michael MD, Dufresne SD, Kahn CR, Goodyear LJ (1999) Exercise modulates postreceptor insulin signaling and glucose transport in muscle-specific insulin receptor knockout mice. J Clin Invest 104(9):1257–1264
11. Funai K, Cartee GD (2009) Inhibition of contraction-stimulated AMP-activated protein kinase inhibits contraction-stimulated increases in PAS-TBC1D1 and glucose transport without altering PAS-AS160 in rat skeletal muscle. Diabetes 58(5):1096–1104. https://doi.org/10.2337/db08-1477
12. Wright DC, Hucker KA, Holloszy JO, Han DH (2004) Ca2+ and AMPK both mediate stimulation of glucose transport by muscle contractions. Diabetes 53(2):330–335
13. Mackrell JG, Cartee GD (2012) A novel method to measure glucose uptake and myosin heavy chain isoform expression of single fibers from rat skeletal muscle. Diabetes 61 (5):995–1003. https://doi.org/10.2337/db11-1299
14. Deves R, Krupka R (1978) Cytochalasin B and the kinetics of inhibition of biological transport. A case of asymmetric binding to the glucose carrier. Biochim Biophys Acta 510 (2):339–348
15. Cartee GD, Arias EB, Carmen SY, Pataky MW (2016) Novel single skeletal muscle fiber analysis reveals a fiber type-selective effect of acute exercise on glucose uptake. Am J Physiol Endocrinol Metab 311(5):E818–E824
16. Wang H, Arias EB, Yu CS, Verkerke AR, Cartee GD (2017) Effects of calorie restriction and fiber type on glucose uptake and abundance of electron transport chain and oxidative phosphorylation proteins in single fibers from old rats. J Gerontol Ser A: Biomed Sci Med Sci 72 (12):1638–1646
17. Pataky MW, Wang H, Yu CS, Arias EB, Ploutz-Snyder RJ, Zheng X, Cartee GD (2017) High-fat diet-induced insulin resistance in single

skeletal muscle fibers is Fiber type selective. Sci Rep 7(1):13642. https://doi.org/10.1038/s41598-017-12682-z

18. Castorena CM, Arias EB, Sharma N, Bogan JS, Cartee GD (2015) Fiber type effects on contraction-stimulated glucose uptake and GLUT4 abundance in single fibers from rat skeletal muscle. Am J Physiol Endocrinol Metab 308(3):E223–E230
19. Wallberg-Henriksson H (1987) Glucose transport into skeletal muscle. Influence of contractile activity, insulin, catecholamines and diabetes mellitus. Acta Physiol Scand Suppl 564:1–80
20. Delp MD, Duan C (1996) Composition and size of type I, IIA, IID/X, and IIB fibers and citrate synthase activity of rat muscle. J Appl Physiol 80(1):261–270
21. Mackrell JG, Arias EB, Cartee GD (2012) Fiber type-specific differences in glucose uptake by single fibers from skeletal muscles of 9- and 25-month-old rats. J Gerontol A Biol Sci Med Sci 67(12):1286–1294. https://doi.org/10.1093/gerona/gls194
22. Castorena CM, Mackrell JG, Bogan JS, Kanzaki M, Cartee GD (2011) Clustering of GLUT4, TUG, and RUVBL2 protein levels correlate with myosin heavy chain isoform pattern in skeletal muscles, but AS160 and TBC1D1 levels do not. J Appl Physiol (1985) 111(4):1106–1117. https://doi.org/10.1152/japplphysiol.00631.2011

Chapter 18

Myoblast Phosphoproteomics as a Tool to Investigate Global Signaling Events During Myogenesis

Fiona K. Jones, Gemma E. Hardman, Samantha Ferries, Claire E. Eyers, and Addolorata Pisconti

Abstract

Protein phosphorylation is a universal covalent chemical modification of amino acids involved in a large number of biological processes including cell signaling, metabolism, proliferation, differentiation, survival/death, ageing, and many more. Regulation of protein phosphorylation is essential in myogenesis and indeed, when the enzymatic activity of protein kinases is distrupted in myoblasts, myogenesis is affected. In this chapter we describe a method to profile the phosphoproteome of myoblasts using mass spectrometry. Phosphate groups are labile and easily lost during the processing of samples for mass spectrometry. Thus, effective methods to enrich for phosphopeptides from protein extracts have been developed. Here, we discuss and present in detail two such methods that we routinely employ. These methods are based on a sample enrichment step performed on titanium dioxide matrices followed by label-free tandem mass spectrometry and semi-quantitation.

Key words Myogenesis, Phosphoproteomics, Myoblasts, Muscle stem cells, Label-free proteomics, Phosphorylation

1 Introduction

Protein phosphorylation is the addition of a phosphate group to a protein and is catalyzed by a family of enzymes called protein kinases. Although the most commonly studied phosphorylation events are those involving serine, threonine, and tyrosine residues, in principle several other amino acids can be phosphorylated, including histidine, arginine, lysine, aspartate, glutamate, and cysteine [1]. Reversible protein phosphorylation is vital for the control of cellular processes and the adaptation of cells to their environment. The addition of a phosphate group to a protein can cause changes in conformation, localized charge, and electrostatic and hydrophobic interactions. Thus, protein phosphorylation modulates inter- and intra-cellular signaling, protein stability, protein binding, and subcellular localization [2, 3].

Sissel Beate Rønning (ed.), *Myogenesis: Methods and Protocols*, Methods in Molecular Biology, vol. 1889, https://doi.org/10.1007/978-1-4939-8897-6_18,

The phosphorylation cascades that are activated in response to extracellular stimuli, dominate the signal transduction landscape that drives and regulates myogenesis [4–7]. Additionally, the function and stability of a plethora of intracellular proteins involved in molecular mechanisms other than signal transduction are regulated by phosphorylation events, including ion channels, transcription factors, adaptor proteins, histones, and many others. Arguably, the most studied protein kinases involved in the regulation of myogenesis are those belonging to the mitogen-activated protein kinase (MAPK) family. MAPKs, such as the four isoforms of p38 and the extracellular signal-regulated kinases (ERKs), are activated by external stimuli such as fibroblast growth factors (FGFs), insulin-like growth factor-1 (IGF-1), hepatocyte growth factor (HGF) etc., and are involved in myogenesis by virtue of phosphorylation of a number of "downstream" kinases, transcription factors, and other regulators of gene expression [5, 8, 9].

Although much is known about the many functions of kinases and their substrates, much still remains to be understood of the global role played by phosphorylation in the process of myogenesis. Phosphoproteomics offers the opportunity to identify and characterize hundreds or even thousands of novel phosphorylation events in a high-throughput, temporal manner [10, 11]. In a mass spectrometry-based phosphoproteomics experiment, proteins are enzymatically digested into peptides, usually with the protease trypsin, which cleaves C-terminal to arginine and lysine residues (except if these are followed by a proline residue). As phosphorylation occurs at sub-stoichiometric levels, phosphorylated peptides must be enriched from a background of predominantly non-phosphorylated peptides to facilate their identification by liquid chromatography-tandem mass spectrometry (LC-MS/MS) [12].

Relative protein quantification using LC-MS/MS-based proteomics data can be achieved using either stable-isotope labeling, whereby individual biological samples are differentially labeled with stable isotopes or isotope-incorporated tags, or using "label-free" approaches [13, 14]. Incorporation of stable isotope labels allows multiple protein samples (e.g., cell extracts) to be combined prior to sample processing and LC-MS/MS analysis. From a single LC-MS/MS run, the sample origin of any given peptide can be differentiated based on the observation of a sample-specific mass (m/z) difference, which is dependent on the isotope label used for each sample. By contrast, in label-free proteomics, each sample is analyzed separately by LC-MS/MS, and protein abundance inferred using the relative signal intensities produced by observed constituent peptide ions. For quantitative phosphoproteomics, the relative abundance of a given phosphopeptide ion from condition-specific samples (e.g., over time and/or in response to external stimuli) is used to determine relative changes in phosphorylation. If this quantitative phosphoproteomics data is overlaid with total

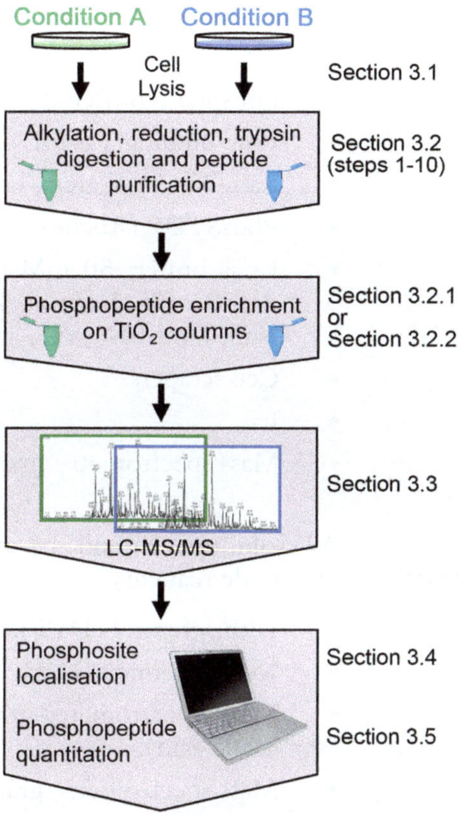

Fig. 1 Schematic workflow of a typical phosphoproteomics experiment. The main steps in a phosphoproteomics experiment are: cell lysis, (Subheading 3.1), sample preparation prior to enrichment (Subheading 3.2, **steps 1–10**); phosphopeptide enrichment (Subheading 3.2.1 or 3.2.2), LC-MS/MS (Subheading 3.3), phosphosite localization (Subheading 3.4), and phosphopeptide quantification (Subheading 3.5)

protein quantification, changes in phosphorylation stoichiometry can then be inferred [15–18].

In this chapter we describe a label-free method to quantitatively profile serine, threonine, and tyrosine phosphorylation events in myoblasts (Fig. 1). The method takes advantage of the high affinity electrostatic interaction of the negatively charged phosphate group with metal ions in metal oxide matrices, in this case titanium dioxide (TiO_2) [19, 20]. Selectivity for phosphopeptide binding is improved by inclusion of additives to reduce interaction of carboxyl groups with the TiO_2 matrix (Fig. 1) [21, 22].

2 Materials

2.1 Lysate Preparation

- Phosphate Buffered Saline (PBS).
- Low-binding 1.5 mL tubes.
- RapiGest™ (Waters, UK).
- PhosSTOP (Roche).
- Lysis buffer: 50 mM ammonium bicarbonate, 0.25% (v/v) RapiGest™. Keep at 4 °C and just before use add 2 mM NaF, 2 mM Na_3VO_4 and 1× PhosSTOP.
- Cell scrapers.
- Ice.
- Mass spectrometry-grade phosphate buffered saline (PBS).

2.2 Sample Preparation for LC-MS/MS

All solutions should be prepared using ultrapure water and MS-grade reagents

- Dithiothreitol (DTT).
- Iodoacetamide (IAA).
- Mass spectrometry-grade trypsin (0.2 μg/μL in 50 mM acetic acid).
- Mass spectrometry-grade trifluoracetic acid (TFA).
- Mass spectrometry-grade bi-distilled water (ddH_2O).
- Mass spectrometry grade methanol.
- 25 mM ammonium bicarbonate (AmBic) made with HPLC grade H_2O.
- 60 mM DTT 9.2 mg/mL prepared in 25 mM AmBic.
- 180 mM IAA 33 mg/mL in 25 mM Ambic—Once prepared keep in the dark.

2.2.1 For TiO_2 Microtips Method

- C18 macro spin column.
- C18 loading/wash buffer: 1% (v/v) TFA in H_2O.
- C18 elution buffer: 80% acetonitrile, 1% (v/v) TFA.
- TiO_2 tip-columns (Titansphere PhosTiO, GL Sciences).
- TiO_2 loading buffer: 60% acetonitrile, 25% (v/v) lactic acid, 0.3% (v/v) TFA.
- TiO_2 wash buffer: 80% (v/v) acetonitrile, 0.4% (v/v) TFA.
- TiO_2 elution buffer 1: 5% (v/v) ammonium hydroxide in H_2O.
- TiO_2 elution buffer 2: 5% (v/v) pyrrolidine in H_2O.
- Solubilization buffer: 97% (v/v) H_2O, 3% (v/v) ACN, 1% (v/v) TFA.

2.2.2 For TiO_2 Bulk Beads Method

- TiO_2 5 μm bulk beads (Titansphere PhosTiO, GL Sciences).
- TiO_2 loading buffer: 80% (v/v) acetonitrile, 5% (v/v) TFA, 1 M Glycolic Acid.
- TiO_2 wash buffer 1: 80% (v/v) acetonitrile, 1% (v/v) TFA.
- TiO_2 wash buffer 2: 10% (v/v) acetonitrile, 0.2% (v/v) TFA.
- TiO_2 elution buffer 1: 1% ammonium hydroxide solution.
- TiO_2 elution buffer 2: 5% ammonium hydroxide solution.
- Solubilization buffer: 97% (v/v) H_2O, 3% (v/v) ACN, 1% (v/v) TFA.

2.3 LC-MS/MS

- Mass spectrometry-grade formic acid.
- Mass spectrometry-grade acetonitrile.
- Solution A: 0.1% formic acid in ddH_2O.
- Solution B: 99.9% acetonitrile, 0.1% formic acid.

2.4 Data Analysis

- PEAKS Studio 8.
- Progenesis QI.

3 Methods

3.1 Lysate Preparation

Culture myoblasts, whether primary myoblasts or cell line, as normal and according to your specific experimental design (*see* **Note 1**). When myoblasts have reached the desired confluence or the desired state (e.g., differentiation time point or stimulus concentration/time point) lyse cells (*see* **Note 2**). Lysates can be produced in two different ways:

3.1.1 Lysis Method 1

Not recommended for myoblasts that require plates coated with a proteinaceous substrate, such as mouse primary myoblasts:

1. Prechill PBS and eppendorf tubes to 4 °C.
2. Aspirate culture medium and move the plates to ice.
3. Wash twice with ice-cold PBS. After the second wash, tilt the plates and wait 10 s for any residual PBS to collect at the bottom of the plate. Aspirate residual PBS and lay the plate again horizontal on ice.
4. Add a small amount of ice-cold lysis buffer to the plate, drop by drop, ensuring that the entire plate is covered. We normally use 200 μL for a 10 cm plate; use smaller or greater amounts of lysis buffer according to plate/flask surface area. Gently rock the plate to further distribute the lysis buffer over the entire surface of the plate.

5. Using a cell scraper, firmly scrape the plate and collect lysis buffer into a prechilled, low-binding tube.
6. Incubate on ice for 30 min, vortexing every 10 min or, alternatively, incubate at 4 °C for 30 min, constantly rotating end-over-end.
7. Further homogenize cell lysates, on ice, by probe sonication for 10 s with amplitude set at 40% of 60 Hz, then rest samples for 30 s. Repeat this cycle three times in total. We use a microtip probe (6 mm) coupled to Sonics Vibracell VCX 130PB.
8. Centrifuge at 13,000 × *g* at 4 °C for 10 min (*see* **Note 3**).
9. Collect supernatant (or whole lysate if not centrifuged as in [8]) and split into two vials: In one collect a small amount (10–15 μL) to use for total protein quantification, in the other one collect the remainder of the lysate.
10. Snap-freeze both aliquots in liquid nitrogen and store at −80 °C until further use.

3.1.2 Lysis Method 2

Recommended for myoblasts that require plates coated with a proteinaceous substrate, such as mouse primary myoblasts:

1. Prechill PBS and eppendorf tubes.
2. Aspirate culture medium.
3. Wash twice with warm (37 °C) PBS. After the second wash, tilt the plates and wait 10 s for any residual PBS to collect at the bottom of the plate.
4. Add trypsin or other protease (we prefer to use collagenase for detachment of primary mouse myoblasts that we normally culture on gelatin-coated plates) and incubate for the required amount of time for that specific protease to induce cell detachment.
5. Add two volumes of culture medium and collect detached cells in a sterile, tissue culture-grade tube.
6. Wash the plate twice with PBS and add each wash to the same tube as in **step 5**.
7. Centrifuge at a speed appropriate for that cell type (we use 500 × *g* for primary mouse myoblasts) and for the appropriate amount of time.
8. Discard the supernatant and resuspend pellet in 1–2 mL of cold PBS (depending on cell numbers).
9. Add 10 mL of cold PBS and mix gently.
10. Repeat **steps 7–9** once.
11. Resuspend pellet in lysis buffer. We normally use 200 μL for a pellet obtained from a 10 cm plate, use smaller or greater amounts of lysis buffer according to plate/flask surface area.

12. Incubate on ice for 30 min, vortexing every 10 min or, alternatively, incubate at 4 °C for 30 min, constantly rotating end-over-end.
13. Further homogenize cell lysates, on ice, by probe sonication for 10 s with amplitude set at 40% of 60 Hz, then rest samples for 30 s. Repeat this cycle three times in total. We use a microtip probe (6 mm) coupled to Sonics Vibracell VCX 130PB.
14. Centrifuge at 13,000 × *g*, 4 °C for 10 min (*see* **Note 3**).
15. Collect supernatant (or whole lysate if not centrifuged as in **step 14**) and split into two vials: in one collect a small amount (10–15 μL) to use for total protein quantification, in the other one collect the remainder of the lysate.
16. Snap-freeze both aliquots in liquid nitrogen and store at −80 °C until further use.

3.2 Sample Preparation for LC-MS/MS

1. Quantify total protein content using your favourite method. We routinely use the bicinchoninic acid assay (BCA), which works well with many types of lysis buffer.
2. Add 200 μg of lysate to a new low-binding tube and dilute to 160 μL with 25 mM AmBic.
3. Add 10 μL of 60 mM DTT (final concentration of DTT is ~3 mM). Vortex, then incubate at 60 °C for 10 min. Cool to room temperature, then spin quickly to return the liquid to the bottom of the tube.
4. Add 10 μL of 180 mM IAA (final concentration of IAA is ~9 mM). Vortex, then incubate at room temperature in the dark for 30 min, then add additional DTT to bring the final DTT:IOA ratio to 1:2—this prevents potential over-alkylation of the sample.
5. Add 20 μL of 0.2 μg/μL MS-grade trypsin (this is a 50:1 protein:trypsin ratio).
6. Incubate overnight at 37 °C.
7. To hydrolyse RapiGest™, add 1 μL tri-fluoracetic acid (TFA) [final concentration of TFA 0.5% (v/v)] and incubate for 2 h at 60 °C.
8. Centrifuge at 17,000 × *g* for 30 min at 4 °C, then carefully move the supernatant to a new low-binding tube.
9. Concentrate supernatants by vacuum centrifugation for approximately 40 min. Volume should be reduced by around half prior to the addition of 150 μL of 1% (v/v) TFA in HPLC-grade H_2O (total sample volume should not exceed 300 μL).
10. Desalt peptides on C18 macro spin columns according to the manufacturer's protocol (we use Harvard Macro Spin Columns, but similar columns from other manufacturers would

be appropriate). Briefly, condition column with 500 μL each of methanol, then 1% (v/v) TFA in H_2O (twice), centrifuging at 110 × *g* to pass liquid through the column. Add the sample to the column, centrifuge as above and add the flow through back onto the column a total of five times. Wash the column with 500 μL 1% (v/v) TFA in H_2O and then elute the peptides with 500 μL 80% (v/v) acetonitrile, 1% (v/v) TFA. Dry samples to completion in a vacuum centrifuge; they can be stored in the freezer at this point prior to TiO_2 enrichment.

The following steps describe two alternative methods for phosphopeptide enrichment that we routinely use. Both enrichment strategies are based on the affinity of the negatively charged phosphate group for TiO_2 but use two different supports for the TiO_2 matrix: either microtips or bulk beads. The method based on microtips is simpler and optimized for small sample volumes. In contrast, the method based on bulk beads is slightly more laborious but also more flexible to be adapted to a range of sample volumes.

3.2.1 Method 1: TiO_2 Microtips

1. Enrich samples with phosphopeptides using columns packed with titanium dioxide [TiO_2, we use Titansphere PhosTiO columns, 200 μL (*see* **Note 4**)]. Place a centrifugal adapter into a tube and insert the tip through the adapter into the tube. Add liquid to the top of the tip in each step, using centrifugation to pass the liquid through the tip.
2. Solubilize dried peptides in 150 μL loading buffer (60% (v/v) ACN, 25% (v/v) lactic acid, 0.3% (v/v) TFA).
3. Prepare the TiO_2 tip by sequentially adding 20 μL wash buffer (80% (v/v) ACN, 0.4% (v/v) TFA) followed by 20 μL loading buffer, centrifuging at 3000 × *g* for 2 min each time, then remove the waste liquid from the tube.
4. Load 150 μL of sample onto the TiO_2 tip, centrifuge at 1000 × *g* for 10 min, then add sample from the tube back to the top of the tip and centrifuge again. Remove the liquid from the bottom of the tube. Phosphopeptides will now be bound to the TiO_2 material in the tip.
5. To wash, add 20 μL loading buffer, centrifuge at 3000 × *g* for 2 min, then add three lots of 20 μL wash buffer, centrifuging at 3000 × *g* for 2 min each time.
6. Transfer the TiO_2 tip and centrifugal adapter to a clean tube. Elute phosphopeptides by the addition of 50 μL of 5% (v/v) ammonium hydroxide (elution buffer 1) followed by 50 μL of 5% (v/v) pyrrolidine (elution buffer 2), with centrifugation at 1000 × *g* for 5 min each time. Dry the samples to completion by vacuum centrifugation. The samples can be stored −20 °C at this point.

7. Solubilize in 20 μL solubilization buffer. Centrifuge at 17,000 × *g* for 15 min, then transfer 10 μL to total recovery vials for LC-MS/MS analysis (*see* **Note 5**).

3.2.2 Method 2: TiO_2 Bulk Beads

1. Weigh 5 mg of TiO_2 beads for each 1 mg sample (we use Titansphere PhosTiO columns, 200 μL).
2. Solubilize dried peptides in 1 mL loading buffer and sonicate for 10 min in a water bath sonicator.
3. Add solubilized sample to the TiO_2 material and agitate on a Thermomixer at 1400 rpm (*see* **Note 6**) for 10 min at room temperature. Centrifuge at 2000 × *g* for 1 min and remove the supernatant.
4. Wash TiO_2 beads with 150 μL loading buffer and agitate on a Thermomixer at 1400 rpm (*see* **Note 6**) for 10 min at room temperature. Centrifuge at 2000 × *g* for 1 min and remove the supernatant.
5. Wash TiO_2 beads with 150 μL wash buffer 1 and agitate on a Thermomixer at 1400 rpm (*see* **Note 6**) for 10 min at room temperature. Centrifuge at 2000 × *g* for 1 min and remove the supernatant.
6. Wash TiO_2 beads with 150 μL wash buffer 2 and agitate on a Thermomixer at 1400 rpm (*see* **Note 6**) for 10 min at room temperature. Centrifuge at 2000 × *g* for 1 min and remove the supernatant.
7. Dry beads in a vacuum centrifuge for 10–30 min.
8. Perform the first elution step by adding 100 μL elution buffer 1 and agitate on a Thermomixer at 1400 rpm (*see* **Note 6**) for 10 min at room temperature. Centrifuge at 2000 × *g* for 1 min, remove the supernatant, and add to a fresh Eppendorf.
9. Perform the second elution step by adding 100 μL elution buffer 2 and agitate on a Thermomixer at 1400 rpm (*see* **Note 6**) for 10 min at room temperature. Centrifuge at 2000 × *g* for 1 min, remove the supernatant, and combine with the first elution solution.
10. Dry samples to completion by vacuum centrifugation at room temperature.

3.3 LC-MS/MS

For phosphoproteomics analysis a nano-HPLC coupled to a high-resolution mass spectrometer is appropriate. We use an Ultimate 3000 RSLC™ nano system coupled to a QExactive™ mass spectrometer (both from Thermo Scientific, Hemel Hempstead). The LC-MS/MS conditions and parameters that we describe here are a suggested starting point that the end user should further optimize on their own instrument.

Table 1
Suggested parameters for analysis of phosphopeptide-enriched samples on a QExactive mass spectrometer

Full MS	
Resolution	70,000
AGC target	1×10^6
Maximum fill time (ms)	250 ms
Scan range	350–2000 *m/z*
MS2	
Resolution	35,000
AGC target	1×10^5
Maximum fill time	100 ms
TopN	10
Isolation window	2 *m/z*
Additional settings	
Charge exclusion	Unassigned, 1, ≥6
Dynamic exclusion	20 s

1. Load 4 μL of the phospho-enriched sample onto the trapping column (we use a Thermo Scientific PepMap100, C18, 300 μm × 5 mm), for 7 min at a flow rate of 9 μL/min with 0.1% (v/v) TFA, 2% (v/v) acetonitrile (*see* **Note** 7).
2. Resolve samples on an analytical column (we use Easy-Spray C18 75 μm × 500 mm, 2 μm column) using a gradient of 97% (v/v) solution A and 3% (v/v) solution B to 60% (v/v) solution A and 40% (v/v) solution B over 90 min at a flow rate of 300 nL/min.
3. The LC column is connected in-line with the nano-electrospray ionization source of the mass spectrometer for analysis in positive ion mode. Suggested parameters for the QExactive mass spectrometer for data-dependent acquisition are detailed in Table 1.
4. Acquire data in data-dependent mode by choosing a program of at least 70,000 resolution full-scan MS scan with automatic gain control (AGC) set to 1×10^6 ions and a maximum fill time of 250 ms.
5. Set the 10 most abundant peaks to be selected for MS/MS using a 35,000 resolution scan (AGC set to 1×10^5 ions with a maximum fill time of 100 ms) and with an ion selection window of 2 *m/z*. To avoid repeated selection of peptides for

MS/MS, it is advisable to set a 20 s dynamic exclusion window, although this should be optimized based on the resolution of the LC separation.

3.4 Phosphopeptide Identification and Phosphosite Localization

Several programmes and workflows have been published for phosphopeptide identification [23], with often the best option dependent on the MS/MS acquisition parameters (MS/MS resolution) and the question to be addressed (*see* **Note 8**). One option is to use PEAKS Studio 8 to search against the species-specific UniProtKB protein database that applies to the samples. In this section we will describe a workflow that utilizes PEAKS Studio 8 for peptide identification and phosphosite localization.

1. In a new PEAKS Studio 8 project add the .raw files and specify the enzyme used, instrument and fragmentation type (In this case: trypsin, Orbi-Orbi and HCD).
2. Proceed to data-refinement. Use default settings to correct precursor mass and charge states between 2 and 5, and filter scans with quality value >0.65.
3. Proceed to identification, with settings as follows. Precursor tolerance: 10 ppm using monoisotopic mass; fragment ion tolerance: 0.01 Da; no nonspecific cleavage; Maximum 2 missed cleavages; fixed modifications: Carbamidomethyl (Cys); variable modification: Oxidation (Met); phosphorylation (Ser, Thr, and Tyr); maximum variable PTMs per peptide: 4.
4. Select the relevant protein database to search against. Tick the box to "estimate FDR with decoy-fusion," then click "Finish."
5. Once processed, filter results to obtain the desired peptide false discovery rate—we recommend selecting a 1% threshold (*see* **Note 9**).
6. The localization probability of all putative phosphorylation sites is reported in PEAKS Studio 8 as the A-score [24]. An A-score of 13 or more is equivalent to $P < 0.05$ or less and should result in the site being localized with 95% certainty or more [25]. An A-score of ≥ 19 ($P < 0.01$) should result in the site being localized with 99% certainty (Fig. 2).
7. Phosphopeptide lists generated by the PEAKS search can be exported for further statistical analysis, and integration with label-free peptide quantification that has been performed using alternative software (such as Progenesis QI) as required.

3.5 Label-Free Peptide Quantification

Although several programmes and workflows have been published for label-free quantification of phosphopeptide abundance, we primarily use Progenesis QI. The general workflow for this type of analysis is described in this section.

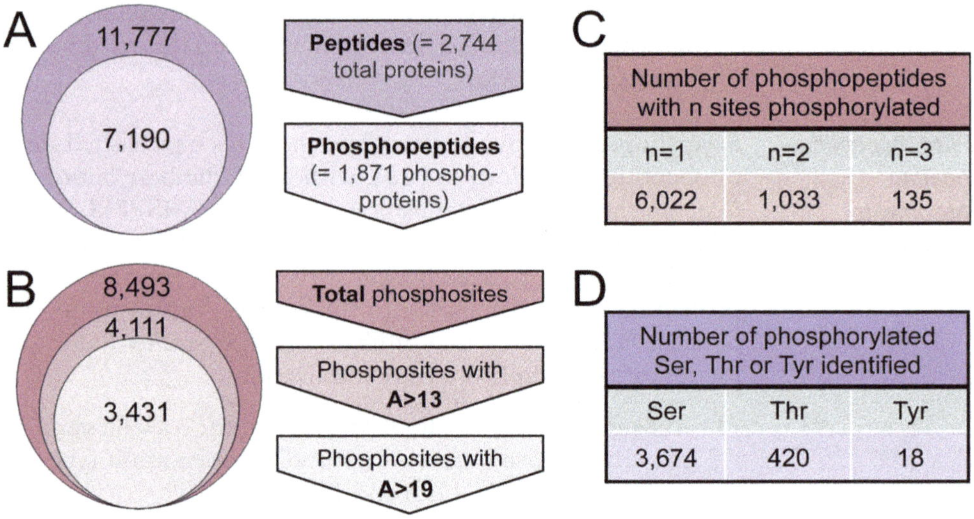

Fig. 2 Levels of phosphopeptide enrichment typically observed in a phosphoproteomic analysis of proliferating C2C12 myoblasts. Proliferating C2C12 myoblasts were lysed as described in Subheading 3.1.1, processed for phosphopeptides enrichment as described in Subheading 3.2.1, enriched lysates ran on a QExactive mass spectrometer as described in Subheading 3.3, analyzed for phosphosite localization as described in Subheading 3.4 and the identified phosphopeptides quantified as described in Subheading 3.5. (**a**) Schematic representation of the fraction of phosphorylated peptides (left: 7190) and phosphorylated proteins (right: 1871) within the total number of identified peptides (left: 11,777) and proteins (right: 2744). In this experiment the enrichment power was approximately 60%. Crucially, in the absence of any enrichment protocol only a very small portion of all the peptides identified by LC-MS/MS in a complex mixture such as a cell lysate is phosphorylated, often as low as 1%. (**b**) Schematic representation of the fraction of phosphosites assigned with good confidence (A-score > 13: 4111 sites) and with great confidence (A-score > 19: 3431 sites) within the total number of identified phosphosites (8493 sites). (**c**) Number of phosphopeptides identified containing 1, 2, or 3 phosphorylation sites. (**d**) Number of phosphorylated serine (Ser), threonine (Thr), and tyrosine (Tyr) residues identified (A-score > 13)

1. Create a “New Experiment”: Select “profile data” and indicate the mass spectrometer type, in this case for data generated with a QExactive™ mass spectrometer this is “Thermo FT-ICR.”
2. Import the .raw files and perform automatic processing. For alignment select the option to “assess all runs in the experiment for suitability.” This will compare the sample chromatograms, deduce which chromatogram is the most similar between all the chromatograms and designate it as reference sample. Alignment scores are considered of good quality when above 80%. Runs can be manually aligned in the software if necessary in order to achieve this score, although this should be done with caution.
3. Perform peak picking with default parameters and select the relevant protein level quantification method, in this case “relative using Hi-N,” where N is the number of the most abundant

peptide ions selected that have been assigned per protein (a value of 3 is usually appropriate).

4. Proceed to "Experiment Design Setup" and select "Between Subject Design." Select the option to "group runs manually" then assign samples to the relevant experimental conditions. Once these pre-processing steps are completed the software will quantify each feature (a given peak identified at a given *m/z* value) by measuring the area under the peak, reporting this as a numerical value.
5. Export the resulting peak list from Progenesis and perform a search with your desired search engine (you could use PEAKS as described above). Import the search results back into Progenesis in order for the software to match the quantified peaks to specific peptide identifications. Further refinement of peptide identifications can be performed in Progenesis, for example filtering based on score, mass error, sequence length, etc. The Progenesis software will provide protein-level quantification information.
6. For interrogation of phosphopeptide quantification data, export the results as a .csv file.
7. Further statistical analysis can be carried out using commercial or open-source statistical packages, or a bespoke-written script. Details on this further analysis exceeds the scope of this chapter, which aims to describe the mass-spectrometry part of a phosphoproteomics experiment, from sample preparation to phosphopeptide quantitation.

In conclusion, the method that we have described here is a powerful tool for studying the role of protein phosphorylation during myogenesis and understanding it at the global level.

4 Notes

1. For example, for routine maintenance of C2C12 myoblasts we use Dulbecco Modified Eagle's Medium (DMEM) supplemented with 10% foetal bovine serum, 1% penicillin/streptomycin, and 2 mM L-glutamine. We always maintain C2C12 proliferating myoblasts in polystyrene uncoated dishes at a confluence between 40% and 70% [26]. A 15 cm petri dish at ~70% confluence yields enough total protein extract for a phosphoproteomics experiment.
2. The choice of the lysis method is heavily dictated by the nature of the myoblasts. Primary myoblasts often require a proteinaceous substrate and it is highly likely that when lysing directly on the plate (or flask), large amounts of the substrate will be removed by vigorous scraping with a detergent-containing

buffer and will be collected in the sample. This will not only affect protein quantification, but is also very likely to be hyper-represented in the sample thus reducing the efficiency of phosphopeptide enrichment and the identification of low-abundance peptides. It is therefore recommended to detach cells and lyse them in a tube to prevent accumulation of coating proteins in the sample. However, cell detachment is normally performed using a protease, which may cleave membrane and pericellular proteins. Consequently, the ability to quantify such proteins may be compromised. Careful cost/benefit considerations for each specific case should be made at the stage of experimental design and preliminary experiments should evaluate the optimal lysis method to determine the final experimental design.

3. If the aim of the study is to profile membrane proteins or pericellular proteins, it is recommended to omit this step as a way to increase representation of membrane and pericellular proteins, which would otherwise be partly lost in the pellet.
4. Care must be taken with the TiO_2 tips to ensure that tip end does not touch the liquid in the bottom of the tube; remove the waste liquid after each step. The PhosTiO_2 kit contains "solution B" which is 100% lactic acid. Prepare solutions for the TiO_2 enrichment just before performing the experiment. Keep the flow-through from the TiO_2 enrichment (store in the freezer) until you are happy with the results of the experiment—the flow-through can be analyzed to check for a successful enrichment or could be enriched again if there were problems with the first enrichment. When possible, it is advisable to pass the flow-through on the column multiple times. Thorough washing of the TiO_2 tip with wash buffer (80% ACN, 0.4% TFA) is vital to remove residual lactic acid used during the binding step. If interfering lactic acid peaks are detected in the phosphopeptide samples increase the number of washes performed.
5. We do not recommend storing phosphopeptide-enriched samples in the freezer if they are not dried—in our experience the number of phosphopeptides identified by MS reduces drastically after even relatively short-term storage of frozen liquid samples. Samples that have been solubilized for MS analysis can be stored temporarily in the fridge in case a repeat injection is immediately required. We also do not recommend storing lysed cell pellets: this has a significant negative impact on phosphopeptide recovery.
6. The mixing speed on the instrument that we use (ThermoMixer, Eppendorf) is only expressed in rpm and cannot be easily converted to g.

7. An injection volume of 4 μL is a good starting point for the analysis of this type of sample, but we advise initially testing different injection volumes in order to choose the right amount of loading for a given experimental setup.
8. Alternative data-processing pipelines can be employed. For phosphopeptide analysis it is important to indicate the degree of confidence in the localization of phosphosites within a given peptide. PEAKS uses A-score for site-localization confidence: an A-score of 13 or more is equivalent to $P < 0.05$, and an A-score of 20 or more is equivalent to $P < 0.01$. Alternative site-localization tools include ptmRS and PTM-score (within MaxQuant) [23].
9. False Discovery Rate (FDR) is an important metric to report for all proteomics data. Calculation of FDR is generally achieved using the target-decoy strategy. According to this approach, the search is repeated with the exact same parameters except the protein database used is one of reversed or randomized sequences, such that there should be no matches between the experimental data set and this false (or "decoy") database. The number of matches obtained against this database is essentially an estimation of the number of false positives that are present in the results for the real (or "target") database search. FDR is thus reported as the number of false positive matches in the decoy database divided by the total number of matches in the target database. 1% FDR is a generally accepted cutoff for data of this type.

Acknowledgments

This work was supported by a PhD studentship to FKJ, GEH and SF from the Biological and Biotechnology Research Council, UK and by a Wellcome Trust ISSF and a Marie Curie IEF to AP.

References

1. Hardman G, Perkins S, Ruan Z, Kannan N, Brownridge P, Byrne DP, Eyers PA, Jones AR, Eyers CE (2017) Extensive non-canonical phosphorylation in human cells revealed using strong-anion exchange-mediated phosphoproteomics. bioRxiv
2. Cohen P (2002) The origins of protein phosphorylation. Nat Cell Biol 4(5):E127–E130. https://doi.org/10.1038/ncb0502-e127
3. Hunter T (2012) Why nature chose phosphate to modify proteins. Philos Trans R Soc Lond Ser B Biol Sci 367(1602):2513–2516. https://doi.org/10.1098/rstb.2012.0013
4. Mashinchian O, Pisconti A, Le Moal E, Bentzinger CF (2018) The muscle stem cell niche in health and disease. Curr Top Dev Biol 126:23–65. https://doi.org/10.1016/bs.ctdb.2017.08.003
5. Olguin HC, Pisconti A (2012) Marking the tempo for myogenesis: Pax7 and the regulation of muscle stem cell fate decisions. J Cell Mol Med 16(5):1013–1025. https://doi.org/10.1111/j.1582-4934.2011.01348.x
6. Pisconti A, Bernet JD, Olwin BB (2012) Syndecans in skeletal muscle development,

regeneration and homeostasis. Muscles Ligaments Tendons J 2(1):1–9

7. Arecco N, Clarke CJ, Jones FK, Simpson DM, Mason D, Beynon RJ, Pisconti A (2016) Elastase levels and activity are increased in dystrophic muscle and impair myoblast cell survival, proliferation and differentiation. Sci Rep 6:24708. https://doi.org/10.1038/srep24708
8. Ghadiali RS, Guimond SE, Turnbull JE, Pisconti A (2017) Dynamic changes in heparan sulfate during muscle differentiation and ageing regulate myoblast cell fate and FGF2 signalling. Matrix Biol 59:54–68. https://doi.org/10.1016/j.matbio.2016.07.007
9. Segales J, Perdiguero E, Munoz-Canoves P (2016) Regulation of muscle stem cell functions: a focus on the p38 MAPK signaling pathway. Front Cell Dev Biol 4:91. https://doi.org/10.3389/fcell.2016.00091
10. Huttlin EL, Jedrychowski MP, Elias JE, Goswami T, Rad R, Beausoleil SA, Villen J, Haas W, Sowa ME, Gygi SP (2010) A tissue-specific atlas of mouse protein phosphorylation and expression. Cell 143(7):1174–1189. https://doi.org/10.1016/j.cell.2010.12.001
11. Zhou H, Di Palma S, Preisinger C, Peng M, Polat AN, Heck AJ, Mohammed S (2013) Toward a comprehensive characterization of a human cancer cell phosphoproteome. J Proteome Res 12(1):260–271. https://doi.org/10.1021/pr300630k
12. Johnson H, Eyers CE (2010) Analysis of post-translational modifications by LC-MS/MS. In: Cutillas PR, Timms JF (eds) LC-MS/MS in proteomics: methods and applications. Humana Press, Totowa, NJ, pp 93–108. https://doi.org/10.1007/978-1-60761-780-8_5
13. Hogrebe A, von Stechow L, Bekker-Jensen DB, Weinert BT, Kelstrup CD, Olsen JV (2018) Benchmarking common quantification strategies for large-scale phosphoproteomics. Nat Commun 9(1):1045. https://doi.org/10.1038/s41467-018-03309-6
14. Hawkridge AM (2014) Chapter 1 practical considerations and current limitations in quantitative mass spectrometry-based proteomics. In: Quantitative proteomics. The Royal Society of Chemistry, Cambridge, pp 1–25. https://doi.org/10.1039/9781782626985-00001
15. Schmutz C, Ahrne E, Kasper CA, Tschon T, Sorg I, Dreier RF, Schmidt A, Arrieumerlou C (2013) Systems-level overview of host protein phosphorylation during Shigella flexneri infection revealed by phosphoproteomics. Mol Cell Proteomics 12(10):2952–2968. https://doi.org/10.1074/mcp.M113.029918
16. Kauko O, Laajala TD, Jumppanen M, Hintsanen P, Suni V, Haapaniemi P, Corthals G, Aittokallio T, Westermarck J, Imanishi SY (2015) Label-free quantitative phosphoproteomics with novel pairwise abundance normalization reveals synergistic RAS and CIP2A signaling. Sci Rep 5:13099. https://doi.org/10.1038/srep13099
17. Humphrey SJ, Yang G, Yang P, Fazakerley DJ, Stockli J, Yang JY, James DE (2013) Dynamic adipocyte phosphoproteome reveals that Akt directly regulates mTORC2. Cell Metab 17(6):1009–1020. https://doi.org/10.1016/j.cmet.2013.04.010
18. Casado P, Alcolea MP, Iorio F, Rodriguez-Prados JC, Vanhaesebroeck B, Saez-Rodriguez J, Joel S, Cutillas PR (2013) Phosphoproteomics data classify hematological cancer cell lines according to tumor type and sensitivity to kinase inhibitors. Genome Biol 14(4):R37. https://doi.org/10.1186/gb-2013-14-4-r37
19. Sano A, Nakamura H (2004) Titania as a chemo-affinity support for the column-switching HPLC analysis of phosphopeptides: application to the characterization of phosphorylation sites in proteins by combination with protease digestion and electrospray ionization mass spectrometry. Anal Sci 20(5):861–864
20. Pinkse MW, Uitto PM, Hilhorst MJ, Ooms B, Heck AJ (2004) Selective isolation at the femtomole level of phosphopeptides from proteolytic digests using 2D-NanoLC-ESI-MS/MS and titanium oxide precolumns. Anal Chem 76(14):3935–3943. https://doi.org/10.1021/ac0498617
21. Thingholm TE, Jorgensen TJ, Jensen ON, Larsen MR (2006) Highly selective enrichment of phosphorylated peptides using titanium dioxide. Nat Protoc 1(4):1929–1935. https://doi.org/10.1038/nprot.2006.185
22. Sugiyama N, Masuda T, Shinoda K, Nakamura A, Tomita M, Ishihama Y (2007) Phosphopeptide enrichment by aliphatic hydroxy acid-modified metal oxide chromatography for nano-LC-MS/MS in proteomics applications. Mol Cell Proteomics 6(6):1103–1109. https://doi.org/10.1074/mcp.T600060-MCP200
23. Ferries S, Perkins S, Brownridge PJ, Campbell A, Eyers PA, Jones AR, Eyers CE (2017) Evaluation of parameters for confident phosphorylation site localization using an orbitrap fusion tribrid mass spectrometer. J Proteome Res 16(9):3448–3459. https://doi.org/10.1021/acs.jproteome.7b00337

24. Beausoleil SA, Villen J, Gerber SA, Rush J, Gygi SP (2006) A probability-based approach for high-throughput protein phosphorylation analysis and site localization. Nat Biotechnol 24(10):1285–1292. https://doi.org/10.1038/nbt1240
25. Paulo JA, McAllister FE, Everley RA, Beausoleil SA, Banks AS, Gygi SP (2015) Effects of MEK inhibitors GSK1120212 and PD0325901 in vivo using 10-plex quantitative proteomics and phosphoproteomics. Proteomics 15(2–3):462–473. https://doi.org/10.1002/pmic.201400154
26. Flamini V, Ghadiali RS, Antczak P, Rothwell A, Turnbull JE, Pisconti A (2018) The satellite cell niche regulates the balance between myoblast differentiation and self-renewal via p53. Stem Cell Rep 10(3):970–983. https://doi.org/10.1016/j.stemcr.2018.01.007

Chapter 19

Preparation and Culturing of Atlantic Salmon Muscle Cells for In Vitro Studies

Tone-Kari K. Oestbye and Elisabeth Ytteborg

Abstract

This chapter outlines methods for isolating myosatellites from Atlantic salmon (*Salmo salar*), how to keep them in culture and differentiate them into mature myocytes. The protocol further describes how to trans-differentiate the myocytes into osteoblasts (bone cells).

Key words In vitro, Atlantic salmon, Muscle cell, Differentiation, Myosatellite, Trans-differentiation

1 Introduction

Fish, as mammals, contain three types of muscle tissue: cardiac muscle, smooth muscle, and skeletal muscle. The skeletal muscle builds up the myotomal muscle, fin muscle, and muscle in head- and gill-region. The myotomal muscle is the most abundant tissue in fish, comprising between 50% and 70% of the body mass in most fish species, and 65% in salmon [1]. The myotomal muscle consists of different fiber types located at specific parts of the fish body, in contrast to other vertebrates, which have the fiber types mixed together within each muscle. The superficial red muscle layers on either side of the fish body are composed of slow fibers, contain a high-volume density of mitochondria and blood capillary network, and are used during slow speed swimming. The white muscles comprise more than 90% of the myotome in most fish species. The white muscles are composed of fast muscle fibers, dependent on anaerobe metabolic pathways. An intermediate pink muscle fiber type, in the zone between the white and red fibers, is recruited as the swimming speed increases and possess intermediate characteristics compared to the red and white muscle fibers. In fish, the muscle grow by both hyperplasia (increase in number of cells) and hypertrophy (increase in size of individual fibers) [2]. This contrasts

Sissel Beate Rønning (ed.), *Myogenesis: Methods and Protocols*, Methods in Molecular Biology, vol. 1889,
https://doi.org/10.1007/978-1-4939-8897-6_19,

with mammalian muscle, where hyperplastic growth is less important after birth.

In fish, three different phases of skeletal muscle development are identified [3]. During the first embryonic phase, muscle fibers and a population of undifferentiated myosatellite cells are formed [4]. These myoblasts are used throughout lifetime for new fiber recruitment and hypertrophy. After the early embryonic phase of myogenesis, muscle fibers are formed from the germinal zone at the dorsal and ventral extremities of the myotome in salmon [5]. This results in a gradation of muscle fiber diameter, which increases away from the germinal zones toward the central part of the myotome [6]. This regionalized process termed "stratified hyperplasia" begins in the late embryo stages [7] and continues until after hatching [5]. The final phase of myogenesis, in both red and white muscles, starts before all the yolk has been internalized and continues until sexual maturity [8]. This phase persists throughout lifetime, and involves both hyperplasia and hypertrophy [1].

A number of regulatory mechanisms and well-orchestrated transcription of genes are important for muscles to grow properly. Members of the MyoD family of muscle regulatory factors (MRFs) together with the myocyte-enhancer factor 2 (MEF2) family members are essential in muscle development, both in the commitment and differentiation of mesoderm cells [9]. The primary MRFs MyoD (myf3) and myf5 are required for myogenic determination, whereas the secondary MRFs myogenin (myf4) and MRF4 (myf6/herculin) are downstream transcription factors involved in differentiation [10]. Mammalian studies have shown that Myf5, MyoD, and MEF2 are all involved in regulation of myostatin, which is a negative regulator of muscle growth [11, 12]. Three functional myostatin genes are cloned from Atlantic salmon, and the differential expression pattern indicates several functional roles of myostatin in salmon than in mammals [13, 14].

Like in other fish species, muscle fibers in Atlantic salmon differentiate from myosatellite cells. These cells are latent in the muscles of the fish and are activated upon growth or repair. The concentration of myosatellite cells is highest above the lateral line in salmon. Myosatellite cells isolated from the white skeletal muscle in this region differentiate into functional muscle fibers in culture in ca 3 weeks at 13 °C (Fig. 1) [15–17]. Temperature has a pronounced effect on fusion, differentiation, and hypertrophic growth of muscle cells [15, 17], and functional amino acids, such as arginine and glutamine, induce gene expression of muscle growth markers [17]. Cultured salmon muscle cells represent well-suited model systems for studies of, for example, lipid metabolism in fish muscle [16].

Mesenchymal stem cells (MSCs) have the capacity to differentiate into a number of cell types, including osteoblasts, chondroblasts, adipocytes, and myocytes [18–20]. The pluripotency of

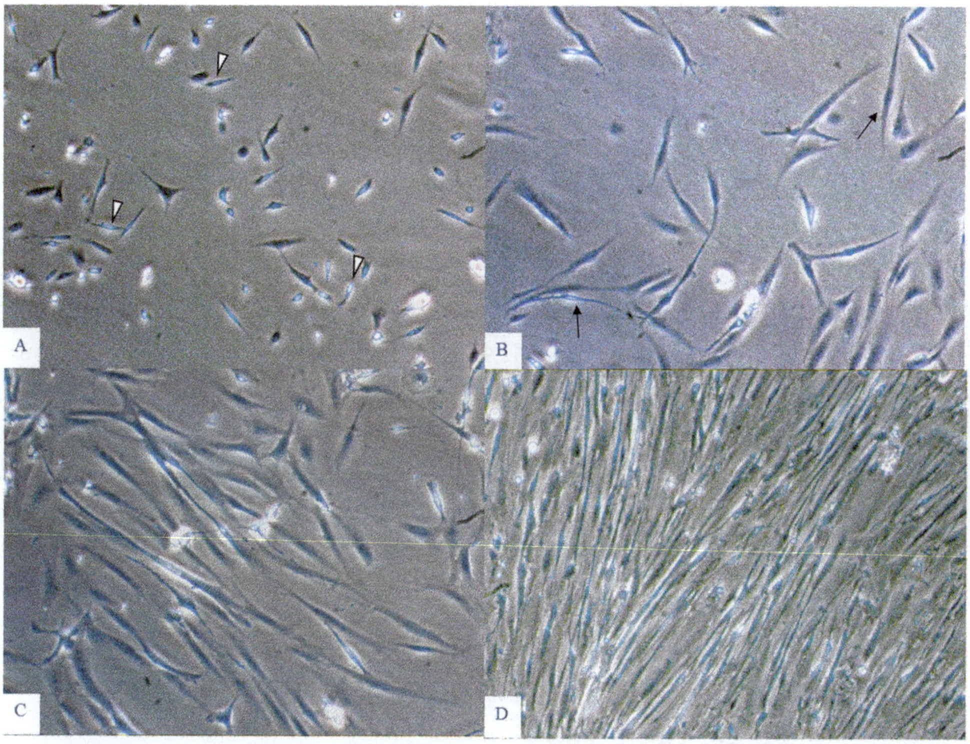

Fig. 1 Myosatellite cells isolated from white salmon skeletal muscle differentiate from myosatellite cells to organized multinuclear myotubes within 3 weeks. The figure shows muscle cells at (**A**) day 1, (**B**) day 4, (**C**) day 8 and (**D**) day 18. Arrowheads: myoblasts, arrows: myotubes

these cells is reflected by the large quantity of regulatory factors controlling their fate. Trans-differentiation is the process where one determined cell type differentiates into another cell type. Trans-differentiation and cellular determination has been suggested to be involved in the disease states in Atlantic salmon, such as vertebral malformations [21, 22], and in healing processes, for example distraction of osteogenesis in rats [23, 24]. For Atlantic salmon, trans-differentiation is described in vitro for both myosatellite cells from muscle and preadipocytes from adipose tissue, differentiating into bone cells [21, 25].

2 Materials

2.1 Atlantic Salmon Primary Muscle Cell Isolation

1. Antibiotic-antimycotic solution.
2. Atlantic salmon (5–7 cm).
3. Aquarium pump.
4. Bright field microscope.
5. Cell culture flasks.

6. Cell culture plates.
7. Cell incubator.
8. Centrifuge.
9. Collagenase.
10. Crushed ice.
11. Erlenmeyer flasks.
12. Ethanol, 70%.
13. Fetal bovine serum (FBS).
14. Filter (100 μM, 40 μM, 20 μM).
15. HEPES.
16. LAF bench.
17. Laminin.
18. L-15.
19. L-15 GlutaMAX.
20. Magnetic stirrer.
21. Pincher.
22. Pipettes (100 μL).
23. Pipette boy.
24. Police rubber.
25. Scalpel.
26. Scissor.
27. Sterile water.
28. Styrofoam box.
29. Thermometer.
30. Tank, 20 L.
31. Trypsin.
32. Tubes, 50 mL.
33. Water bath.
34. Prepare all sterile solutions in clean benches.
35. Autoclave all glassware and equipment, and use autoclaved water for all solutions used in the experiment.
36. L-15, 90%: Prepare a 90% dilution of L-15 by adding 45 mL of sterile water and 5 mL antibiotic-antimycotic solution (100×) to 450 mL L-15. Store at 4 °C until use.
37. Collagenase, 0.0005 U/mL: Use 5 mL collagenase solution (0.0005 U/mL) per gram muscle tissue. The collagenase solution must be prepared just before use. Weigh out 0.134 U collagenase using sterile equipment. Add 250 mL L-15, 90% containing antibiotics-antimycotic solution.

38. Trypsin, 0.1%: Use 5 mL trypsin (0.1%) per gram muscle tissue (*see* **Note 1**). Prepare the trypsin solution just before use by adding 4 mL trypsin solution (25 g/L in 0.9% NaCl) to 100 mL L-15, 90% added antibiotic-antimycotic solution.
39. FBS, 25%, in L-15: Prepare the FBS, 25%, by adding 75 mL FBS to 220 mL L-15 and 5 mL antibiotic-antimycotic (*see* **Notes 1** and **2**).
40. Growth media, 10% FBS: Prepare growth media by adding 50 mL FBS, 5 mL HEPES and 5 mL antibiotic-antimycotic (100×) to 440 mL L-15-GlutaMAX. Make fresh growth media every week.
41. Coating of culture ware: Precoat the culture ware (flasks, wells) with laminin.

2.2 Culturing of Atlantic Salmon Muscle Cells

1. Antibiotic-antimycotic.
2. Cell incubator.
3. FBS.
4. Growth media.
5. HEPES.
6. LAF bench.
7. L-15 GlutaMAX.
8. Growth media, 10% FBS: Prepare growth media by adding 50 mL FBS, 5 mL HEPES, and 5 mL antibiotic-antimycotic (100×) to 440 mL L-15 GlutaMAX. Make fresh growth media every week.

2.3 Trypsination of Atlantic Salmon Muscle Cells

1. Cell culture flasks.
2. Growth media.
3. LAF bench.
4. L-15 GlutaMAX.
5. Phosphate buffer saline (PBS).
6. Trypsin-EDTA.
7. Tubes, 50 mL.

2.4 Stimulation of Atlantic Salmon Muscle Cell Differentiation

1. Antibiotic-antimycotic.
2. Cell incubator.
3. HEPES.
4. Horse serum (HS).
5. L-15 GlutaMAX.
6. Differentiation media, 2% HS: Prepare differentiation media by adding 10 mL HS, 5 mL HEPES, 5 mL antibiotic-antimycotic 100× to 480 mL L-15 GlutaMAX L-15.

2.5 Trans-Differentiation of Atlantic Salmon Myosatellite Cells

1. Antibiotic-antimycotic solution.
2. Aquarium pump.
3. Atlantic salmon (5–7 cm).
4. B-glycerophosphate.
5. Bright field microscope.
6. $CaCl_2$.
7. Cell culture flasks.
8. Cell culture plates.
9. Cell incubator with CO_2.
10. Centrifuge.
11. Collagenase.
12. Crushed ice.
13. Dexamethasone.
14. DMEM.
15. 1.25-dihydroxyvitamin D3.
16. Erlenmeyer flasks.
17. FBS.
18. Filter (100 μM, 40 μM, 20 μM).
19. HEPES.
20. LAF bench.
21. Laminin.
22. L-ascorbic acid.
23. L-15.
24. L-15 Glutamax.
25. Magnetic stirrer.
26. Pincher.
27. Pipettes (100 μL).
28. Pipette boy.
29. Police rubber.
30. Scalpel.
31. Scissor.
32. Sterile water.
33. Styrofoam box.
34. Tank, 20 L.
35. Thermometer.
36. Trypsin.
37. Tubes, 50 mL.

38. Water bath.
39. Growth media: 435 mL DMEM added 50 mL FBS, 5 mL antibiotic-antimycotic solution (100×), 5 mL HEPES.
40. Bone differentiation media: DMEM growth media added 4 mM CaCl2,10 mM β-glycerophosphate, 150 μM L-ascorbic, 1 μM 1,25-dihydroxyvitamin D3, 10 nM dexamethasone.

3 Methods

3.1 Atlantic Salmon Primary Muscle Cell Isolation

1. Coating of culturing ware: Coat all plastic or glassware for cell culturing with laminin to facilitate attachment of myosatellite cells (*see* **Note 3**). Add 1.6 μL laminin per cm^2 culture area, typically 40 μL for 25 cm^2 cell flask.
2. Isolation of myosatellite cells: Use 80–100 salmon (approximately 5–7 cm) for one experiment with 24 wells (9.6 cm^2) of cells (*see* **Note 4**).
3. Go through **step 2–5** with one fish at a time.
4. Euthanize one fish at a time by a sharp blow to the head.
5. Remove mucus on the skin by a scalpel and dip the fish in ethanol.
6. Remove the skin from the gills and above the lateral line using scalpel and tweezers.
7. Cut out the epaxial white muscle and transfer to pre-weighed 50 mL tubes with L-15 with antibiotic-antimyocotic solution.
8. Cut the tissue into small pieces using scissors.
9. Centrifuge at 300 × *g* and 4 °C for 5 min (*see* **Note 5**).
10. Remove the supernatant.
11. Wash the tissue by addition of L-15 with antibiotic-antimycotic solution and centrifugation at 300 × *g* and 4 °C for 5 min.
12. Remove the supernatant.
13. Add collagenase solution (0.2%) to the tissue and transfer to a suitable Erlenmeyer flask with a magnet.
14. Place the flask in a Styrofoam box containing water at 11–13 °C. Monitor the temperature and adjust by adding ice to the sounding water if the temperature increases.
15. Put the Styrofoam box on a magnetic stirrer. Adjust the speed of the magnetic stirrer to keep a slow but constant movement of the solution. Incubate for 90 min.
16. Transfer the solution to 50 mL tubes and centrifuge at 300 × *g* and 4 °C for 5 min.
17. Remove the supernatant.

18. Wash the tissue twice by the addition of 30 mL L-15, 90%, to each tube and by gentle mixing inverting the tubes or by pipetting.
19. Centrifuge at 300 × *g* and 4 °C for 5 min.
20. Remove the supernatant.
21. Resuspend the pellet in trypsin solution (0.1%, 5 mL/g tissue) and transfer to a suitable Erlenmeyer flask with a magnet.
22. Place the flask in a Styrofoam box containing water at 11–13 °C. Monitor the temperature and adjust by adding ice to the surrounding water if the temperature increases.
23. Put the Styrofoam box on a magnetic stirrer. Adjust the speed of the magnetic stirrer to keep a slow but constant movement of the solution. Incubate for 30 min.
24. Centrifuge the cell suspension at 300 × *g* and 4 °C for 1 min to remove cell debris. The cells are now in the supernatant.
25. Stop the collagenase and trypsin treatment by diluting the supernatant containing the cells with FBS, 25%, in L-15 containing antibiotic-antimycotic. Add 15 mL FBS, 25%, to 30 mL cell suspension and mix by inverting the tubes.
26. Centrifuge the cell suspensions at 300 × g and 4 °C for 15 min.
27. Carefully remove the supernatant and resuspend the cell pellet in L-15 containing antibiotic-antimycotic solution.
28. Filter the cell suspension stepwise through 100 μm filter, 40 μm filter, and finally 20 μm filter.
29. Centrifuge the cell suspensions at 300 × *g* and 4 °C for 15 min.
30. Carefully remove the supernatant and resuspend the cell pellet in growth medium.
31. Repeat the last step by centrifugation of the cell suspensions at 300 × *g* and 4 °C for 15 min and removing of the supernatant.
32. Resuspend the cell pellet in desired volume of growth medium and seed out in cell culture ware (flasks, wells), e.g., add 20 mL growth medium and seed out in four cell flasks (25 cm^2) (*see* **Note 6**).
33. Incubate the cells at 13 °C overnight without CO_2.
34. Wash the cells from this first seeding carefully with L-15 containing antibiotic-antimycotic solution and add fresh growth media.
35. Save the growth medium containing un-attached cells from the first seeding for a second seeding by transferring the cell suspension to a 50 mL tube.
36. Centrifuge the cells at 300 × g and 4 °C for 15 min, remove the supernatant, and resuspend in 10 mL fresh growth media.

37. Seed out in two new cell flasks (25 cm^2) and incubate for 6 h.
38. Wash the cells from carefully with L-15 containing antibiotic-antimycotic solution and add fresh growth media.

3.2 Culturing of Atlantic Salmon Muscle Cells

1. Culture the muscle cells at 13 °C without CO_2. The isolated myoblasts will go through a proliferative phase followed by differentiation to mature multinuclear myotubes.
2. Let the myoblasts grow by proliferation for typically 1–-4 days before trypsination (*see* **Note 6**).
3. Evaluate the cells every day in microscope.
4. Wash the cells approximately every third day in L-15 and add fresh growth media.

3.3 Trypsination of Muscle Cells (see Note 6)

1. Remove the growth media and wash twice in PBS to remove FBS.
2. Add 0.5 mL trypsin-EDTA (for example Sigma T3924) to a cell flask (25 cm^2) and let it cover the whole cell culture area.
3. Follow the process in microscope and transfer the trypsinated cells to a tube containing growth medium to stop the trypsin (after ca. 2–5 min).
4. Repeat the process by adding 0.5 mL trypsin, follow the process in microscope, and transfer to growth medium.
5. Pool the trypsinated cells from all flasks and seed out in new cell culture ware.
6. Incubate the cells at 13 °C overnight without CO_2. Change the medium to fresh growth medium the next day.

3.4 Stimulation of Muscle Cell Differentiation

1. Wash the cells in L-15 (*see* **Note 7**).
2. Add fresh differentiation media (Fig. 1).

3.5 Trans-Differentiation of Atlantic Salmon Myosatellite Cells

1. Isolate myosatellite cells as described in Subheading 3.1.
2. Allow the myosatellite cells to adhere overnight in DMEM-based growth media at 13 °C with 5% CO_2.
3. Trypsinate the cells using 1 mL of stock solution, concentration 2.5 g/L, and re-seeded (day 0) in order to get rid of cell debris and to obtain cleaner cell cultures.
4. When the cells are 80–100% confluent, wash the cells twice in DMEM and add specific bone differentiation media.
5. Wash the cells every third day twice in growth media and add fresh bone differentiation media.
6. Observe the cells microscopically. Clusters of cells will form prior to mineralization (*see* **Note 8**).

4 Notes

1. When opening bottles with trypsin, FBS, HEPES, and antibiotics, aliquot in appropriate tubes and store at correct temperatures (4 °C or 20 °C).
2. Thaw FBS cold and in the darkness.
3. Allow the laminin to thaw at 4 °C before use. The coating should be done in the morning the same day as the isolation takes place.
4. Keep the fish in the lab using the same water as they were farmed in and with sufficient oxygen supply using an aquarium pump. Try to cover the fish so that they are in the dark and do not get too stressed.
5. Perform all centrifugation steps at 300 × *g* and 4 °C.
6. It is difficult to know exactly the yield of muscle cells, and to obtain a homogenous density of cells and cells in the same developmental stage in all the cell culture wares, all cells must be trypsinated after a period of proliferation. The myoblasts must be trypsinated and transferred to new cell culture ware before the cells have too many contact points and start to differentiate and fuse to multinuclear myotubes. When the cells are trypsinated they are developmentally put back.
7. The muscle cells will spontaneously start to mature and fuse to multinuclear myotubes when they grow and reach each other in the culture. However, differentiation of the muscle cells can be stimulated by changing to differentiation medium.
8. Trans-differentiated cells will start depositing extracellular matrix after approximately 3 weeks and start mineralizing the matrix after approximately 4 weeks. Matrix may be stained by alizarin red S (Fig. 2).

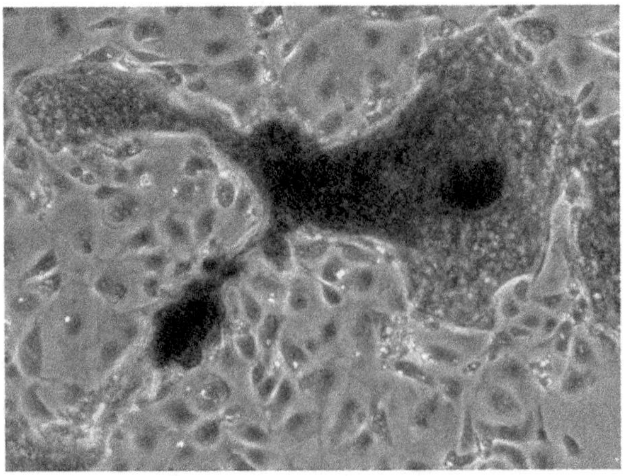

Fig. 2 Alizarin stained extracellular matrix in trans-differentiated bone cell culture

References

1. Johnston IA (2001) Genetic and environmental determinants of muscle growth patterns. Fish Physiol 18:141–186
2. Stickland NC (1983) Growth and development of muscle fibres in the rainbow trout (Salmo gairdneri). J Anat 137(Pt 2):323–333
3. Johnston IA (1999) Muscle development and growth: potential implications for flesh quality in fish. Aquaculture 177(1–4):99–115
4. Koumans J, Akster H, Dulos G, Osse J (1990) Myosatellite cells of Cyprinus carpio (Teleostei) in vitro: isolation, recognition and differentiation. Cell Tissue Res 261(1):173–181
5. Johnston IA, McLay HA (1997) Temperature and family effects on muscle cellularity at hatch and first feeding in Atlantic salmon (Salmo salar L.). Can J Zool 75(1):64–74
6. Rowlerson A (2001) Cellular mechanisms of post-embyonic muscle growth in aquaculture species. Fish Physiol 18:102–140
7. Stickland N, White R, Mescall P, Crook A, Thorpe J (1988) The effect of temperature on myogenesis in embryonic development of the Atlantic salmon (Salmo salar L.). Anat Embryol 178(3):253–257
8. Johnston IA, McLay HA, Abercromby M, Robins D (2000) Phenotypic plasticity of early myogenesis and satellite cell numbers in Atlantic salmon spawning in upland and lowland tributaries of a river system. J Exp Biol 203 (Pt 17):2539–2552
9. Watabe S (2001) Myogenic regulatory factors. Fish Physiol 18:19–41
10. Rudnicki MA, Jaenisch R (1995) The MyoD family of transcription factors and skeletal myogenesis. BioEssays 17(3):203–209
11. Spiller MP, Kambadur R, Jeanplong F, Thomas M, Martyn JK, Bass JJ, Sharma M (2002) The myostatin gene is a downstream target gene of basic helix-loop- helix transcription factor MyoD. Mol Cell Biol 22 (20):7066–7082
12. Salerno MS, Thomas M, Forbes D, Watson T, Kambadur R, Sharma M (2004) Molecular analysis of fiber type-specific expression of murine myostatin promoter. Am J Physiol Cell Physiol 287(4):C1031–C1040
13. Østbye TK, Galloway TF, Nielsen C, Gabestad I, Bardal T, Andersen Ø (2001) The two myostatin genes of Atlantic salmon (Salmo salar) are expressed in a variety of tissues. FEBS J 268(20):5249–5257
14. Østbye T-KK, Wetten OF, Tooming-Klunderud A, Jakobsen KS, Yafe A, Etzioni S, Moen T, Andersen Ø (2007) Myostatin (MSTN) gene duplications in Atlantic salmon (Salmo salar): evidence for different selective pressure on teleost MSTN-1 and-2. Gene 403 (1):159–169
15. Matschak T, Stickland N (1995) The growth of Atlantic salmon (Salmo salar L.) myosatellite cells in culture at two different temperatures. Experientia 51(3):260–266
16. Vegusdal A, Østbye TK, Tran TN, Gjoen T, Ruyter B (2004) Beta-oxidation, esterification, and secretion of radiolabeled fatty acids in cultivated Atlantic salmon skeletal muscle cells. Lipids 39(7):649–658
17. Østbye TK, Ruyter B, Standal IB, Stien LH, Bahuaud D, Dessen JE, Latif MS, Fyhn-Terjesen B, Rørvik KA, Mørkøre T (2018) Functional amino acids stimulate muscle development and improve fillet texture of Atlantic salmon. Aquac Nutr 24(1):14–26
18. Mizuno H (2009) Adipose-derived stem cells for tissue repair and regeneration: ten years of research and a literature review. J Nippon Med Sch 76(2):56–66
19. Pittenger MF, Mackay AM, Beck SC, Jaiswal RK, Douglas R, Mosca JD, Moorman MA, Simonetti DW, Craig S, Marshak DR (1999) Multilineage potential of adult human mesenchymal stem cells. Science 284 (5411):143–147
20. Pittenger M, Mosca J, McIntosh K (2000) Human mesenchymal stem cells: progenitor cells for cartilage, bone, fat and stroma. In: Lymphoid organogenesis. Springer, New York, pp 3–11
21. Ytteborg E, Vegusdal A, Witten PE, Berge GM, Takle H, Østbye TK, Ruyter B (2010) Atlantic salmon (Salmo salar) muscle precursor cells differentiate into osteoblasts in vitro: polyunsaturated fatty acids and hyperthermia influence gene expression and differentiation. BBA-Mol Cell Biol L 1801(2):127–137
22. Helland S, Denstadli V, Witten PE, Hjelde K, Storebakken T, Skrede A, Åsgård T, Baeverfjord G (2006) Hyper dense vertebrae and mineral content in Atlantic salmon (Salmo salar L.) fed diets with graded levels of phytic acid. Aquaculture 261(2):603–614
23. Yasui N, Sato M, Ochi T, Kimura T, Kawahata H, Kitamura Y, Nomura S (1997)

Three modes of ossification during distraction osteogenesis in the rat. J Bone Joint Surg 79 (5):824–830

24. Choi IH, Chung CY, Cho T-J, Yoo WJ (2002) Angiogenesis and mineralization during distraction osteogenesis. J Korean Med Sci 17 (4):435

25. Ytteborg E, Todorcevic M, Krasnov A, Takle H, Kristiansen IØ, Ruyter B (2015) Precursor cells from Atlantic salmon (Salmo salar) visceral fat holds the plasticity to differentiate into the osteogenic lineage. Biol Open 4 (7):783–791

Chapter 20

RNA Interference Screening for Genes Regulating *Drosophila* Muscle Morphogenesis

Aynur Kaya-Çopur and Frank Schnorrer

Abstract

RNA interference (RNAi) is the method of choice to systematically test for gene function in an intact organism. The model organism *Drosophila* has the advantage that RNAi is cell autonomous, meaning it does not spread from one cell to the next. Hence, RNAi can be performed in a tissue-specific manner by expressing short or long inverted repeat constructs (hairpins) designed to target mRNAs from one specific target gene. This achieves tissue-specific knock-down of a target gene of choice. Here, we detail the methodology to test gene function in *Drosophila* muscle tissue by expressing hairpins in a muscle-specific manner using the GAL4-UAS system. We further discuss the systematic RNAi resource collections available which also permit large scale screens in a muscle-specific manner. The full power of such screens is revealed by combination of high-throughput assays followed by detailed morphological assays. Together, this chapter should be a practical guide to enable the reader to either test a few candidate genes, or large gene sets for particular functions in *Drosophila* muscle tissue and provide first insights into the biological process the gene might be important for in muscle.

Key words *Drosophila*, Muscle, Sarcomere, RNAi, GAL4-UAS, Genetic screens, Off-target effects

1 Introduction

RNA interference (RNAi) offers a simple method to test for gene function in a developing organism by knocking down target mRNAs [1–3]. In contrast to the situation in *C. elegans*, the RNAi machinery acts cell autonomously in *Drosophila* [4]. This enables the researcher to perform tissue-specific knock-down of chosen target mRNAs using the inducible GAL4-UAS system. The availability of transgenic genome-wide UAS inverted repeat collections (UAS-RNAi, UAS-IR, UAS-hairpin) in *Drosophila* [5, 6] allows to test the function of almost any gene. The combination of these UAS-IR fly lines with the wide variety of GAL4 lines available in *Drosophila* facilitates tissue or cell-type specific knock-down at different developmental stages [3, 7].

Sissel Beate Rønning (ed.), *Myogenesis: Methods and Protocols*, Methods in Molecular Biology, vol. 1889, https://doi.org/10.1007/978-1-4939-8897-6_20,

Here, we provide all practical details to knock-down mRNAs in developing and adult *Drosophila* muscles. Depending on the selected GAL4 line this can be done either in all body muscles or in specific muscle subgroups only. The addition of the GAL4 repressor GAL80-temperature-sensitive (GAL80-ts) allows an even more precise timing of the RNA knock-down onset during development [8]. We give practical tips how to most effectively plan and perform large-scale screens in muscle, similar to those done successfully before [9–13], in particular how to effectively test for gene function by a combination of high-throughput assays and detailed but more time consuming muscle and sarcomere morphology and function assays. Together, this should enable the reader to quickly test any given gene for a function during *Drosophila* muscle development and also provide a starting point to identify the biological process the gene might be important for during muscle development.

2 Materials

Apart from the standard fly husbandry (fly food in small tubes and large bottles, stereomicroscope with CO_2 to sort flies, temperature and humidity controlled incubators to grow flies, brushes) we describe all the specific reagents and materials below to test gene function in *Drosophila* muscle by RNAi.

2.1 Fly Strains and Large Fly Collections

1. Large UAS-RNAi fly line collections exist at three different stock centers, the Vienna *Drosophila* Resource Center (VDRC) in Vienna (https://stockcenter.vdrc.at), the Bloomington *Drosophila* Stock Center (BDSC) in Bloomington, Indiana (https://bdsc.indiana.edu/), and the NIG collection in Kyoto (https://shigen.nig.ac.jp/fly/nigfly/). Stock centers require the users to register with a laboratory account and then ship stocks to any location in the world for a small processing fee. The VDRC equips each fly vial with a barcode, which can simplify fly handling in large-scale screens. For a comprehensive comparison of the RNAi libraries in terms of library preparation, coverage, logistics, and pricing *see* [3]. The Vienna and Kyoto collections contain mainly long inverted repeats (300–400 base pairs), whereas the TRiP collection found in Bloomington contains a large amount of short hairpins, based on the endogenous miR-1 scaffold [6]. The UAS-IR constructs of the TRiP collection are found in different UAS backbone vectors (Valium). Generally, the UAS-t-based vectors (Valium 10 and Valium 20) do work better than the UAS-p-based ones (Valium 22) for zygotic knock-down in muscle.

In our experience, it is not clear a priori if a short or a long hairpin does result in more efficient knock-down of the target mRNAs. Thus, we recommend testing all available hairpins for a given gene, if the function of this gene should be investigated.

2. Large scale fly line GAL4 collections are also available at the BDSC and the VDRC. Various muscle-specific GAL4 lines are available from these sources, however some other useful GAL4 lines for muscle-specific RNAi must be requested from specific laboratories. Table 1 provides a comprehensive list of useful muscle-specific GAL4 lines, including their expression characteristics and sources from where to potentially acquire them.
3. Transgenic or knock-in fluorescent fly marker strains and their sources as listed in Table 2. These lines are useful markers to visualize muscle components when assaying gene function after RNAi knock-down either using live imaging or also after fixation as antibodies against the tags can easily be used for visualization of the protein.

2.2 Reagents for Functional Muscle Tests

1. Flight test:
   (a) One meter long Plexiglas tube with a diameter of 8 cm, with 5 indicated subdivisions (20 cm each).
   (b) Water at the bottom in a Petri dish.
   (c) A funnel inserted at the top of the cylinder to drop down flies [36].
2. Spontaneous muscle contractions during development: a standard spinning disc or scanning confocal microscope is required for imaging [37].

2.3 Reagents for Muscle Morphology Assays

1. Live muscle protein markers listed in Table 2.
2. Antibodies against muscle and sarcomere proteins listed in Table 3.
3. PBS and PBT (PBS with 0.3% Triton-X).
4. Normal goat serum (NGS).
5. Relaxing solution: 20 mM phosphate buffer, pH 7.0; 5 mM $MgCl_2$; 5 mM EGTA, 5 mM ATP; 0.3% Triton X-100. (needs to be freshly prepared).
6. Fixing solution: 4% Paraformaldehyde (PFA) in PBS with 0.3% Triton X-100 or 4% PFA in relaxing solution.
7. Mounting medium for fixed samples, optionally with DAPI.
8. 50% glycerol for mounting larvae and live samples.
9. Microscope slides and cover slips.
10. Dissection materials:
    (a) Sharp scissors (Fine Science Tools, No. 15000-02),

Table 1
GAL4 drivers

GAL4 drivers	Expression	Source and type of line	Reference
1151-GAL4	Adult myoblasts; not very strong during late muscle differentiation	Enhancer trap (on X, close to *ct*)	Roy and Raghavan [14]
24B-GAL4	Muscles and embryonic tendons	Bloomington stock: 1767; enhancer trap (on 3rd, in *how*)	Brand and Perrimon [15]
Act79B-GAL4	Jump muscle (after fusion)	Enhancer transgene (on X)	Bryantsev et al. [16]
Act88F-GAL4	Very strong in differentiating indirect flight muscles (after fusion, from 14 h APF on)	Enhancer transgene (on X)	Bryantsev et al. [16]
duf-GAL4	Founder cells of embryonic muscles; muscles and tendons during pupal stages.	Bloomington stock: 66682; enhancer trap (on X, in *duf/kirre [rP298]*)	Menon and Chia [17]
him-GAL4	Adult myoblasts (AMPs, turns off after fusion)	Enhancer transgene (on 2nd or 3rd)	Spletter et al. [18]
Mef2-GAL4	All myoblasts and muscles	Bloomington stock: 27390; enhancer transgene (on 3rd)	Ranganayakulu et al. [19]
Mhc-GAL4	Very strong in all differentiating muscles (not in myoblasts)	Bloomington stocks 55132, 55133 (on X and 3rd); enhancer transgene (same enhancer as in *Mhc*-Tau-GFP)	Klein et al. [20]
salm-GAL4	Indirect flight muscles after fusion; also in many non-muscle cells	Enhancer trap (on 2nd, in *salm*)	Schönbauer et al. [21]
sns-GAL4	Fusion competent myoblasts (FCMs) of embryonic muscles (GAL4 will be present in myotube after fusion)	Enhancer transgene	Kocherlakota et al. [22]
sr-GAL4	All embryonic and larval tendon cells, developing tendons of pupal thorax	Bloomington stock: 26663; enhancer trap (on 3rd, in *stripe*); homozygous lethal	Usui et al. [23]

Table 2
Live markers

Live protein marker	Expression	Subcellular localization	Source and type of line	References
Endogenously tagged proteins				
Ilk-GFP (ZCL3192 and ZCL3111)	All muscles	Myotendinous junction	Protein-GFP trap; DGRC, Kyoto stocks: 110962 and 110951	Morin et al. [24]
Mhc[weeP26]-GFP	All muscles; in flight muscle labels only early isoforms of Mhc	Thick filaments	Protein-GFP trap	Orfanos and Sparrow [25]
Sls-GFP (G00053 and ZCL2144)	All muscles	Z-disc	Protein-GFP trap; DGRC, Kyoto stocks: 110573 and 110855. Homozygous Sls-GFP shows some sarcomere defects	Orfanos et al. [26]
Zasp52-GFP (G00189)	All muscles	Z-disc	Protein GFP-trap; DGRC, Kyoto stock: 110621	Katzemich et al. [27]
Zasp66-GFP (ZCL0663)	All muscles	Z-disc	Protein-GFP trap; DGRC, Kyoto stock: 110740	Katzemich et al. [27]
βPS-Integrin-GFP	All muscles	Myotendinous junction	C-terminal GFP fusion, knock-in	Klapholz et al. [28]
Protein expressed under large genomic or tissue-specific enhancer				
Act88F-GFP	Very strong in differentiating indirect flight muscles (after fusion, from 14 h APF on)	Thin filaments	VDRC, Fosmid transgene (fTRG78 and fTRG10028 on 3rd)	Sarov et al. [29]
Dlg1-GFP	Myoblasts and all body muscles	Membranes	VDRC, Fosmid transgene (fTRG502 on 3rd)	Sarov et al. [29]
Kettin-GFP	All muscles	Z-disc	VDRC, Fosmid transgene (fTRG569 on 3rd)	Sarov et al. [29]

(continued)

Table 2
(continued)

Live protein marker	Expression	Subcellular localization	Source and type of line	References
Mf-GFP	All muscles (Mf-Iso A, G, N)	Thick filaments	VDRC, Fosmid transgene (fTRG501 on 3rd)	Sarov et al. [29]
Obscurin/Unc89-GFP	All muscles	M-line	VDRC, Fosmid transgene (fTRG1046 on 3rd)	Sarov et al. [29]
Prm-GFP	All muscles	M-line or thick filaments	VDRC, Fosmid transgene (fTRG475 on 3rd)	Sarov et al. [29]
Talin-GFP (Rhea-GFP)	All muscles	Myotendinous junction	VDRC, Fosmid transgene (fTRG587on 3rd)	Sarov et al. [29]
TpnC25D-GFP	Leg and visceral muscles (not in flight muscles)	Thin filaments	VDRC, Fosmid transgene (fTRG1257 on 3rd)	Sarov et al. [29]
βTub60D-GFP	All body muscles	Microtubules	VDRC, Fosmid transgene (fTRG958 on 3rd)	Sarov et al. [29]
ap^{ME}-NLS::dsRED, *ap*^{ME}-GFP, *ap*^{ME}-NLS::eGFP	LT1-4 muscles in embryonic muscles	Cytoplasmic/nuclear	P-element transgene; marker under *apterous* mesodermal enhancer 680	Richardson et al. [30]
GFP-Talin (GFP-Rhea)	All muscles	Myotendinous junction	P-element transgene	Klapholz et al. [28]
him-GFP-Gma	Adult myoblasts (AMPs, turns off after fusion)	Actin binding	P-element transgene; marker under *him* enhancer (on 2nd and 3rd)	Spletter et al. [18]
Mhc-Tau-GFP	Very strong in all differentiating muscles (not in myoblasts)	Microtubules	P-element transgene; marker under Mhc enhancer (on X)	Chen and Olson [31]

(continued)

Table 2
(continued)

Live protein marker	Expression	Subcellular localization	Source and type of line	References
twist-GFP-Actin	All FCs and FCMs at embryonic stage (possibly also adult myoblasts)	Actin	P-element transgene; marker under twist enhancer	Richardson et al. [30]
Vinculin-GFP, Vinculin-RFP	All muscles	Myotendinous junction	P-element transgenes	Klapholz et al. [28]
UAS driven transgenes				
UAS-Cherry-Gma	GAL4 dependent	Actin binding	P-element transgene (on 3rd);	Millard and Martin [32]
UAS-GFP-Gma	GAL4 dependent	Actin binding	P-element transgene; Bloomington stocks: 31775 (on 2nd), 31776 (on 3rd)	Dutta et al. [33]
UAS-Lifeact-GFP	GAL4 dependent	Actin binding	Transgene; Bloomington stock: 35544 (on 2nd)	Hatan et al. [34]
UAS-Lifeact-Ruby	GAL4 dependent	Actin binding	Transgene; Bloomington stock: 35545 (on 2nd)	Hatan et al. [34]
UAS-mCD8-GFP	GAL4 dependent	Membranes	P-element transgene; multiple Bloomington stocks, e.g. 5137 (on 2nd)	Lee and Luo [35]

(b) Blunt forceps, sharp forceps (Dumont #5, Fine Science Tools),

(c) Glass embryo plates with lids,

(d) Silicon-filled Petri dish, insect pins,

(e) Double-sided tape,

(f) Microtome blade (PFM C35, No. 207500003),

(g) Transparent nail polish;

11. Custom-made imaging slide with a groove in the middle to fit the pupae.
12. Barcode reader and printer.
13. A standard point scanning confocal microscope for imaging.

3 Methods

3.1 Planning of Experiments—Testing Positive and Negative Controls

It is well known that RNAi can result in unwanted unspecific effects, mainly knock-down of other genes than the anticipated gene. These off-target effects strongly depend on the GAL4 line, the temperature, and the assay conditions used. Thus, it is recommended to choose a set of negative control genes, which have no role in muscle [10] to estimate how many of these would score positive in the assay chosen. This can be used to calculate the false positive rate (number of negative control genes scoring positive divided by all negative control genes tested). If the false positive rate is too large, the assay conditions need to be adjusted.

The false negative rate can be estimated by testing a set of positive control genes, which are known to play a role in the tested process (number of positive control genes scoring negative divided by all positive control genes tested). In particular for a large scale screen these numbers are important to calculate the false discovery rate (false positive rate divided by total hit rate) [3]. If the false negative rate is high, the co-expression of UAS-Dcr2 can be used to enhance RNAi activity [5]. However, this usually also increases the false positive rate. Together, these gene test sets can be used to optimally adjust RNAi knock-down efficiencies to the used GAL4 line and the particular process studied.

3.2 Fly Crosses for Muscle-Specific RNAi

All fly crosses should be performed in an incubator at defined temperature and humidity (about 70%). Note that GAL4 activity is higher at higher temperature. Thus, raising the temperature from the standard 25 °C to 27 °C or even 29 °C will increase the level of knock-down.

1. Collect virgin females from a chosen muscle-specific GAL4 line (*see* Table 1), which in addition may contain a fluorescently-tagged marker protein if in addition microscope-based evaluation is planned or UAS-Dcr2 if RNAi efficiency should be enhanced (*see* **Notes 1** and **2**).
2. Cross the collected virgin females with males from the UAS-IR line against the gene of interest. Use 3–6 virgin females per small tube and about 20 virgin females per large bottle with a similar amount of males.
3. Incubate in the incubator until the desired age and then perform the desired assay at larval, pupal, or adult stages (*see* **Note 3**).

3.3 High-Throughput Assays After RNAi-Based Gene Knock-Down: Lethality, Locomotion, and Flight Tests

If a gene is essential for muscle function, an efficient knock-down in muscle results in lethality. If the gene is only required for flight muscle function, its knock-down results in flightlessness. These simple assays can be easily used to test the function of many hundred genes in muscle [10].

1. Perform an RNAi cross as described in Subheading 3.2. Use barcodes if a high-throughput screen is done.
2. Assay after 10 days at 27 °C or after 14 days at 25 °C if adult progeny carrying GAL4 and UAS-IR construct are present (viable) or not (*see* **Note 4**). If not, check the vial for dead larvae or pupae to determine the lethality stage. If a lethality stage was identified, low-throughput assays (*see* Subheadings 3.4 and 3.5) can be used to investigate the muscle phenotype of the respective stage in more detail (*see* **Note 3**).
3. Locomotion assay:
   (a) If progeny of the cross are viable collect 30 males (*see* **Note 4**) and incubate them for another 7 days at 27 °C or 25 °C.
   (b) To roughly estimate locomotion, bang the flies down in the vial and estimate the speed of climbing compared to wild-type. Obvious locomotion defects can be easily identified by this high-throughput assay.
4. Flight test:
   (a) Use the same flies from **step 3** and drop them into the flight cylinder through the funnel at the top.
   (b) Score the zone in which the flies are landing. Wild-type flies will land in the upper 2 zones of the cylinder. Flightless animals will drop into the water at the bottom of the tube (*see* **Note 5**).
   (c) If more than 70% of the flies tested end in the water the flight muscle have a major functional defect and should be investigated in more detail by antibody stainings.
5. If a high-throughput screen is done all results should be collected using an electronic database, such as Filemaker or Microsoft Access. The use of barcodes and barcode scanners ensures the error-free entry of the fly line data.
6. Retest the line blindly if it scored positive in the high-throughput assay by feeding it again into the screen (**step 1**).
7. If an RNAi line scores reproducibly positive, additional independent RNAi lines targeting the same gene should be tested in the same assay. Ideally, at least 2 independent lines should score positive before starting a detailed phenotypic analysis (*see* Subheadings 3.4 and 3.5). This quality check efficiently eliminates unspecific off-target hits (*see* **Note 6**).

3.4 Low-Throughput Fixed Assays After RNAi-Based Gene Knock-Down: Whole Mount, Muscle Dissections, and Antibody Stainings

3.4.1 Larval Muscles

1. Visualize the larval muscle morphology in whole mount larvae with a fast heat fixation protocol. This requires having a fluorescent muscle marker in the genetic background (*see* Table 2) such as ZASP66-GFP (Protein-GFP trap), labeling Z-discs, or *Mhc*-Tau-GFP, labeling all the larval body wall muscles.
2. Depending on the lethality stage use blunt forceps to collect larvae after 2–5 days at 27 °C (L2 and L3 larvae are easier to handle).
3. Immobilize them by placing them into hot water (65 °C) for a few seconds.
4. Then mount them in 50% glycerol on a glass slide. Since the larvae are thick use coverslips as spacers between the slide and the coverslip located on the larvae.
5. Image the larvae using a low magnification objective (10× or 20×) of a fluorescent microscope or with a point scanning confocal at high resolution. More detailed muscle morphology and also neuromuscular junctions can be visualized by preparing larval filets and performing antibody stainings using either sarcomeric antibodies such as Mhc, α-Actinin or postsynaptic density antibodies such as Dlg1 [47–49].

3.4.2 Pupal or Adult Muscles

The histological dissection followed by antibody stainings for pupal or adult flight and abdominal muscles have been described in detail elsewhere [49]. Here, we briefly describe how to dissect and stain indirect flight muscles at pupal and adult stages. These assays can be used for example to visualize the actin cytoskeleton of sarcomeres (using phalloidin), the nuclei (using DAPI) together with other proteins of interests using antibodies listed in Table 3. These assays allow analysis of not only gross developmental defects such as muscle atrophy or attachment defects but also detailed myofibril and sarcomere quantification to identify putative defects after RNAi-mediated knock-down (*see* **Note** 7).

Dissections of flight muscles at pupal stages (works best from 8 h to 50 h after puparium formation, APF):

1. Set up the RNAi cross (*see* Subheading 3.2).
2. To stage pupae of the correct age, collect young white pre-pupae (0–1 h APF old) with a wet brush and place them on a slide with a stripe of double-sided tape on it. Put the slide into a Petri dish that contains a wet filter paper to avoid drying of pupae.
3. Incubate the pupae until the desired stage.
4. Under the dissection microscope use blunt forceps to remove all the parts of the pupal case that are not facing the slide. Since the pupae are placed on a double-sided tape the bottom of the

Table 3
Antibodies

Publicly available antibodies (clone name)	Expression	Subcellular localization	Host species	Notes	Source	References
Sarcomeric components						
Kettin (MAC155) (KIg16)	All muscles	Z disc	Rat	Epitope near the middle of the molecule	Babraha m Institute	Lakey et al. [38]
Mhc (3E8-3D3)	All muscles	Thick filaments	Mouse		DSHB	Saide et al. [39]
Mhc (MAC147)	All muscles	Thick filaments	Rat		Babraha m institute	Qiu et al. [40]
Projectin (*bent*) (MAC150)	All muscles	Z disc	Rat		Babraha m institute	Lakey et al. [38]
Tropomyosin (MAC141)	All muscles	Thin filaments	Rat		Babraha m institute	Bullard et al. [41]
α-Actinin (MAC276)	All muscles	Z disc	Rat	Works best with MeOH fixation	Babraha m Institute	Lakey et al. [38]
Myotendinous junction						
Integrin αPS1 (*mew*) (DK.1A4)	All muscles	Myotendinous junction	Mouse	Aldehyde sensitive epitope; not suitable for immunoblotting	DSHB	Wilcox et al. [42]

(continued)

Table 3 (continued)

Publicly available antibodies (clone name)	Expression	Subcellular localization	Host species	Notes	Source	References
Integrin αPS2 (*if*) (CF.2C7)	All muscles	Myotendinous junction	Mouse	Aldehyde sensitive epitope; not suitable for immunoblotting	DSHB	Brower et al. [43]
Integrin βPS (*mys*) (CF.6G11)	All muscles	Myotendinous junction	Mouse		DSHB	Brower et al. [43]
Talin (A22A and E16B)	All muscles	Myotendinous junction	Mouse	For immunostainings, a cocktail of A22A and E16B works best	DSHB	Brown et al. [44]
Other components						
Dlg1 (4F3)	Flight muscles	All membranes	Mouse	Also labels neuromuscular junctions in larval muscles	DSHB	Razzaq et al. [45]
Futsch (22C10)	Early flight muscles and abdominal adult founder cells	Microtubule associated protein	Rabbit	Also labels neurons	DSHB	Atreya and Fernandes [46]

pupal case will remain sticking to the slide. Using a wet brush gently lift the pupa and transfer to a silicon-filled Petri dish filled with ice-cold PBS. Until all the pupae from the same genotype are collected keep this dish on ice. This should not take more than 15 min to decrease developmental variation among different pupae.

5. Orient the pupae with the ventral side facing up. To fix the position of the pupa use two fine insect pins and push them through the posterior part of the abdomen until the pins are inserted into the silicon.
6. Since the flight muscles are located on the dorsal side of the thorax, the ventral part needs to be removed. Using fine scissors make a small hole in the head of the pupa. Using this hole as an entry point for the scissors cut the pupa on both sides laterally until the middle of the abdomen. Then remove the ventral side by cutting it on the posterior side.
7. On the remaining dorsal part there are other tissues such as gut, trachea, and fat. Using blunt forceps remove these tissues. Avoid touching the muscles underneath.
8. Using fine scissors cut the thorax into two along the dorsal midline until the position insect needles were fixed. Then cut each thorax half at the posterior end and transfer into a glass embryo dish filled with ice-cold PBS.
9. Once all the thorax halves from all the pupae are collected remove PBS and add cold 4% PFA in PBT (200 μL suffice to cover all the samples). Fix for 15 min at room temperature. Remove the fixative and add PBT.
10. Once the tissue is fixed standard staining procedures can be used (*see* **Note 8**). In all subsequent steps use PBT as the medium to improve antibody penetration. In general, block the tissues for 1 h at room temperature using NGS diluted 1/30 in PBT, incubate in the primary antibody overnight at 4 °C and in the secondary antibody for 2 h at room temperature. In between the antibody steps and after the secondary antibody incubation wash minimum three times, 10 min each, with PBT. To visualize actin cytoskeleton use Rhodamine-phalloidin (1/500 in PBT) added to the secondary antibody solution. Mount half thoraces using Vectashield (optionally with DAPI to visualize nuclei) as the mounting medium. Immobilize the coverslip using a transparent nail polish.
11. Image the muscles using a scanning confocal microscope.

Dissections for adult flight muscles:

1. On the fly pad with CO_2 supply anesthetise flies and use fine scissors to remove the wings, legs, abdomen and head.

2. Transfer the thorax to a glass embryo dish filled with ice-cold PBS. After all the thoraces from the same genotype are collected in the embryo dish remove the PBS and add 4% PFA in relaxing solution (*see* **Note 9**). Fix for 15 min at room temperature. Remove the fixing solution and wash two times with relaxing solution.
3. To visualize the dorsal-longitudinal flight muscles (DLMs), the thorax needs to be cut sagittally. Thus, place the thorax on a slide with a stripe of double-sided tape on it. The anterior most part of the thorax should be facing the slide. Using a microtome blade cut the thorax in sagittal plane, ideally in one movement to avoid disrupting the myofibrils with the movement of the blade. Transfer the half thoraces into a glass embryo dish and proceed with standard staining procedure, as described for early pupa (*see* **Note 8**).

3.5 Low-Throughput Live Assays After RNAi-Based Gene Knock-Down: Myoblast Fusion, Myotube Attachment, Myofibrillogenesis, and Muscle Twitching

A detailed video protocol how to image developing adult flight or abdominal muscles at pupal stages has recently been published [37].

1. Set up RNAi cross (*see* Subheading 3.2) using the appropriate live marker in addition to the GAL4 line and the UAS-IR, for example Mef-GAL4, UAS-GFP-Gma to image myoblast fusion or myotube attachment, Mhc[weeP26]-GFP to image myofibrillogenesis or Talin-GFP to image spontaneous muscle twitching.
2. To stage pupae of the correct age, collect young white pre-pupae (0–1 h old) with a wet brush and place them on a slide. Put the slide into a Petri dish that contains a wet filter paper to avoid drying of pupae.
3. Incubate the pupae at 27 °C until the desired age, for example 12 h for imaging myoblast fusion, 18 h to image myotube-tendon attachment or 30 h to image myofibrillogenesis or 30 h to 48 h to quantify spontaneous muscle twitching.
4. Use sharp forceps to open an observation window by cutting away parts of the pupal case on the dorsal side above the thorax or above the abdomen depending on which muscles should be observed (*see* [37] for a video).
5. Place the pupae on a custom-made slide with a deeper groove that fits a large part of the pupae, with the observation window facing up (*see* **Note 10**). Place a coverslip on each side of the pupae as a space holder. Put a drop of 50% glycerol on a third coverslip for each pupa mounted, respecting the correct spacing of all pupae used. Place this coverslip on the pupae to cover the observation window of each pupa with the glycerol to avoid drying out. Fix the coverslip with sticky tape.

6. Depending on which structure should be observed this preparation is suitable for fast imaging, including laser micro-lesions followed by imaging with a spinning disc confocal [50], long-term imaging over several hours observing myoblast fusion or attachment, or high-resolution imaging observing myofibrillogenesis using a two-photon microscope [29, 50]. It can also be used to quantify spontaneous muscle contractions using a regular scanning confocal microscope with a fast scanning option [18].

4 Notes

1. In the GAL4 stock, using a Y-heat-shock-hid virginizer chromosome facilitates virgin collection [51].
2. If an RNAi library with long dsRNAs are used (VDRC, NIG-FLY, and first generation of TRiP libraries), using Dicer-2 together with the GAL4 driver increases knock-down efficiency. However, this also increases the risk of off-targets, which needs to be controlled carefully.
3. If RNAi results in lethality at early stages, however later developmental stages are of interest, use *Tub*-Gal80-ts to overcome the early lethality (TARGET) [8]. Alternatively to overcome early lethality fly crosses can initially be kept at low temperatures and then shifted to higher temperatures during late larval or pupal stages.
4. Since RNAi works more efficiently in males [5] use males for scoring.
5. If flight tests are used for a high-throughput screen, it is practical not to place a water-filled Petri dish at the bottom but simply count the flies in the lower zone. This enables the simple cleaning of the flight tube with a vacuum cleaner before the next flight assay.
6. An even better specificity test than testing a second independent RNAi line is to rescue the RNAi phenotype by either expressing an RNAi insensitive construct from a related fly species [52] or to over-express the target gene using a UAS construct. This often results in inefficient knock-down and thus rescue of the phenotype.
7. The ImageJ plugin MyofibrilJ (https://imagej.net/MyofibrilJ) can be used for detailed analysis of sarcomere phenotypes as it allows measuring sarcomere length and width automatically using Phalloidin labeled-muscles [18].
8. After fixation, half thorax samples can be kept in PBT at 4 °C for a few days until the staining steps will be performed.

9. For the mature muscles, which can contract, it is important to use relaxing solution instead of the PBT to make sure that the muscles are in the relaxed state.
10. Turn the pupae carefully that the observed structures are closest to the surface. The deeper the muscles are located, the less fluorescent light can be detected.

References

1. Mohr SE, Perrimon N (2011) RNAi screening: new approaches, understandings, and organisms. WIREs RNA 3:145–158. https://doi.org/10.1002/wrna.110
2. Boutros M, Ahringer J (2008) The art and design of genetic screens: RNA interference. Nat Rev Genet 9:554–566. https://doi.org/10.1038/nrg2364
3. Kaya-Copur A, Schnorrer F (2016) A guide to genome-wide in vivo RNAi applications in Drosophila. Methods Mol Biol 1478:117–143. https://doi.org/10.1007/978-1-4939-6371-3_6
4. Roignant J-Y, Carré C, Mugat B, Szymczak D, Lepesant J-A, Antoniewski C (2003) Absence of transitive and systemic pathways allows cell-specific and isoform-specific RNAi in Drosophila. RNA 9:299–308
5. Dietzl G, Chen D, Schnorrer F, Su K-C, Barinova Y, Fellner M et al (2007) A genome-wide transgenic RNAi library for conditional gene inactivation in Drosophila. Nature 448:151–156. https://doi.org/10.1038/nature05954
6. Ni J-Q, Zhou R, Czech B, Liu L-P, Holderbaum L, Yang-Zhou D et al (2011) A genome-scale shRNA resource for transgenic RNAi in Drosophila. Nat Methods 8:405–407. https://doi.org/10.1038/nmeth.1592
7. Mohr SE, Smith JA, Shamu CE, Neumüller RA, Perrimon N (2014) RNAi screening comes of age:improved techniques andcomplementary approaches. Nat Rev Mol Cell Biol 15:591–600. https://doi.org/10.1038/nrm3860
8. McGuire SE, Le PT, Osborn AJ, Matsumoto K, Davis RL (2003) Spatiotemporal rescue of memory dysfunction in Drosophila. Science 302:1765–1768. https://doi.org/10.1126/science.1089035
9. Pospisilik JA, Schramek D, Schnidar H, Cronin SJF, Nehme NT, Zhang X et al (2010) Drosophila genome-wide obesity screen reveals hedgehog as a determinant of brown versus white adipose cell fate. Cell 140:148–160. https://doi.org/10.1016/j.cell.2009.12.027
10. Schnorrer F, Schönbauer C, Langer CCH, Dietzl G, Novatchkova M, Schernhuber K et al (2010) Systematic genetic analysis of muscle morphogenesis and function in Drosophila. Nature 464:287–291. https://doi.org/10.1038/nature08799
11. Handler D, Meixner K, Pizka M, Lauss K, Schmied C, Gruber FS et al (2013) The genetic makeup of the Drosophila piRNA pathway. Mol Cell 50:762–777. https://doi.org/10.1016/j.molcel.2013.04.031
12. Reim G, Hruzova M, Goetze S, Basler K (2014) Protection of armadillo/β-catenin by armless, a novel positive regulator of wingless signaling. PLoS Biol 12:e1001988. https://doi.org/10.1371/journal.pbio.1001988
13. Neumüller RA, Richter C, Fischer A, Novatchkova M, Neumüller KG, Knoblich JA (2011) Genome-wide analysis of self-renewal in Drosophila neural stem cells by transgenic RNAi. Cell Stem Cell 8:580–593. https://doi.org/10.1016/j.stem.2011.02.022
14. Roy S, VijayRaghavan K (1998) Patterning muscles using organizers: larval muscle templates and adult myoblasts actively interact to pattern the dorsal longitudinal flight muscles of Drosophila. J Cell Biol 141:1135
15. Brand AH, Perrimon N (1993) Targeted gene expression as a means of altering cell fates and generating dominant phenotypes. Development 118:401–415
16. Bryantsev AL, Duong S, Brunetti TM, Chechenova MB, Lovato TL, Nelson C et al (2012) Extradenticle and homothorax control adult muscle fiber identity in Drosophila. Dev Cell 23:664–673. https://doi.org/10.1016/j.devcel.2012.08.004
17. Menon SD, Chia W (2001) Drosophila rolling pebbles: a multidomain protein required for myoblast fusion that recruits D-Titin in response to the myoblast attractant Dumbfounded. Dev Cell 1:691–703
18. Spletter ML, Barz C, Yeroslaviz A, Zhang X, Lemke SB, Bonnard A et al (2018) A transcriptomics resource reveals a transcriptional transition during ordered sarcomere morphogenesis

in flight muscle. elife 7:1361. https://doi.org/10.7554/eLife.34058

19. Ranganayakulu G, Zhao B, Dokidis A, Molkentin JD, Olson EN, Schulz RA (1995) A series of mutations in the DMEF2 transcription factor reveal multiple functions in larval and adult myogenesis in Drosophila. Dev Biol 171:169–181. https://doi.org/10.1006/dbio.1995.1269
20. Klein P, Müller-Rischart AK, Motori E, Schönbauer C, Schnorrer F, Winklhofer KF et al (2014) Ret rescues mitochondrial morphology and muscle degeneration of Drosophila Pink1 mutants. EMBO J 33:341–355. https://doi.org/10.1002/embj.201284290
21. Schönbauer C, Distler J, Jährling N, Radolf M, Dodt H-U, Frasch M et al (2011) Spalt mediates an evolutionarily conserved switch to fibrillar muscle fate in insects. Nature 479:406–409. https://doi.org/10.1038/nature10559
22. Kocherlakota KS, Wu J-M, McDermott J, Abmayr SM (2008) Analysis of the cell adhesion molecule sticks-andstones reveals multiple redundant functional domains, protein-interaction motifs and phosphorylated tyrosines that direct myoblast fusion in Drosophila melanogaster. Genetics 178:1371–1383. https://doi.org/10.1534/genetics.107.083808
23. Usui K, Pistillo D, Simpson P (2004) Mutual exclusion of sensory bristles and tendons on the notum of dipteran flies. Curr Biol 14:1047–1055. https://doi.org/10.1016/j.cub.2004.06.026
24. Morin X, Daneman R, Zavortink M, Chia W (2001) A protein trap strategy to detect GFP-tagged proteins expressed from their endogenous loci in Drosophila. Proc Natl Acad Sci U S A 98:15050–15055. https://doi.org/10.1073/pnas.261408198
25. Orfanos Z, Sparrow JC (2013) Myosin isoform switching during assembly of the Drosophila flight muscle thick filament lattice. J Cell Sci 126:139–148. https://doi.org/10.1242/jcs.110361
26. Orfanos Z, Leonard K, Elliott C, Katzemich A, Bullard B, Sparrow J (2015) Sallimus and the dynamics of sarcomere assembly in drosophila flight muscles. J Mol Biol 427:2151–2158. https://doi.org/10.1016/j.jmb.2015.04.003
27. Katzemich A, Liao KA, Czerniecki S, Schöck F (2013) Alp/enigma family proteins cooperate in Z-disc formation and myofibril assembly. PLoS Genet 9:e1003342. https://doi.org/10.1371/journal.pgen.1003342.s007
28. Klapholz B, Herbert SL, Wellmann J, Johnson R, Parsons M, Brown NH (2015) Alternative mechanisms for talin to mediate integrin function. Curr Biol 25:847–857. https://doi.org/10.1016/j.cub.2015.01.043
29. Sarov M, Barz C, Jambor H, Hein MY, Schmied C, Suchold D et al (2016) A genome-wide resource for the analysis of protein localisation in Drosophila. elife 5:e12068. https://doi.org/10.7554/eLife.12068
30. Richardson B, Beckett K, Nowak S, Baylies M (2007) SCAR/WAVE and Arp2/3 are crucial for cytoskeletal remodeling at the site of myoblast fusion. Development 134:4357
31. Chen E, Olson E (2001) Antisocial, an intracellular adaptor protein, is required for myoblast fusion in Drosophila. Dev Cell 1:705–715
32. Millard TH, Martin P (2008) Dynamic analysis of filopodial interactions during the zippering phase of Drosophila dorsal closure. Development 135:621–626. https://doi.org/10.1242/dev.014001
33. Dutta D, Bloor JW, Ruiz-Gómez M, VijayRaghavan K, Kiehart DP (2002) Real-time imaging of morphogenetic movements in Drosophila using Gal4-UAS-driven expression of GFP fused to the actin-binding domain of moesin. Genesis 34:146–151. https://doi.org/10.1002/gene.10113
34. Hatan M, Shinder V, Israeli D, Schnorrer F, Volk T (2011) The Drosophila blood brain barrier is maintained by GPCR-dependent dynamic actin structures. J Cell Biol 192:307–319. https://doi.org/10.1083/jcb.201007095
35. Lee T, Luo L (1999) Mosaic analysis with a repressible cell marker for studies of gene function in neuronal morphogenesis. Neuron 22:451–461
36. Spletter ML, Barz C, Yeroslaviz A, Schönbauer C, Ferreira IRS, Sarov M et al (2015) The RNA-binding protein Arrest (Bruno) regulates alternative splicing to enable myofibril maturation in Drosophila flight muscle. EMBO Rep 16:178–191. https://doi.org/10.15252/embr.201439791
37. Lemke SB, Schnorrer F (2018) In Vivo imaging of muscle-tendon morphogenesis in Drosophila pupae. J Vis Exp:e57312–e57312. https://doi.org/10.3791/57312
38. Lakey A, Labeit S, Gautel M, Ferguson C, Barlow DP, Leonard K et al (1993) Kettin, a large modular protein in the Z-disc of insect muscles. EMBO J 12:2863–2871
39. Saide J, Chin-Bow S, Hogan-Sheldon J, Busquets-Turner L, Vigoreaux J, Valgeirsdottir K et al (1989) Characterization of components of Z-bands in the fibrillar flight muscle of Drosophila melanogaster. J Cell Biol 109:2157

40. Qiu F, Brendel S, Cunha P, Astola N, Song B, Furlong E et al (2005) Myofilin, a protein in the thick filaments of insect muscle. J Cell Sci 118:1527
41. Bullard B, Leonard K, Larkins A, Butcher G, Karlik C, Fyrberg E (1988) Troponin of asynchronous flight muscle. J Mol Biol 204:621–637
42. Wilcox M, Brower DL, Smith RJ (1981) A position-specific cell surface antigen in the drosophila wing imaginal disc. Cell 25:159–164
43. Brower DL, Wilcox M, Piovant M, Smith RJ, Reger LA (1984) Related cell-surface antigens expressed with positional specificity in Drosophila imaginal discs. Proc Natl Acad Sci U S A 81:7485–7489
44. Brown NH, Gregory SL, Rickoll WL, Fessler LI, Prout M, White RAH et al (2002) Talin is essential for integrin function in Drosophila. Dev Cell 3:569–579
45. Razzaq A, Robinson I, McMahon H, Skepper J, Su Y, Zelhof A et al (2001) Amphiphysin is necessary for organization of the excitation--contraction coupling machinery of muscles, but not for synaptic vesicle endocytosis in Drosophila. Genes Dev 15:2967
46. Atreya K, Fernandes J (2008) Founder cells regulate fiber number but not fiber formation during adult myogenesis in Drosophila. Dev Biol 321:123–140
47. Schmid A, Sigrist SJ (2008) Analysis of neuromuscular junctions: histology and in vivo imaging. Methods Mol Biol 420:239–251. https://doi.org/10.1007/978-1-59745-583-1_14
48. Budnik V, Gorczyca M, Prokop A (2006) Selected methods for the anatomical study of Drosophila embryonic and larval neuromuscular junctions. Int Rev Neurobiol 75:323–365. https://doi.org/10.1016/S0074-7742(06)75015-2
49. Weitkunat M, Schnorrer F (2014) A guide to study Drosophila muscle biology. Methods 68:2–14. https://doi.org/10.1016/j.ymeth.2014.02.037
50. Weitkunat M, Kaya-Copur A, Grill SW, Schnorrer F (2014) Tension and force-resistant attachment are essential for myofibrillogenesis in Drosophila flight muscle. Curr Biol 24:705–716. https://doi.org/10.1016/j.cub.2014.02.032
51. Starz-Gaiano M, Cho NK, Forbes A, Lehmann R (2001) Spatially restricted activity of a Drosophila lipid phosphatase guides migrating germ cells. Development 128:983–991
52. Langer CCH, Ejsmont RK, Schönbauer C, Schnorrer F, Tomancak P (2010) In vivo RNAi rescue in Drosophila melanogaster with genomic transgenes from Drosophila pseudoobscura. PLoS One 5:e8928. https://doi.org/10.1371/journal.pone.0008928

Index

Sissel Beate Rønning (ed.), *Myogenesis: Methods and Protocols*, Methods in Molecular Biology, vol. 1889, https://doi.org/10.1007/978-1-4939-8897-6, © Springer Science+Business Media, LLC, part of Springer Nature 2019

T

Z

Lightning Source UK Ltd.
Milton Keynes UK
UKHW051017311018
331460UK00004B/179/P

9 781493 988969